Letters of Gold

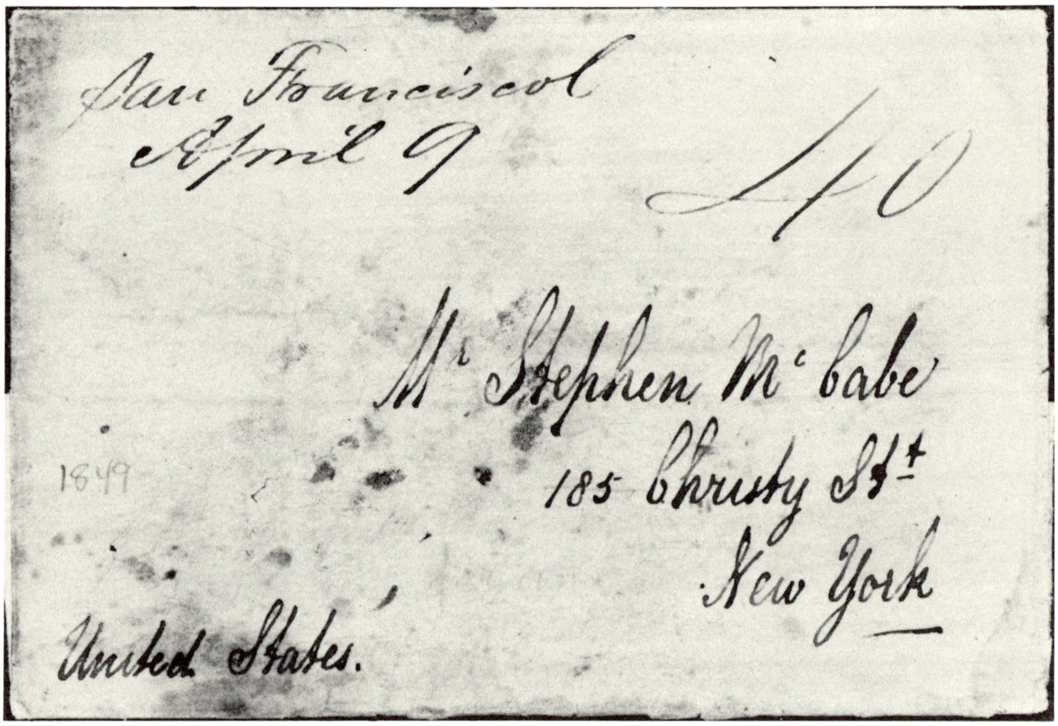

Manuscript "San FranciscoC/April 9" (1849), to New York. One of the very few surviving letters from the first U.S. contract mail from California. Carried on the *Oregon* from San Francisco on 12 April 1849.

Letters of Gold

California Postal History Through 1869

Jesse L. Coburn
Professor Emeritus, Santa Barbara City College

The U.S. Philatelic Classics Society, Inc.
The Philatelic Foundation, Inc.
1984

Designed by Susan M. McDonald
Composed in Century Schoolbook and printed by
Standard Printing Co., Canton, Ohio

Published by the U.S. Philatelic Classics Society, Inc.
Canton, Ohio

Library of Congress Catalog Card No. 84-050922

ISBN 0-9603548-1-6

Acknowledgements

This book was accomplished only through the assistance of many people — far too many to acknowledge individually, lest, to paraphrase St. John: "Of the writing of this book there is no end!" Yet, special mention must be made of Marc Haas, philatelic scholar and collector who insisted that the book must be written and then financed three years of my research, and made his vast collection of Western Americana available to me. My introduction through him to postal history as the graduate division of philately, and his generosity in both funding and expertise are gratefully acknowledged. When unavoidable delays caused me to put the book manuscript "back on the shelf," the insistent encouragement to complete the work came from Oscar Salzer. Without the unselfish help of these two men, the book would never have been written.

Grateful acknowledgement must also be made to the many members of the Western Cover Society. The late Dr. W. Scott Polland, the patriarch of western postal history, gave me so many hours of unstinting guidance. Basil C. Pearce was both mentor and critic. David Jarrett suggested the book's format. Irwin Vogel and Stanley Piller allowed me to photograph the rarities from their collections.

So many others shared both their knowledge and material with me that it would be impossible to express individual gratitude, so my thanks to them all. Their names are written in the cloistered recesses of my heart.

And I certainly would be remiss if I did not thank Thomas J. Alexander, President of the U.S. Philatelic Classics Society, and my editor, Susan McDonald, who both insisted that postal history is different from general history — and then showed me how to do it correctly.

It was my wife, Marcia, who was my patient amanuensis and who for five years cataloged my notes, typed the many drafts of the text, made valuable stylistic suggestions, and never complained during my many research absences that a writer is also a husband with family obligations. To her I dedicate this book.

J. L. C.

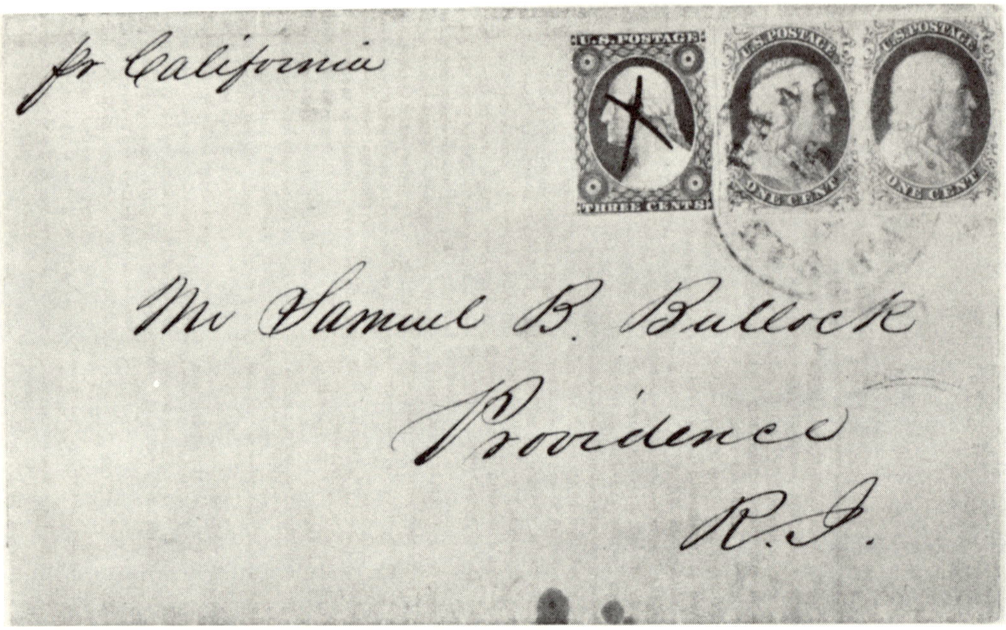

A circular postmarked at San Francisco January 16, 1852, "pr California." The very rare 5¢ circular rate for over 3,000 miles paid by stamps of the 1851 issue.

Table of Contents

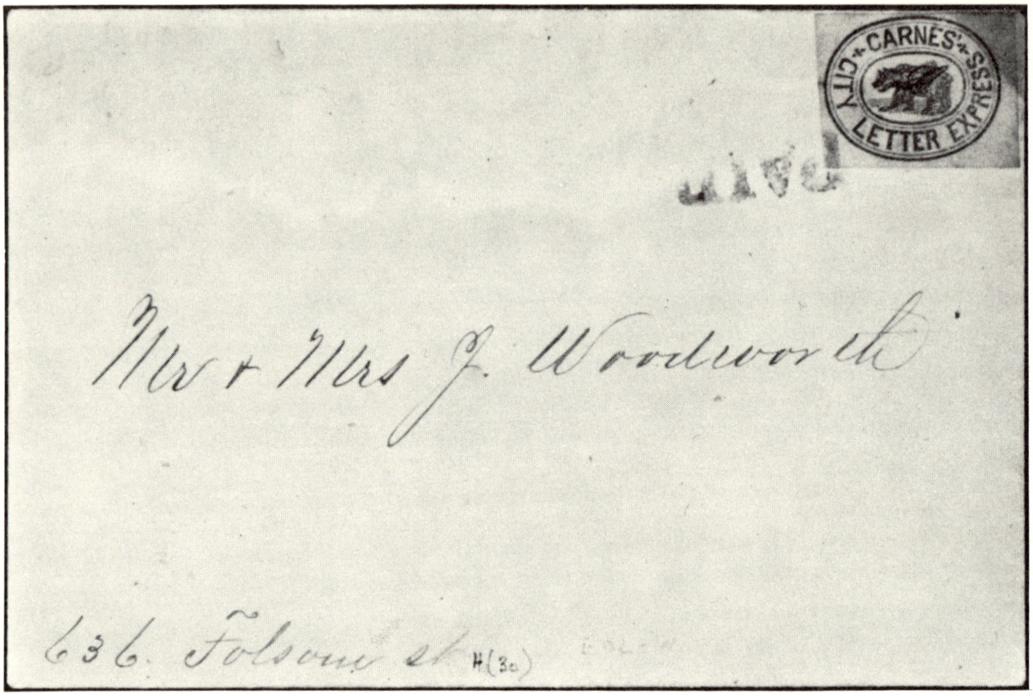

A San Francisco letter carried by Carnes' City Letter Express and bearing the adhesive stamp of that local before "Carnes'" was removed.

Preface

From the advent of civilization and the development of written language, always somebody has been communicating with somebody by writing letters. The archaeologists and scholars explore the debris of the past, seeking not only artifacts but archives of letters. They find them impressed on clay tablets, inscribed on papyrus rolls, scratched on copper sheets, written on vellum pages. From urns, rooms, and tombs they come, making the past live again. Sumerian, Akkadian, Egyptian, Babylonian, Assyrian, Persian, Greek, and Roman history is known most intimately through its letters.

St. Paul's letters to churches and friends are enshrined in sacred literature. The spread of Christendom is recorded in the letters of the Church Fathers. The actual format of the Bible itself was determined by the Easter pastoral letters of the Christian bishops 1,500 years ago.

In the 13th century Genghis Khan had established relays of riders to carry messages throughout the Mongol Empire from China to Western Europe. Royal couriers bearing mail were sacrosanct as they ran along the stone roads of the Inca Empire in the 15th century, or rode down the Grand Trunk road of the Mogul emperors in India in the 16th century. When the mail coaches of Europe sounded their posthorns, the roads were cleared before them. Somebody was sending mail to somebody!

Postal history is a vital part of the story of civilization. The United States is a newcomer to the pages of history, and California is a very recent chapter. Yet we study California's postal history because it is a story of tragedy and romance, often poignant, occasionally comic. It tells of hopes denied and dreams fulfilled. In its totality and with homely detail, it records the decay of the Hispanic empire and the expansion of a vigorous new nation as it spread across a continent from the Atlantic to the Pacific shores. It is a story of people venturing into the unknown yet eager to keep in touch with their roots, and today we know the intimate details of the adventure because people wrote letters to people.

The historian values a letter for its content, for the information it gives about its times, culture, civilization, and the personality of its writer. The postal historian not only respects the content, but also studies the postage stamps and postal markings: the manuscript or handstamped markings and cancellations which directed the passage of the letter from its writer to its addressee.

The collector of postal material places an additional value upon it, namely, the monetary cost of such material in the marketplace. The value of a cover is determined by its rarity, its condition, and the demand by collectors. The physical condition of an envelope (cover) or the address-portion of a folded letter is secondary to its scarcity. Every collector dreams of finding neat undamaged covers with clearly struck or written postal markings. Realistically, such are rare. The manuscripts, straight lines, cancellations, and handstamps too often are crude or partially illegible. Many of the covers became "shopworn" as they were carried in the rough pockets of the senders, or were carelessly opened by the receivers, or traveled under adverse conditions. Only an old-fashioned almost reverent respect for a letter has preserved as many as we now see in good condition. Therefore rarity must outweigh aesthetic appeal.

Obviously the postal history of California began when an inhabitant first sent a letter to someone else. Our primary concern, however, is the two decades between 1849 and 1869. This 20-year period encompassed the fumbling attempts of the United States Post Office to solve problems which it did not seem to understand, the local solutions offered by private agencies, the role played by expressmen, stage lines, and steamships in the transmission of mail, and the postal *curiosa* which were part of California's entry into national history.

The author's purpose in writing this book has been to correlate and synopsize the material which has existed in scores of books, segmented in hundreds of philatelic articles, isolated in numerous library collections, and carried in the minds of collectors who generously have shared their knowledge with the author. As may be said of many historians, there is little that is completely original information herein. The field of western postal history has been thoroughly "mined" during the last fifty years by many scholars with specialized interests. In a sense this book is a compendium, gathering into one volume the work of many men. In the light of newer knowledge and recent discoveries, it attempts to correct some of the errors of older research.

It is difficult to record the events and factors of the 20-year period, 1849 to 1869, in a strict chronological sequence to achieve a smooth flow of narration. Too many postally-related things were happening at the same time. The California postal system was being established while the expresses were operating. The expresses and stage lines co-existed with the mail steamers. The supplementary mail distributors such as the locals and letter bag operators overlapped the other activities. The development of the overland mails was an integral part of the same scene. The author, therefore, has been forced to treat the various strands of this historical tapestry as somewhat isolated factors, meanwhile trying to indicate the interrelationship of each to the other.

This book terminates its account in 1869, the year in which the transcontinental railroad was completed. This historically climactic event largely ended the postal role of other transportation agencies in California. By then, also, the government post office had surmounted most of the problems which had vexed its western users, and which had created the unique postal history of the period. A fascinating chapter of postal history had ended, one which had spanned from missions to miners, from packmules to railroads.

Chapter 1

Prologue

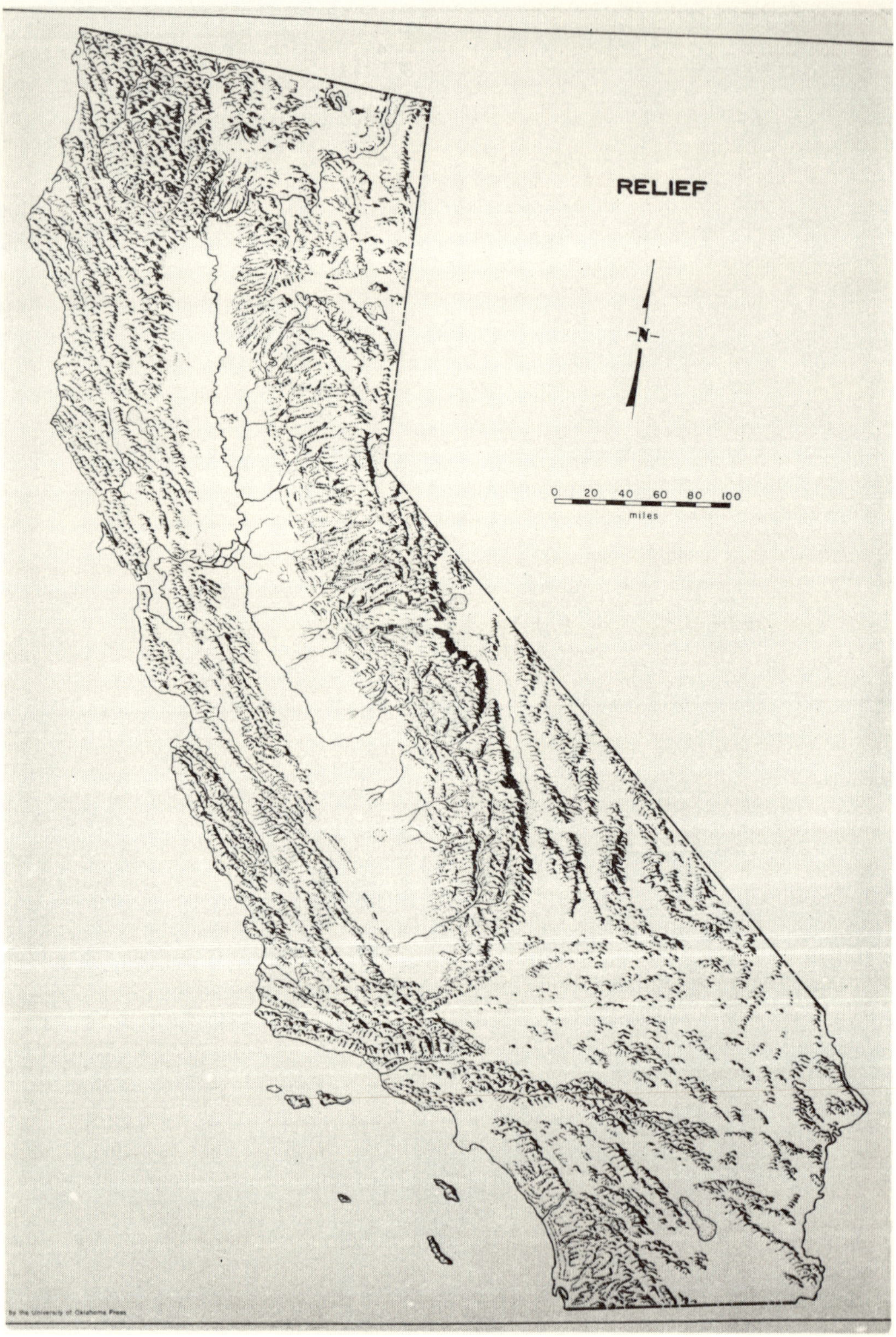

Relief map of California.

Prologue

California was long a stepchild among the lands of North America. Virtually ignored by Spain, neglected by Mexico, it finally had to force its statehood upon a reluctant nation. It was an isolated area until in the Horatio Alger fashion it went from rags to riches, to be known as the Golden State, entered through a Golden Gate.

Topography and Geology

Its isolation is explained by a glance at a topographical map. Less than 300 miles wide, and a bit over 700 miles long, with a coastline of over 750 miles, it has only three natural landlocked harbors. On the ocean side are the Coast Ranges, more than a score of mountain ranges running north and south, from 20 to 40 miles in width, intersected by longitudinal valleys. From the Mexican border they range north to join the precipitous Klamath Mountains which fill the entire northwestern part of the state. The Klamath Range in the east joins the Cascade Range, with only the volcanic Modoc plateau offering a dubious relief in the northeast corner. At Mt. Lassen the Cascades merge into the great eastern wall of the Sierra Nevadas which are 40 to 80 miles broad and extend south for 430 miles, then to join the transverse range of the Tehachapis which go west to become part of the Coast Ranges. Almost the entire southeastern portion of the state is desert: the Mohave and Colorado Deserts, an extension of Mexico's great Sonoran Desert, interspersed with scores of short barren mountain groups. In the center of the state is the great Central Valley, 450 miles long, forming a deep trough between the Coast Ranges and the Sierra Nevadas. On the map it looks like, and for centuries was, an isolated and inhospitable land. To reach it from the western edge of the United States along the Mississippi and Missouri Rivers was a five months' journey, across the great treeless prairie, over the vast Rockies, through the almost waterless American desert and over the mighty barrier of the Sierras. From the Atlantic seaboard it was a five to eight months' voyage by sailing ship around treacherous Cape Horn.

It was little wonder that for two and one-half centuries after its discovery California seemed to rest in isolation, as if it were slowly recovering from the cataclysmic geological convulsions which had brought it into physical existence thousands of millennia earlier.

To understand the subsequent history of California one must know the geology of the area, for the story of California begins with the formation of its physical features during the Jurassic, Cenozoic, Tertiary, Miocene, Pliocene, and Pleistocene ages.

More than one hundred million years ago

a vast sea covered the present land, lapping against the Rockies. Then the earth's mantle, always in motion, began to crack and fault, to fold and wrinkle. Great masses of molten granite invaded the folded strata from below and slowly crystallized into hard rock. Mountain chains formed. In turn these ancestral sierras were eroded by mountain streams. Perhaps sixty to eighty million years in the past, the earth shivered again. Once more its crust wrinkled and faulted and molten rock poured upward, creating a low mountain ridge against whose western base the Pacific Ocean now sent its tides. The forces of nature, however, are continuous if not constant and again erosion began. Thus, about fifty million years ago a series of tectonic pulsations pushed up the sea bottom, and in a pattern of upheaval, subsidence, and erosion the Coast and Klamath Ranges began to appear, the Sierras rose higher and a great trough ran for hundreds of miles between them. This central valley was a catch basin for the Tertiary rivers that cut their way down the slopes of the mountain ranges. During the Miocene and Pliocene epochs, from sixteen to thirty million years ago, the pattern of activity and quiescence continued. Twice the Sierras rose and fell, broad floods of volcanic mud poured from the fissures that reached down to the molten core, and buried valleys and blocked streams, forcing them into new channels as they once more cut their way through the hardening lava flows and the eroded debris.

Perhaps fifteen million years in the past, the final uplift of the Sierras began to occur. Apparently the earth's crust here was too rigid and brittle to bend and fold in response to disturbances below. Instead, a huge massive block of the earth's crust simply tilted upward, from ten to twenty thousand feet in height, with its eastern edge an almost sheer drop-off. Then the last of the geological epochs, the Pleistocene, beginning about one and one half million years ago, occurred. Four successive "ice ages" ensued, three of which covered California with deep sheets of ice. Their great glaciers gouged out valleys on the mountain slopes and as each ice age retreated, the softened rock eroded down to the hard granite beneath, and the melting ice fed the rivers that cut deep channels along the fault lines of the mountains. The final ice age retreated perhaps twenty-five thousand years ago, leaving the topography of California much as it is now.

It was this geological activity of the earth's crust faulting, allowing volcanic action to force molten rock containing minerals of almost every kind to emerge upon the earth's surface, that produced the golden treasure which would affect California's history. Gold is one of the heaviest of natural metals, and as the molten rock thrust upward, the gold was deposited in veins along the fissures. No geologist knows all of nature's secrets, therefore there is no explanation of why gold veins did not form everywhere. But they did form in the Klamath and Cascade Mountains, and in lesser amounts in the transverse ranges and in the small ranges of the southeast desert. The largest deposits were in the Sierras. Between the Yuba River on the north and the Mariposa River on the south, a long ribbon of molten gold ore had flowed upward into the fissures of the granite rock, forming a series of quartz veins in an almost continuous belt about 120 miles long and varying in width from a few hundred feet to two miles. This was the famed "Mother Lode." The Tertiary rivers had cut through it, tearing away pieces and particles of it. Erosion supplemented the process and washed the particles into the rivers and creeks where they moved downward to lodge behind sandbars and stones or in potholes in the streambeds, to later be known as "placer" gold. For millions of years it lay there: perhaps a billion dollars worth in the placer deposits and in the higher elevated quartz veins waiting for the late-comer called "man" to discover and harvest over a period of fifty years.

The Coming of Spain

California's man-made history began with the Spanish explorers and conquistadores. Columbus in the 1490s claimed the Caribbean for Spain. Balboa explored Central America, crossing the Isthmus of Panama and was the first European to see the Pacific Ocean. Pizarro discovered Peru and the Inca Empire where gold had been mined and venerated for centuries. Hernando Cortez, the conqueror and looter of the Aztec Empire of Mexico, first sent ships north along the Pacific coast and in 1533 made landings on the peninsula of Lower or Baja California. Successors to Cortez explored the Gulf of California and reached the mouth of the Colorado River in 1540. Spanish expeditions traveled the length of California's great interior valley, and Spanish sailing ships explored California's coast as far north as Oregon. Most of these explorations were motivated by the explorer's simple desire to be where no one else had been, and other than mapping Spain's California possessions, they accomplished little of lasting value.

From 1602 to 1765 California was left to its Indians and its isolation. Much of its coast was dangerous; winds and currents made sailing north difficult; land travel was even more forbidding as it crossed waterless deserts up to mountain chains where passes were few. And for what purpose? Centuries earlier the Spaniards had found Indian empires in Mexico and South America where gold and silver had long been mined and treasured. But in the Californias the Indians had no acquaintance with any metal whatsoever. *Ipso facto:* precious metals did not exist in the Californias. It was enough simply to know that the Californias were Spanish possessions.

The Missions

In 1765 the picture changed. An energetic Visitor General, Jose de Galvez, arrived in Mexico City as a special deputy of the King of Spain. Soldiers of the Cross, dedicated to the spiritual conquest of the souls of the native Indians, already had moved into Baja California to establish missions. Soldiers of the Sword had accompanied the padres, under orders to protect the priests and by their presence to insure the physical possession of the land as Spanish territory. Originally the Spanish American missions had belonged to the Jesuit order, but in 1767 Carlos III of Spain ordered the expulsion of all Jesuits from all Spanish dominions. Their missions and their missionary activities were given over to the Franciscan order. Galvez turned the Baja missions and pueblos over to Captain Gaspar de Portola militarily, and to Father Junipero Serra religiously. Galvez also was the first Spanish American to be interested seriously in the Californias. He knew that the Russians and the British were encroaching

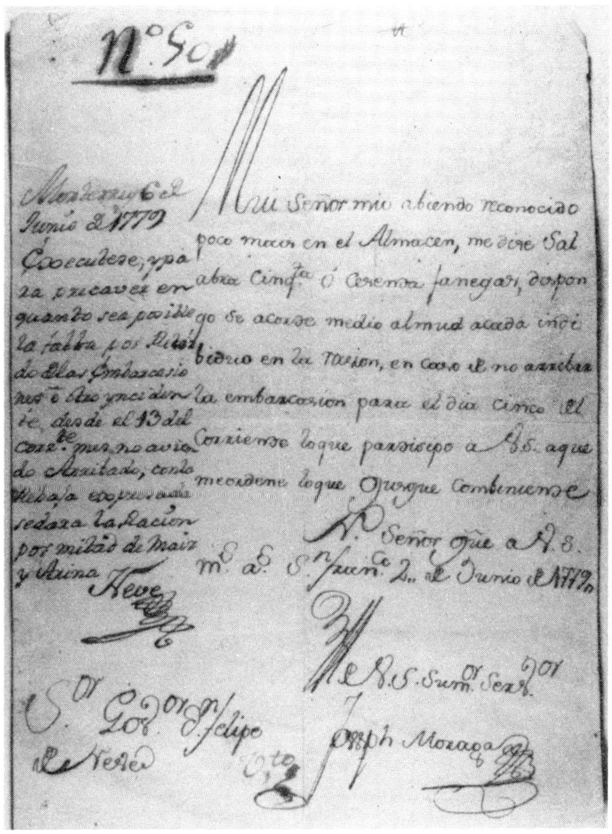

A very early California letter, dated 2 June 1779, from Joseph Moraga, commandant of the San Francisco presidio, to Monterey to Gov. Neve, whose reply is at left.

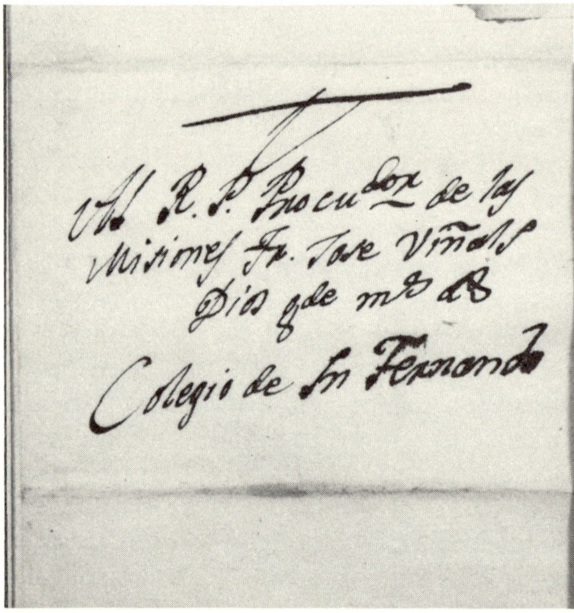

Folded letter dated Oct. 27, 1806, from Mission San Jose to the Apostolic College of San Fernando by courier.

Letter dated Oct. 31, 1807, to San Fernando by courier. Straightline ACAPULCO.

from the north into territory claimed by Spain. So in 1769 he ordered Portola and Serra to explore and Christianize Upper or Alta California. Early in that year, from the new seaport of San Blas, the expedition moved north by land and sea.

During the next 60 years, 21 missions, four military presidios, and three pueblos or towns had been established between San Diego and San Francisco, all in the coastal valleys or on the coastal plain. In the process a trail was blazed, the *El Camino Real,* a bridle path and mule track which followed the passes through the 13 primary mountain ranges which separated the missions. The missions, which combined churches, agricultural communities, and vocational schools, by 1830 owned more than 1,470,000 acres of the most fertile land, on which they grazed over 400,000 cattle, 60,000 horses, and 300,000 sheep, goats, and hogs. Close by the missions were the 20 private land-grant ranchos which the Spanish crown had given to soldiers when they retired from garrison duty.

From the first establishment of the missions in Upper and Lower California, there was continuous mail service between the missions themselves as well as to their headquarters at the Apostolic College of San Fernando in Mexico City. Mounted military couriers twice weekly made the trip from Monterey near San Francisco to Loreto in Baja California. They carried letters and reports, both military and religious, as well as mail from the rancheros, which from Loreto went by ship to San Blas and then overland to Mexico City.

Mexican Rule

In 1821 after a 10-year struggle Mexico achieved its independence from Spain, and California became part of the Mexican Republic. During the next 25 years of Mexican rule, however, political turmoil ensued in California. The office of Governor changed hands 12 times. The ranchero families, long envious of the prosperity of the missions, forced the secularization of the missions in 1834 and divided their holdings among themselves. Between 1822 and 1846 more than 800 land grants were made, totalling 13,000,000 acres,

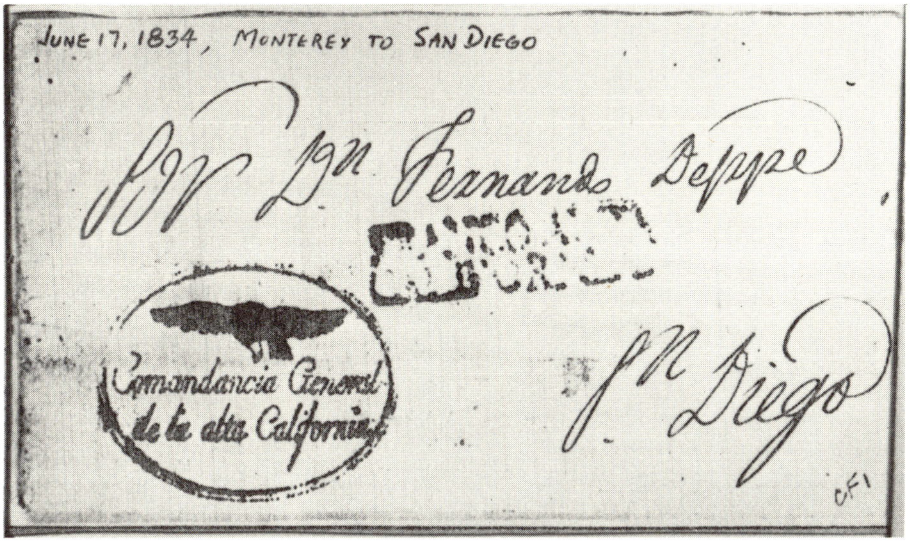

Earliest known handstamped California cover: FRANCO ALTA/CALIFORNIA. Monterey to San Diego June 17, 1834, by military courier. Oval "Comandancia General de la alta California."

42 percent of the land grants going to non-Mexicans, mostly Americans who had become "naturalized Mexican citizens." In 1851 the United States Land Act appointed a commission to adjudicate conflicting land-grant claims and approved only 553, totalling 8,850,000 acres. The losers were primarily Mexican families who could not furnish proper documentation establishing their claims.

These "naturalized Mexican citizens" in most cases were American immigrants who had come to California seeking land or to act as commercial agents for eastern mercantile houses. While much has been said of the isolation of California, that isolation was not total. From the late 1700s the sailing ships of England, France, Holland, Russia, Portugal, and the United States had plied the Pacific Ocean, engaging in trade with the Orient, the East Indies, the Hawaiian Islands, and Australia. The seals, sea otters, and whales brought New England ships to the harbors of California. Many of these ships used the harbors to obtain water and provisions, and to establish trade centers. The "hide and tallow" trade attracted more Americans as the Californias slaughtered over a million cattle to trade the hides and tallow for manufactured goods. Yankee merchants placed agents who were men of means and ability in the California port villages to buy and sell, until finally American enterprise had made an economic conquest of California.

Also during the early 1800s the adventurous trappers of the American Fur Company and the Rocky Mountain Fur Company were a vanguard of American expansion. They discovered the passes across the northern mountain ranges and through the mighty Sierras. When they returned to their homes their tales of the fertile valleys of Oregon and the great central valley of California aroused a land hunger in the East. The trappers shared their knowledge of the trails, deserts, and mountain passes, and guided immigrant trains westward.

These immigrants must be distinguished from the trapper-explorers and the merchant-agents. The immigrants set forth westward to find new homes and to settle down there. Their orientation was primarily agricultural. At first most of them went into Oregon, following the Methodist missionaries of 1836. Then many began leaving the Oregon Trail near Salt Lake at Fort Hall to cross the Humboldt Desert and to make their way over the Sierras into California. Some wagon trains turned south at Santa Fe and followed the old

Letter datelined Monterey Oct. 6, 1843, to Boston. Endorsed "pr Bark 'North America'." Received at New London April 4, 1844, rated SHIP 14½ due.

San Francisco Oct. 19, 1845. "Politeness of Mate of whaleship Sarah." Received as ship letter at New Bedford, Mass., May 25, 1846, rated 7¢ due.

Spanish Trail into southern California. The composition of one immigrant party led by John Rowland and William Workman which entered California in 1841 is indicative of the character of this westward movement: there were two physicians, four carpenters, a tailor, a naturalist, a mineralogist, a musician, an engineer, a gunsmith, a blacksmith, a cooper,

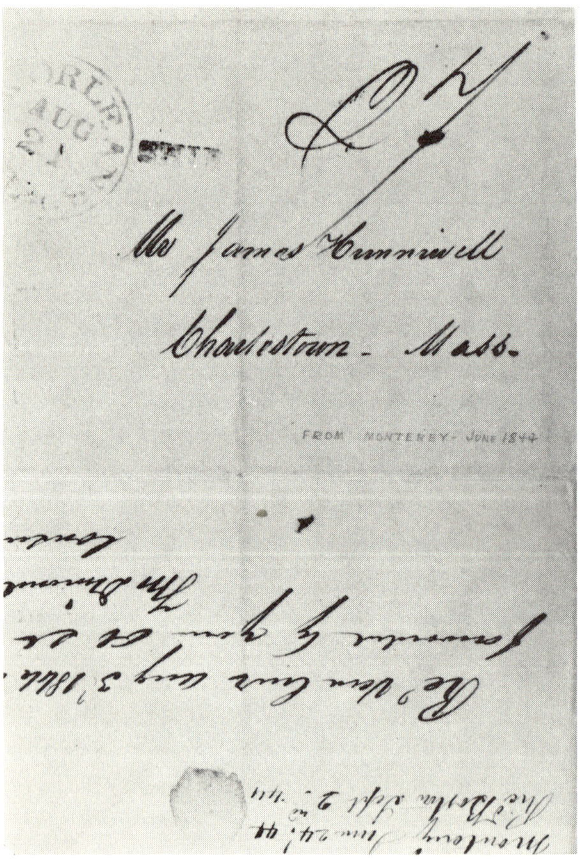

Monterey June 24, 1844, to Mass. Forwarded by F. M. Dimond, U.S. consul at Vera Cruz. Landed at New Orleans Aug. 21 as ship letter rated 27¢ due (2¢ ship and 25¢ for over 400 miles).

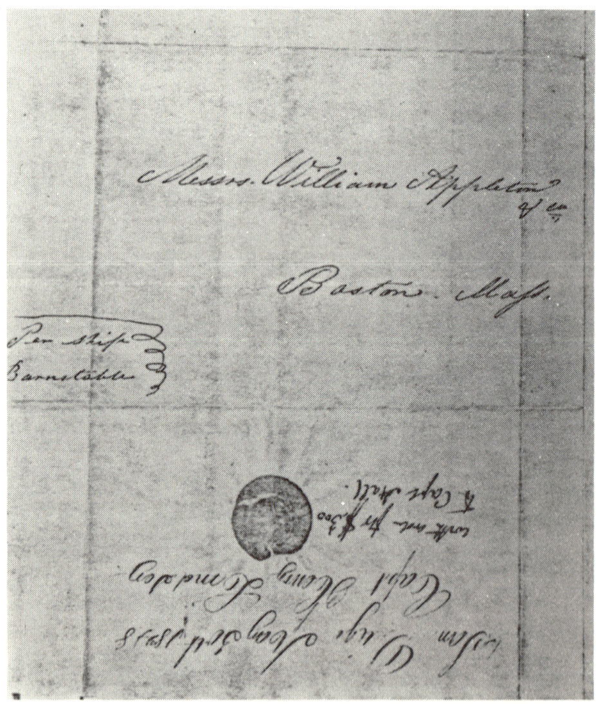

San Diego, May 30, 1848, to Boston. "Per ship Barnstable." Carried by favor entirely outside mails.

Granville, Ohio, Aug. 26 (1848), to Monterey. Addressed in care of postmaster at Independence, Missouri, to be forwarded "by the first opirtunity," which was nearly 10 months coming. Independence postmark June 21 (1849). Rated 10¢ Ohio to Mo.; 40¢ Mo.-Cal., for total of 50¢ due.

and 11 farmers, all with their families. These were people who had come to stay. By 1845 more than 600 American pioneers had traveled overland to settle in California.

As the American population increased with successive immigrant trains, more letters began seeking routes homeward. There was a need for correspondence with business associates, friends, and relatives they had left in the United States. Merchants and businessmen usually made three copies of their correspondence and sent them out by differ-

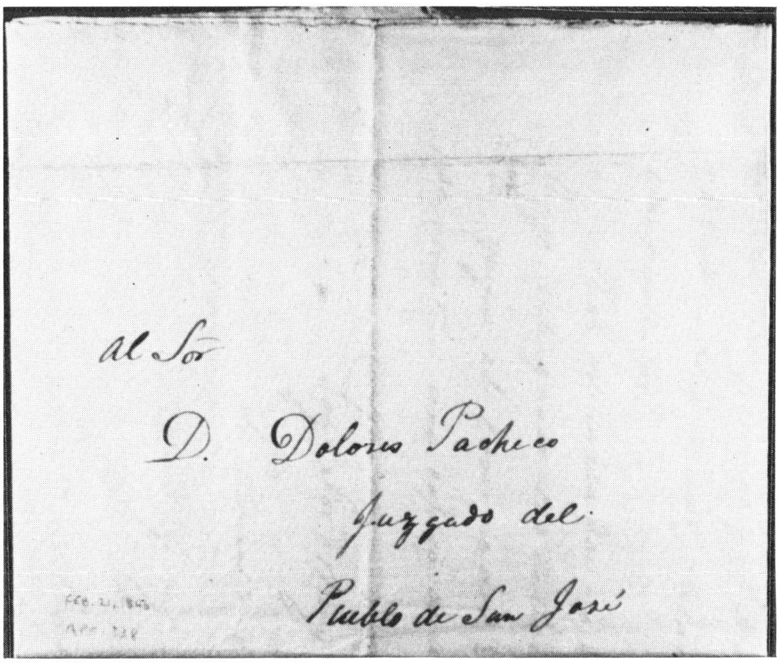

Letter dated Feb. 21, 1846, at encampment near Santa Cruz, from John C. Frémont, Capt. U.S. army, to the mayor of San Jose. Probably by military courier.

Bridgeport, Conn., to Hawaii, Dec. 6, 1842, in care of forwarder at Mazatlan. East to west example of via Mexico route. Boxed VERA CRUZ/ABRIL 24 and "4" Mexican charge.

ent ships, and trusted to the whims of tide and weather. Others sought out returning ships' officers also, or entrusted their letters to the trapper-mountain men returning to act as guides for westbound groups, the letters to be posted at Fort Leavenworth or some Missouri-Mississippi River frontier post office. Answers to these letters followed the same uncertain paths.

Perhaps the most popular mail route between 1842 and 1846, other than around Cape Horn, was across Mexico from Mazatlan to Vera Cruz. Ships from the Atlantic ports and from New Orleans would leave mail with the American Consul in Vera Cruz. He would place it in the Mexican mail service to be carried via Guadalajara and Durango to Mazatlan where the American Consul would

West to east via Mexico, mailed at Chihuahua July 17, 1849. Crown circle PAID AT VERA CRUZ (Brit. p.o.) 1/-; then to Mobile, rated 10¢ due to Texas.

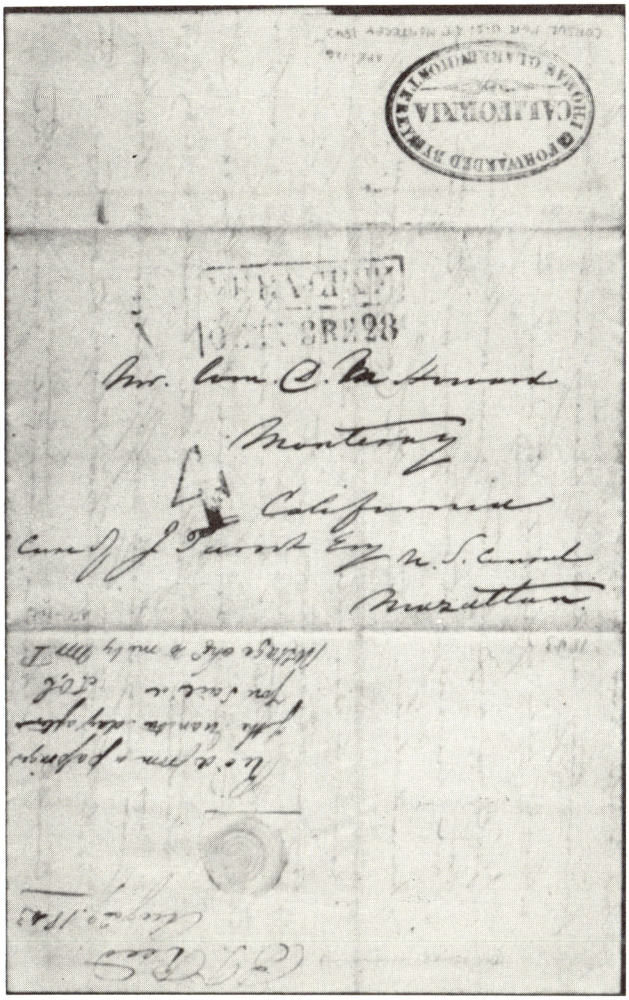

Cover to Monterey via Vera Cruz, Oct. 28, 1843, addressed in care of J. Parrott, U.S. consul at Mazatlan. Forwarding handstamp of Thomas O. Larkin, Monterey, and Larkin's ms. note: "Rec'd from a passenger of the Juanita day after you sail'd. T.O.L. Postage chgd. to me by Mr. P."

Via Mexico route.

send it on to Monterey by ship. Normally mail traveling by this route in either direction took about 80 days, although if there were internal political disruptions in Mexico it might be in transit for six months or longer.

War with Mexico, and Statehood

The next climactic event in California's history was the war between the United States and Mexico. Early in the 1840s the United States began to indicate that its "Manifest Destiny" required that its western border should be the Pacific Ocean, a belief which made conflict with Mexico inevitable. Early in September 1842 Commodore Thomas Jones, commander of the American Pacific Naval Squadron, received a false report that the United States and Mexico were at war. He promptly sailed from Callao, Peru, and on 18 October entered Monterey harbor and demanded the peaceful surrender of the Mexican garrison. Two days later Jones raised the American flag and proclaimed that California was under American occupation. On 21 October, after examining official documents, Jones realized his blunder and sent a letter of apology to Governor Micheltorena in Los Angeles. The one-day occupation had ended, but it was a portent of events to come.

President James K. Polk, a disciple of Andrew Jackson, had been elected in 1844 on a platform of Manifest Destiny. Two days before his inauguration on 4 March 1845 Congress had approved the annexation of the Re-

Immigrant routes to the west coast.

public of Texas. On coming to office Polk began efforts to acquire California. He issued secret instructions through Secretary of State James Buchanan to Thomas O. Larkin, the American Consul in Monterey, to encourage insurrection among the Americans residing in California and to aid them to secede from Mexico and seek American protection, but warned Larkin that the United States could not officially sponsor the secessionist movement unless Mexico should begin hostilities. Thomas Larkin immediately began conspiring with prominent American landholders and merchants.

Mexico had never recognized the independence of Texas, and was especially bitter when Texas proclaimed that its southern boundary was the Rio Grande rather than the Nueces River. President Polk sent John Slidell to Mexico City as a special envoy to nego-

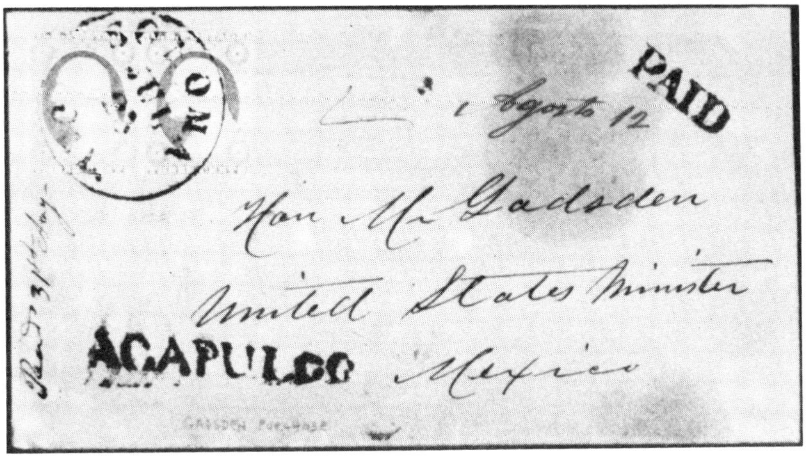

Cover from Sonoma (1853-56), addressed to James Gadsden, U.S. Minister to Mexico.

tiate the dispute and to offer the Mexican government $40 million for the purchase of Upper California and New Mexico. The Mexican president refused to receive Slidell. In reply, on 8 April 1846, President Polk ordered General Zachary Taylor to cross the Nueces River with American troops and to occupy the disputed Texas territory. Several Americans were killed in the resultant armed skirmishes. With this excuse in hand, Polk asked Congress to declare that a state of war existed, and Polk signed the declaration on 13 May 1846.

It was a shortlived conflict. On 14 June a group of 30 American pioneers descended on Sonoma, compelled the surrender of General Mariano Vallejo, and proclaimed California an independent republic under "the bear flag." The Bear Flag Republic disappeared into California's mythology three weeks later when on 2 July Commodore John D. Sloat anchored the Pacific Squadron on Monterey Bay and on 7 July raised the American flag above the Custom House and declared that California was now part of the United States. In the meantime for almost a year John C. Frémont had been ranging Upper California with a band of 100 American riflemen and Delaware Indian scouts under the mysterious pretext of making a "scientific survey." On 28 December his force occupied Santa Barbara. Earlier, in June, at Fort Leavenworth, General Stephen W. Kearney had been ordered to move the Army of the West into New Mexico to subdue the garrison at Santa Fe and then to proceed by the old Spanish Trail to California

San Francisco to Los Angeles, Dec. 27, 1847, by military courier. SF markings in magenta ink.

to complete its occupation. In early December Kearney reached southern California and in joint action with Frémont's riflemen and Commodore Stockton's naval troops forced the final capitulation of all Mexican forces in California on 13 January 1847 following the final battle of Cahuenga Pass. In the meantime an American army under General Winfield Scott had invaded Mexico through Vera Cruz, and at the battle of Chapultepec had defeated the army of Mexico. In the subsequent Treaty of Guadelupe Hidalgo, signed on 2 February 1848, much of what are now the states of Arizona, Nevada, New Mexico, and California was ceded to the United States. (Today's Mexico-United States border was finalized by the Gadsden Purchase on 30 December 1853 of 30,000 square miles of present day southern Arizona and New Mexico for $10 million.)

Significant in the subsequent history of California were two American groups who came west to fight in California's war for independence but arrived too late to participate in the subjection of California. A battalion of Mormons, 360 strong, had marched through Santa Fe and Tucson and had arrived in San Diego in January 1847. They had marked a wagon road from Santa Fe to San Diego which would be used by later immigrants and gold rush argonauts. Also, in the summer of 1846, a regiment of 250 men had been recruited in New York by Jonathan Stevenson to assist in the conquest of California. Sailing around Cape Horn, they arrived in San Francisco on 6 March 1847. Most of them remained in California to become influential in the state's future history.

Oddly, between the actual acquisition of California in 1847 and its being granted statehood in 1850, it was never officially designated as a "territory." It was governed by a series of military officers, some of whom signed official correspondence as territorial governor and some who did not. It was an interregnum period of confusion. Seven Army and Navy officers in succession governed California in less than three years, the last of them being General Bennett Riley who took command on 13 April 1849.

General Riley was confronted by a civilian populace which was weary of military government, and which impatiently demanded an end to military rule, so he wisely decided to direct the movement toward statehood rather than oppose it. Without consulting the desires of Congress, he ordered the election of delegates to a convention in Monterey to frame a state constitution. On 13 November 1849, in a general election, Peter H. Burnett was elected governor and John C. Frémont and William M. Gwin were elected as United States Senators. Quite literally, the independently-minded Californians had created a State. Presented with a *fait accompli,* the United States Congress on 9 September 1850 admitted California as a "free" State into the Union.

The postal historian must note the establishment of a rudimentary postal system within California under the interregnum period of military government. In March 1847 General Kearney began a semi-monthly mail service to run regularly between San Francisco and San Diego, to be carried by soldiers on horseback. They followed the old *Camino Real* which had been utilized by the mission courier mail service decades earlier. Though the primary purpose of this service was the carrying of military dispatches, Kearney ordered his military postmen also to carry such letters and newspapers as the American civilians should place in their possession. General Kearney's successors carried on this mail service until early in 1849. Such was the beginning of a mail communication system within California prior to the establishment of post offices.

Although unaware of the future, California was about to experience a momentous event which would change its character from quietude to incredible confusion and chaos. The current of western history was to enter a new channel paved with gold.

Chapter 2

Gold!

"Miners' Coat of Arms" mailed from Goodyears Bar to Ill. The handstamp PAID/3 is struck twice to show payment of 6¢ transcontinental rate.

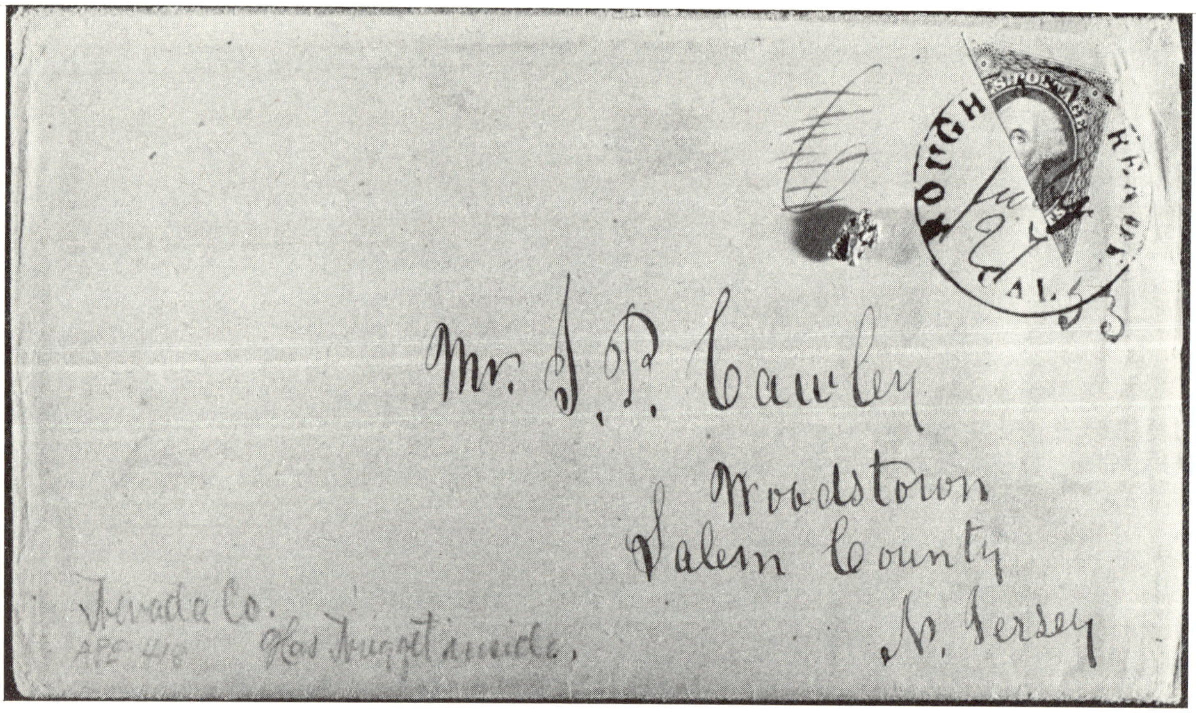

Rough and Ready to N.J., July 27/53, tying a diagonal bisect of the 1851 12¢. The gold nugget mailed in this envelope is shown left of the postmark.

Gold!

In the dramatic history of California no other event has equaled the explosive influence of the Gold Rush. Almost overnight a remote and relatively unknown corner of North America became the focus of the world's attention.

As early as March 1842 Francisco Lopez, a Mexican ranchero, while pulling wild onions had found gold particles clinging to the roots. This was on Placerita Creek, about 35 miles northwest of Los Angeles. It was a small placer deposit, quickly exhausted, and attracted little attention. Abel Stern, a Los Angeles merchant-landowner, purchased 20 ounces of this gold and forwarded it through Alfred Robinson to the United States Mint. Still, it was considered a one-time happenstance and, though talked about, it created no excitement.

The crucial discovery occurred on 24 February 1848 when John Marshall, building the Sutter sawmill on the South Fork of the American River about 45 miles northeast of Sutter's Fort at a place called Coloma, found extensive gold particles in the millrace. Although Captain Sutter sought to keep the news secret, the families of his mill employees gossiped, and spent their spare time successfully digging out gold deposits with primitive instruments. They invited friends to share in their good fortune. Yet the news spread slowly and for several months little attention was paid to it, and even that attention was greeted with wide skepticism. The press was dubious, for on 15 March the *Californian* noted on its back page that gold had been found, but it downplayed its importance by inaccurately adding that "gold has been found in every other part of the country."

On 18 March San Francisco's and California's only other weekly newspaper, the *California Star,* similarly noted the discovery and similarly disparaged its importance. On 25 March the *Star* reported that gold had become "an article of traffic" and was being used to make purchases at New Helvetia, but the editor added that agriculture was of more importance. A week later a *Star* article implied that the new discovery was similar to the Los Angeles placero of 1842 and would not amount to much.

Nevertheless, here and there a few individuals were slipping away to the Coloma area, to discover that some men were finding significant amounts of gold in their diggings. And on 12 May 1848 Mormon Elder Samuel Brannan infected California with "gold fever."

Elder Sam Brannan had arrived in San Francisco on the ship *Brooklyn* on 31 July 1846 as the leader of a group of 136 Mormons from New York. Brannan was a man of energy and business acumen, and as he received the tithes of his fellow-religionists (10 percent of a Mormon's gross income), he refused to send them to Brigham Young, the supreme head of the Mormon faith. It is said that when Brigham Young tried to recover the tithes, Brannan replied "I'll give up the Lord's money when Young sends me a receipt signed by the Lord, and no sooner." The tithes, combined with his own money, permitted Brannan in less than two years to become a leading businessman in the sleepy little town of San Francisco. By 1848 he owned land lots, a hotel, a store, and a newspaper, the *California Star,* printed on a press he had brought around the Horn, and another store at Sutter's Fort in New Helvetia (Sacramento). In March 1848 Brannan heard that customers at the New Helvetia store were paying for purchases with gold dust and nuggets. During the first week of April Brannan investigated, going to New Helvetia and Coloma. Realizing the potential, Brannan spent the next month quietly buying every article of merchandise he could find in northern California that would be needed by gold-seekers, and hoarding these articles in his stores. Then on 12 May, holding aloft a quinine bottle filled with gold dust, he strode through the streets of San Francisco, shouting in stentorian tones: "Gold! Gold! Gold from the American River!"

California went mad! Within two weeks San Francisco lost half of its population as men hurried to "the diggings." Within a month business was at a standstill, newspapers suspended publication, sailors deserted their ships, and soldiers deserted their garrisons. The fever spread through the state and by mid-summer most Californians were in the gold fields.

When the news of the gold discovery had become believable, many of the settlers owning ranches decided to try their luck at gold digging, assisted by their employees and by friendly Indians. John Sinclair in his first five weeks took $16,000 from along the American River. Claude Chana discovered rich deposits in the area to be known as Auburn, where men took out as much as $15,000 in a single day. Captain Charles M. Weber, owner of the French Camp Ranch, a grant of 48,000 acres on the San Juaquin River, and William Daylor, a ranch owner on the Consumnes River, found rich deposits around what would become Placerville. Weber not only profited from gold digging, but also laid out the town of Stockton on his ranch property. William Knight also decided to combine gold digging and commerce: he built a trading post and a ferry where the traffic crossed the Stanislaus River. From his 4,000-acre property at Santa Clara, John M. Murphy crossed the Central Valley to find gold. Landowners J. H. Carson and James Woods gave their names to camps and creeks where gold was found. John Bidwell, owner of a large grant near the Feather River, struck it rich and laid out the town of Chico. Patrick McChristian, Jasper O'Farrell, and Jacob Leese left their property in Sonoma to take out $75,000 in three months along the Yuba River. Major Pierson B. Reading, who owned the most northerly of the Mexican grants, found gold on the Trinity River. From Los Angeles, Antonio Coronel and Dolores Sepulveda led a party of 30 Mexicans to rich strikes in what is now called Sonora.

This name review, which could be much lengthier, calls attention to an interesting characteristic of 1848. These men were largely people who knew each other, who were the pioneer settlers and native Californians from the ranchos and pueblos.

Within weeks the contagion spread to Oregon and Hawaii. By autumn it had reached Mexico, Peru, and Chile. It spread to Australia. By the end of the year, probably 10,000 men were digging for gold. (Statistics for 1848 can be only approximate, which perhaps is why historians persist in publishing different figures.) They were the merchants, lawyers, tradesmen, farmers, and landowners of California; they were two-thirds of the able-bodied men of Oregon; they were a few of the

New York ad for sailings to California.

substantial citizens of Hawaii; they were some experienced miners from Mexico, Peru, and Chile. In the early months of 1848 gold-seeking was a California monopoly, and by its end still largely a Pacific monopoly. However, by the end of 1851 most of the miners from Chile, Peru, and the provincc of Sonora had taken their profits and gone home.

Although the California State Mining Bureau years later would maintain that the gold yield of 1848 was only $245,301, this statistic ignores the obvious fact that no accurate record is available. Most of this early gold was not deposited in record-keeping banks or sent to mints for coinage. It was used in its raw state as an exchange medium, or crudely smelted and retained; the lucky settlers kept the amounts secret and took it home with them. A reasonable estimate is that the total yield of 1848 approached $1 million.

The Mining Bureau figures for subsequent years may be more accurate: 1849, $10,151,360; 1850, $41,273,106; 1851, $75,938, 272; 1852, $81,294,700; 1853, $67,613,487. Yet again, these figures are merely approximations, based on accounts and receipts for treasure shipments. Only after 1854, as placer mining ended and quartz hardrock mining began, and a mint was established in San Francisco, do statistics become more exact. Gold production peaked at over $81 million in 1852, but California remained the world's largest gold producer until 1900. This gave the United States its first great source of precious metals for both coinage and currency support. It was the world's first major infusion of gold since the Spanish conquistadores looted the Inca and Aztec Empires in the 1500s.

The gold fever was somewhat slower to infect the eastern United States. Although in September 1848 Lieutenant Edward F. Beale had arrived in Washington by way of Hawaii, Peru, across Panama, and via Jamaica, with reports of the gold discoveries and with some gold specimens sent by Thomas Larkin, they were received with considerable skepticism. Larkin, who had been the U.S. Consul to Mexican California and was now the U.S. Naval Agent in Monterey, had kept the State Department well advised on affairs in California. On 1 June and 21 June he had written letters, carried by Lieutenant Beale, describing the placer mining and accurately predicting its importance. The letters were ignored. It was not until 5 December 1848 that President Polk, in his Annual Message to Congress, confirmed the news, and appended a lengthy, detailed report of Colonel Mason, military governor of California, who had spent six weeks touring the mining areas with his aide, Lieutenant William Tecumseh Sherman.

Almost overnight the rumor became confirmed belief. Lieutenant Lucien Loeser had sailed from Monterey by schooner to Payta, Peru, then a British steamer took him to Mazatlan; he crossed Mexico to Vera Cruz and shipped to New Orleans; from there he proceeded by stagecoach arriving in Washington on 7 December with the famous tea caddy crammed with 230 ounces of gold, which Colonel Mason had dipped into civil funds to purchase from miners.

The War Department placed it on display, and the newspapers excitedly published the confirmation of Polk's Message. The belief now became a mania for sudden wealth. From the Atlantic seaboard to the Mississippi and Missouri River frontiers, the epidemic of hunger for gold spread like a virulence. No social class was immune: doctors, lawyers, merchants, tradesmen, farmers, all were infected. Men sold their businesses and farms, quit their jobs, borrowed money, mortgaged their possessions — and "took off" for the West. "A man can become rich in a week in California. How do I get there?" Before the Gold Rush ended, about 10 percent of the 22 million people in the United States had gone to the California gold fields — the young, the adventurous, the failures, and the dreamers. Some would remain to become the state's leading citizens; many more would return to the East, some prosperous, many penniless.

Travel Routes to the Goldfields

How to get to California? That was the first question for the victims of the gold fever. Both geography and history provided answers. From 1790 American ships, fur and hide traders, whalers, and naval squadrons had rounded Cape Horn. From 1820 to the 1840s first the trappers, then the immigrant trains, had trekked overland. Then there was a third option: the Isthmus of Central America, via Mexico, Nicaragua, or Panama. Official mail had crossed Mexico by Vera Cruz, Mexico City, and Mazatlan for years. Lieutenants Beale and Loeser had brought confirmation of the gold discovery via Tehuantepec and Panama. In addition, the government had just authorized the U.S. Mail Steamship Company and the Pacific Mail Steamship Company to provide steamer service to Panama and thence to San Francisco.

Overland

The most difficult and yet, to the majority of Americans, the first choice was the overland trails across the great plains. In many respects it was the hardest journey of all: it meant months of plodding through clouds of choking dust, crossing the burning alkali deserts of Utah and Nevada, and finally surmounting the massive wall of the Sierra Nevadas. It had to be traversed between the grass of spring and the snow of winter.

For many it was the first choice simply because they lived on the American frontier along the Mississippi Valley, and could start from home, already owning their wagons and stock and provisions. The standard equipment on any farm could be converted into a trail outfit. Also the overland routes had been marked earlier. The Oregon Trail was a known artery. Brigham Young and his Mormons had built a road to Salt Lake City. The majority of argonauts followed the trail of the earlier immigrants along the Platte River, through the South Pass of the Rockies into the Humboldt Sink and crossed the Sierras by the Donner or Carson Passes. The Old Spanish Trail to Santa Fe and on to San Diego and Los Angeles, then northward across the Tehachapi Mountains was the choice of some. There were those who tried to avoid the Sierras, and thus Death Valley earned its name. Small contingents went far north and crossed the Sierras by the Lassen Pass, to follow the Pit River southward.

All of these were overland routes, but few of their travelers were realistic about the physical hardships involved. Regardless of the particular overland trail chosen, it soon was littered with the carcasses and skeletons of dead animals, broken wagons, abandoned equipment, and the graves of the drowned and the cholera and dysentery victims. Only the strong survived, and few reached California with more than the ragged clothing on their bodies. Yet before the first phase of the Gold Rush ended, more than twice as many gold-seekers would go the overland route in preference to the sea passages. Approximately 50,000 adventurers in the first migration chose the 60 to 90 days' route by horseback or three to five months by wagon. About 5,000 died on the trail.

Cape Horn

The second most popular choice in 1849 was the hazardous sea passage around Cape Horn or through the Straits of Magellan. It was the long way around, from 17,000 to 18,000 nautical miles and four to six months in probable traveling time, but it too was a well-known route. Hundreds of ships had charted its dangers, currents, and winds. It was the most disease-free way to California. In 1849 it allowed men to leave New England and New York and other Atlantic ports without delay. There were merchant ships loaded with cargo for California on which they could book passage, or sign on as crew members. Thousands of argonauts chose to organize stock companies and buy ships, sometimes anything that would float, for this way they could take larger amounts of baggage, clothing, provisions, and equipment with them. They mistakenly believed that they could sail directly to the gold fields, or sell their ships on arrival and thus recoup the expenses of their passage. Perhaps 20,000 men survived the long trip around Cape Horn; by July more than 600 ships had joined the ghost fleet in San Francisco Bay, many abandoned and left to decay, and some to become waterfront hotels and warehouses.

Via Panama

The route across the Isthmus of Panama was the third choice of the argonauts. It was in the province of New Granada, Colombia, and in 1846 a treaty had given the United States rights to utilize a trade route across Panama. It also was a well-known route. For more than two centuries the Spaniards had used it in transporting treasure, so that their galleons could avoid the dangerous passage around the Horn. It was easily accessible on the eastern side by ships from New Orleans or the Atlantic ports. Also, the U.S. Mail Steamship Company and the Pacific Mail Steamship Company had contracted to provide steamer service from New York to Panama to San Francisco. Geographically viewed, it offered potentially the most rapid transit to the gold fields and could be used all year around. Theoretically, it was a five to eight weeks' journey. The other side of the coin was that it was disease-ridden and subject to torrential rains up to 126 inches from May to December.

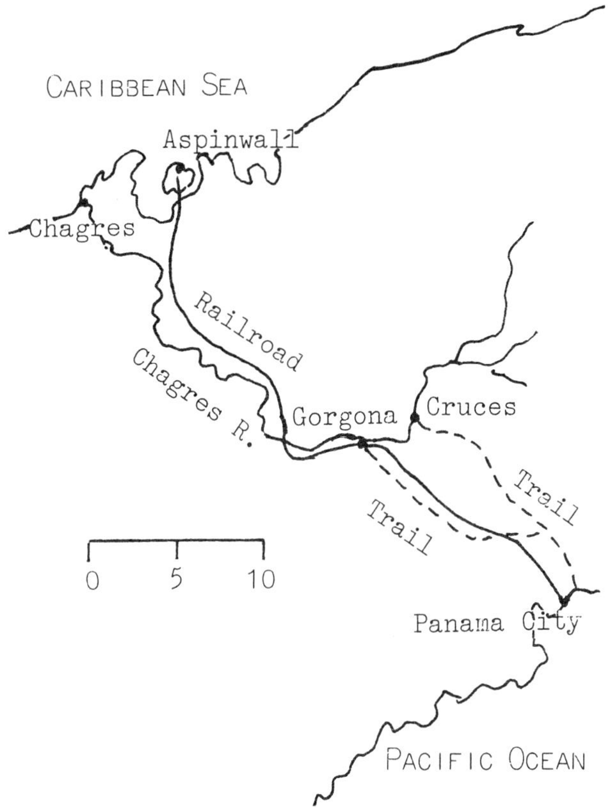

Map of Panama route.

Though the Isthmus is only 28 miles wide at its narrowest point, the terrain forced a wandering route of about 70 miles. It was the traveler's responsibility to bargain for passage, often with dishonest boatmen and baggage carriers. Literally dumped ashore on the eastern side at Chagres or Colon (Aspinwall), the argonauts would proceed up the meandering Chagres River in dugout boats to Gorgona or Cruces. From there they would bargain again for riding animals or pack mules to complete the 70-mile trip over a narrow, muddy track to Panama (City). There they would hope to find ship transportation to San Francisco, but instead they too often would find disappointment and frustration. If they had survived the yellow fever, blackwater fever, cholera, amoebic dysentery, and malaria, they now piled up by the hundreds scanning the horizon for non-existent ships.

When the first Pacific Mail steamer *California* arrived on 17 January 1849, there were 700 passengers trying to board a vessel with a rated 250 capacity. When the second steamer *Oregon* came in on 23 February, there were 1,200 clamoring claimants, of which she could take only 250. The *Panama* arrived on 18 May to pick up 209 of the 2,000 awaiting passengers. In desperation the argonauts, many of whom had already paid for through tickets, sought passage on any ships they could persuade to sail north. The coal collier *Philadelphia* which had unloaded coal from England was pressed into service. The brig *Belfast* took on 76 passengers. The whaler *Equator* emptied her tanks of sperm oil and turned north with passengers. The whaler *Niantic* carried 300 more. The *Humboldt* sailed with 300, the *Phoenix* with 60, the *Two Friends* with 164. Any ship available was utilized. These sailing vessels, because of prevailing winds and currents, often had to go out to the Hawaiian Islands before making landfall at San Francisco. The supposedly short trip sometimes took from three to five months, almost as long as the voyage around Cape Horn. Until the improvement of steamer service and the completion of the Panama Railroad in 1855, this route decidedly was a third choice. Nevertheless, perhaps 10,000 men survived the diseases and bandits of the Panama crossing in 1849.

The Gold Rush

The year of 1849 marked the beginning of the real Gold Rush when the "outsiders" began to arrive. By the end of 1849 the population of California had reached 100,000. Though the majority of these were Americans, there were also thousands who came from Europe, Latin America, Hawaii, the South Pacific, and Asia. Most of them proceeded immediately to the proven gold fields of the Mother Lode area, to widen the frantic search for placer deposits.

Moving up the two great waterways that drained the Sierras, the Sacramento River and the San Juaquin River, they fanned out along the rivers that emptied into the two waterways: the Feather, the Yuba, the Bear, the American, and the Consumnes in the north; the Mokelumne, the Calaveras, the Stanislaus, the Tuolumne, and the Merced in the south. Stopping at every sandbar and bend (there were 51 large bars of gravel on the Yuba River alone between Marysville and Downieville, each rich with placer gold), exploring every creek that was a tributary, following every canyon ravine, building dams to divert waterways in order to reach the streambeds beneath, they skimmed off the surface gold.

At first they panned, then the rocker and cradle came into use, then the long tom or extended sluice box. Later in 1852 hydraulic mining would be used first at Yankee Jims. Each in turn was an improved though still crude method of separating gold dust and flakes and nuggets from the dirt and gravel. They discovered that the gravel hills could be rich in gold, that under thick layers of earth

Scenes from a miner's life.

How the mines are worked.

debris and down at bedrock were the ancient beds of Tertiary rivers. In pairs and groups, they dug and wheelbarrowed and washed the tons of earth to recover the flakes and particles of precious metal. It was wet, unhealthful, and backbreaking work and they often sickened or died of land scurvy, cholera, malaria, dysentery, pneumonia, mine cave-ins, or were robbed and murdered for their "pokes of dust." They said rightly: "The finding of gold is easy; the mining of gold is mighty hard."

As one writer described it: "Gold mining meant swinging a pick, shovel or crowbar in the burning sun with temperatures up to 100 degrees. It meant digging through sand and gravel until one's back ached and his shoulders cried out in rebellion. It meant slogging about in cold muddy water until one's feet were numb, and his boots were a soggy useless wreck. One might have to spend hours kneeling or stooping over while he scraped out crevices, and anxiously washed out the dirt in his pan. Or he watched the daylight hours pass while he monotonously rocked the cradle to and fro and shoveled in dirt and poured in water."

Realistically, it must be recognized that the majority of argonauts simply did not "strike it rich." Mining was largely a matter of luck. The few lucky ones received wide publicity; the less fortunate dreamed of "next time" as they labored and ached and scratched in their flea-and-lice-infested clothes and blankets. Probably not more than one in ten was successful. After 1849 an average miner's "take" was six to ten dollars per day, if he was fortunate. With high prices for everything they purchased, most miners merely existed. Many went home in despair, many who remained took up their old trades or turned to farming, or became teamsters. A letter from a digging near Auburn expresses the honest feeling of many "There is, I judge, not more than 100 at work in this place who are averageing perhaps $10 per day. Prices high . . . I would gladly warn my brethern and friends against comeing to this place of Torment."

Other miners' letters are equally despairing, cynical and sardonic: "In regard to emigrating to California I say in one word a man is a fool." "No married man should ever come to California for this is a land of disappoint-

Primitive mining methods.

ment to thousands of men. Some go crasy, some take to drinking and gamboling and horing and so on . . . there are thousands who came in forty-nine cant get home and never will." "Men who have famileys and interests at home and wishing to come to this country and make a fortune and return home within a year, I say they had better cut there throats then theyr friends can attribute the act to insanity and honor them with a decent burial."

These, of course, were the voices of the nine out of ten who "mucked" for gold and did not find it. It must be remembered that half of those who came to California in the Gold Rush days never became miners. They settled in the fertile valleys between the coastal ranges where the earlier Franciscan missions had developed agriculture, and they too planted orchards and vineyards. They quickly found that the great interior valley was virgin topsoil for growing grain and produce. These immigrants sent for their families and friends to join them in a land which they described in Biblical language as a "place of milk and honey."

Nevertheless it was the gold-searchers who would symbolize the era and give it the romanticism of legend. The dreamers still dreamed even as the easy-to-get gold began to disappear. With a primitive geological understanding, they reasoned that all this placer gold must have eroded and washed down from some higher sources. They began a search of the higher Sierra foothills into the actual mountain chains, and found the quartz-imbedded gold ore veins. This led to hardrock mining, the digging and timbering of deep shafts and the use of blasting powder. Thousands of tons of hard rock must be crushed into powder and the gold separated from it. This required capitalization and practically ended individual effort.

In this brief, perhaps over-simplified, account an entire decade of the Gold Rush has been covered. The easier surface placer mining and the nuggets which could be dug out with a jackknife were a phase that had ended by 1853. Not completely, of course, for the millions of acres of gold-bearing soil simply were too extensive for even 100,000 prospectors to cover every foot of it. Occasional finds occurred for years and stimulated new rushes to new areas. The gold of the northwest along the Klamath, Salmon, Pit, Scott, and Trinity Rivers lay in almost impassable mountain ranges, and mining exploration there did not occur extensively until the placer deposits of the Mother Lode area had been exhausted.

The story of the Gold Rush is also the account of "men on the move." Men would find a rich digging and within hours or days hundreds of other men would be there staking out claims. A tent and shack community would

Cooking the evening meal.

come into being and a store-saloon would appear. Then word or rumor of a rich strike

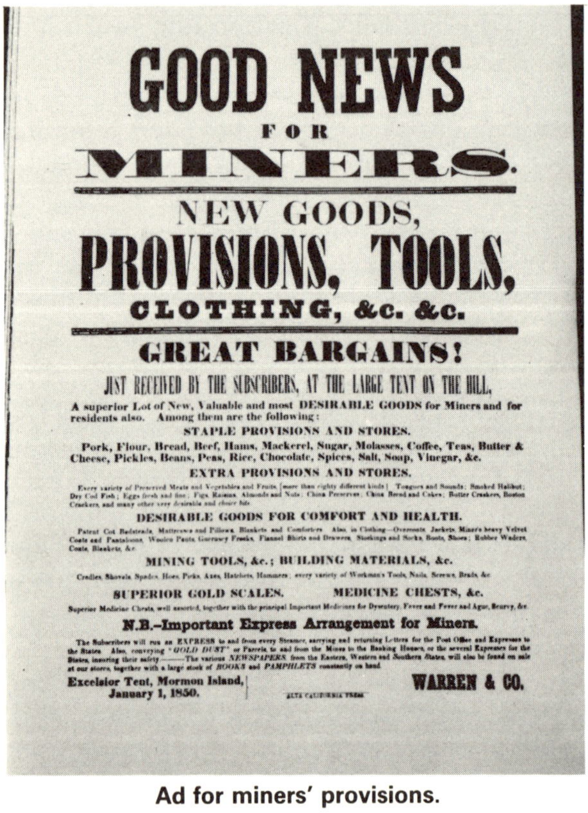

Ad for miners' provisions.

somewhere else would circulate and many of the men would take off for a new El Dorado. Often another wave of newcomer argonauts would arrive to work the abandoned claims, until they too and in turn followed Dame Rumor to new areas, to form new communities with exotic or comic or tragic names. The names of the mining camps reveal much about those who named them: Ophir, Bidwells Bar, Poverty Bar, Murderers Bar, Missouri Bar, Mormons Bar, Drunkards Bar, Dead Mans Bar, Rattlesnake Bar, Spanish Flats, French Corral, Angels Camp, Coyote Diggings, Hangtown, Whiskeytown, Yankee Jims, Mad Mule Gulch, Bottle Hill, Rough and Ready, Bed Bug, You Bet, Forlorn Hope, Loafers Hill, Poker Flat, Port Wine, Red Dog Camp — the list is almost endless as communities became identified with founders, events, hopes, and disappointments. Hundreds of these camps came into existence from the Kern River in the south to the Oregon border in the north. Some became villages that still exist, some became supply centers and developed into towns, many disappeared under weeds and brush.

Covers of towns with picturesque names: Whiskey Creek, Yankee Jims, Mountain Springs, Forlorn Hope, Rabbit Town, Pleasant Springs, Spanish Flat, Drytown, Angels Camp, Fiddletown.

More unusual town names: Goodyears Bar, Johnson's Ranch, Gibsonville, Chinese, Chinese Camp.

As mentioned earlier, the miners had to be fed, clothed, and occasionally entertained. They were too frantically involved in their digging to spare the days or weeks to go down to San Francisco, or even to Sacramento or Stockton, to "stock up on vittles and tools." Fortunately there were men who saw that more money could be made by mining the miners — men who were merchants at heart, or unfitted for the rigors of mining life, or disappointed miners themselves. They built the stores in the more permanent camps, organized the express companies that freighted the supplies and carried the mail to every camp and digging in the hundreds of miles of the gold areas, and carried back the gold dust and bullion for shipment east or for safekeeping in the larger cities.

The ranchers had a vast new market for their cattle, sheep, and hogs, and fresh meat brought incredibly high prices in the gold camps. Many persons quickly realized that a patch of garden vegetables could be more profitable than a mining claim. Wheat and barley were planted over thousands of acres, often by men who deserted the mines and reverted to their original vocation of farming. Lumber was at a premium in the mines and mining communities and scores of lumber sawmills were built, especially among the extensive redwood groves. By 1853 California was practically self-sufficient.

If there was any emotion which exceeded the greed of the miners, it was their homesickness. Most were young men away from their families and friends for the first time. They wanted to hear from their loved ones. Living squalid lives in the gold camps, they felt that everything good and clean and decent was back home. They wrote long and frequent letters, then anxiously waited for replies, and complained bitterly about the slowness of mail service. Interestingly, many of the miners were articulate men of a letterwriting, journal-keeping generation. Educated or semiliterate, these argonauts had

"Sunday in Mining Camp" from Nahl drawings.

carried writing paper, notebooks, and steel pens with them. And they used them! From a letter by William Miller come familiar words: "I have looked for letters until I got tired of asking the postmasters. This is the 15th letter I have rote to you since I left home and receive 4 letter."

Almost every historian quotes the words of Enos Christman describing the delivery of mail by an expressman in Sonora: "A huge fellow stepped up. A beard almost covered his face, and a large heavy revolver in his belt gave him the appearance of a ruffian without a tender spot in his composition. A handsomely enveloped letter was handed him. After weighing out his two dollars for it he stepped aside, broke open the seals and commenced reading it. In a few moments this same burly, stern looking man was in tears. He had heard from loved ones and memory carried him back to happier scenes . . . Another man was told that no letters had come for him. He raised his

Overcome by news from home.

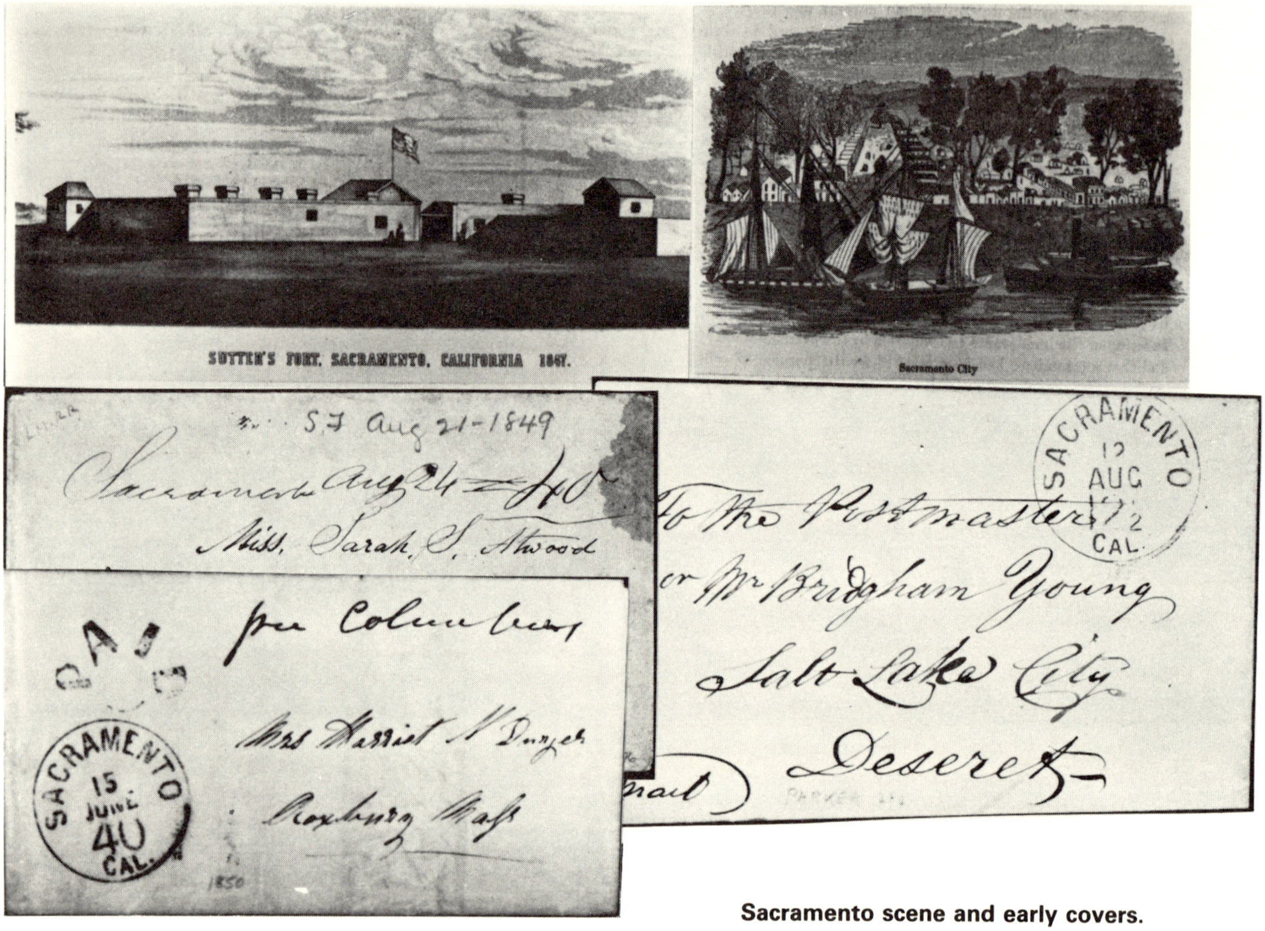

Sacramento scene and early covers.

voice in blasphemy and swore his friends had forgotten him and cared nothing about him."

The flood of argonauts continued through successive years. By January 1850 California had 107,000 inhabitants, compared with 20,000 in 1848. In 1852 a state-conducted census showed a population of 223,000. By the end of the decade, the first accurate federal census of 1860 placed California's population at 379,800. This census is particularly interesting because it contains a national or ethnic breakdown which reveals that 40 percent of the people in the state were not American-born. There were, in round figures, 4,000 Negroes, 35,000 Chinese, 33,000 Irish, 22,000 Germans, 16,000 British, 9,000 Latin Americans, 8,000 French, 5,000 Canadians, and 3,000 Italians. California was not only the most populous state west of the Mississippi River, but it also was the most cosmopolitan in the Union.

This increase in population was accompanied by a mushroom growth of towns in the interior valley. For example, in November 1848 Sacramento consisted of Sutters Fort and a surveyor laying out a town site. One year later the population of Sacramento was 12,000 people. With tens of thousands of miners in the foothills, certain communities because of their geographical locations became natural supply centers. The three base towns of Sacramento at the joining of the American and Sacramento Rivers, Marysville on the Feather River, and Stockton on the San Joaquin River, all connected to California's main point of entry, San Francisco, by river steamers. During high water large boats could reach Oroville on the Feather,

Stockton as it appeared in 1849; some early covers at 40¢ and 12½¢ rates.

Red Bluff on the Sacramento, and Fresno on the San Joaquin. These towns became the transshipment points between the steamers and the wagon train and muleback expresses which served the mining camps. Shasta, Yreka, and Crescent City developed to serve the far northern mines. Salinas, Weaverville, and Colusa were agricultural centers, providing food for the transient camps. Often overlooked by romantic writers is the fact that about one half of all the immigrants remained in the towns and cities, or rapidly returned there after viewing the hardships of mining.

Marysville ms. postmark, to Mass., 1850; early circle, 1851.

Marysville street scene.

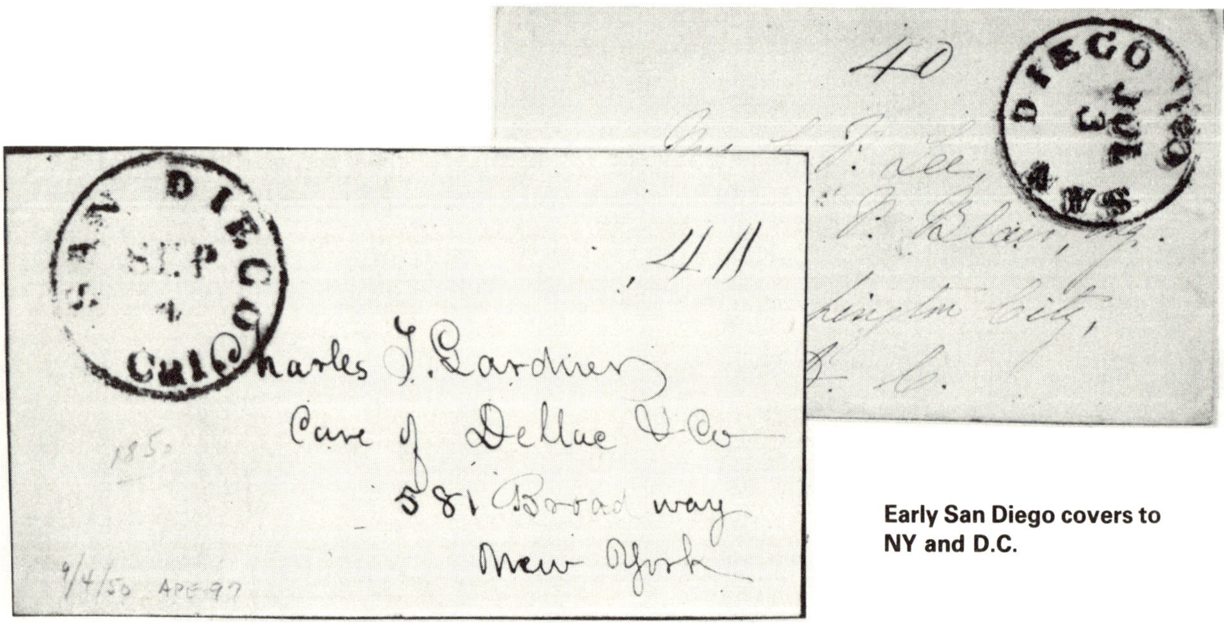

Early San Diego covers to NY and D.C.

Farther into the hills, there were hundreds of small towns along every river, creek, and ravine on the western watershed of the Sierras and within the Klamath Mountains. Some of them grew larger and became in turn supply centers: Nevada City, Coloma, Grass Valley, Placerville, Jackson, Mokelumne Hill, and Sonora. Also Downieville, whose development serves as a good example: Major William Downie found gold in the winter of 1849 at the junction of the North Fork of the Yuba River. By spring 1850 numerous shanties and cabins were there. In 1851 it was a town with stores, saloons, blacksmiths, a pharmacy, a bank, and a church. Almost every present county seat in northern and eastern California from the Oregon border to the Merced River began as a mining and supply center.

A succession of Downieville postal markings.

Covers from early mining and supply centers: Nevada City, ms. and circle; Culloma, Grass Valley, Placerville, Mokelumne Hill, rough circle and ms., Sonora ms. and straightline, Jackson on cover to England, overpaid 1¢.

Chapter 3

Establishment Of California Postal System

Two consecutive letters from the same correspondence to Vermont. Above: San Francisco straightline, July 2 (1849), on unpaid folded letter datelined June 28. Rated 40¢ due. Docketed "No. 8." Below: circular San Francisco postmark AUG 1 on double letter, due 80¢, datelined Sacramento River and docketed "No. 9." Carried by *California*.

Establishment of California Postal System

Both the academic historian and the postal historian have a common interest in the development of California's postal system. To the ordinary historian it is part of the drama of a restless people expanding the borders of a young nation. The postal historian is a specialist who interprets mail as a harbinger of culture and civilization. His interest in postal markings is like that of the archaeologist who scrutinizes every shard for seals and names and markings which allow him to date his artifacts and add knowledge to the story of human progress. (The particular and individual postal markings upon California mail will be noted in detail in the chapter on Early California Postal Markings.)

Early California Mail

The movement of mail to and from California began when Americans first became resident there from the 1830s onward. Before 1 July 1847 all United States mail was stampless, except for the semi-official stamps used in the New York City carrier service starting in 1842 and the comparatively rare use of locally printed postmaster provisional stamps, none of which were ever used in California. Most letters were simply sheets of paper, folded and sealed, with an outside area used for addressing. The postal markings on these folded letters, or occasionally handmade envelopes, either by manuscript nota-

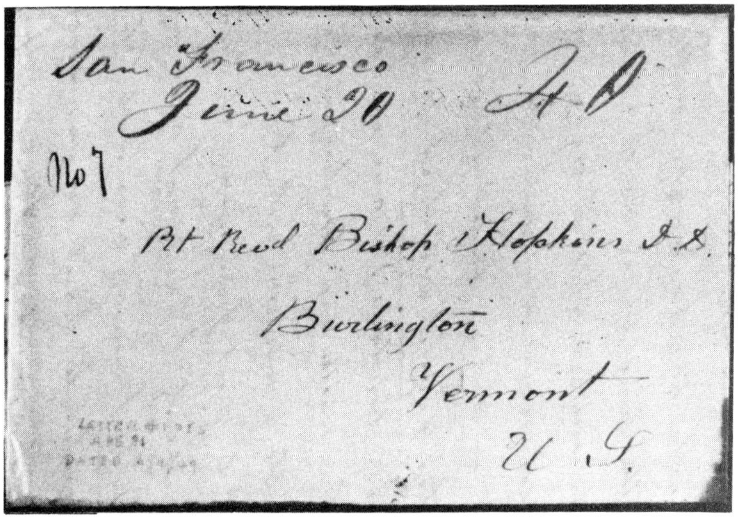

Folded letter with ms. San Francisco (June 20, 1849) to Vt. Unpaid rated 40¢ due.

Letter addressed to Brooklyn, N.Y. Received from private ship at SF Nov. 1 (1849) and there rated "Ship 42:" 40¢ transcontinental postage, 2¢ ship.

tion or by handstamp, indicated the town and state of origin, the day and month of entry into the postal system, the postage rate and whether or not the letter had been prepaid. These markings were required by the Postal Laws and Regulations from the beginning of the postal system under the Federal government.

Prior to February 1849 there was no government mail service in the far west. A letter writer made individual arrangements with a ship's officer to carry his letters. In the case of eastbound mail, the ship's officer was required to deliver these letters to the post office at his eastern port of entry, at which place the letters received their first postal markings. For this service the ship's officer was paid a ship letter fee of 2¢, provided the ship was American owned. Unless such mail had been handled through a forwarding agent, there was no way to prepay the U.S. postage, thus this mail was sent collect. If delivered locally at the port of entry, the ship letter rate was 6¢; if delivered beyond the port of entry, it was the inland postage rate from the port of entry to the letter's destination plus 2¢. On most westbound mail, postage to the port of departure was prepaid, and a ship's officer usually delivered a letter to a west coast custom's officer or a commercial agent to await pickup

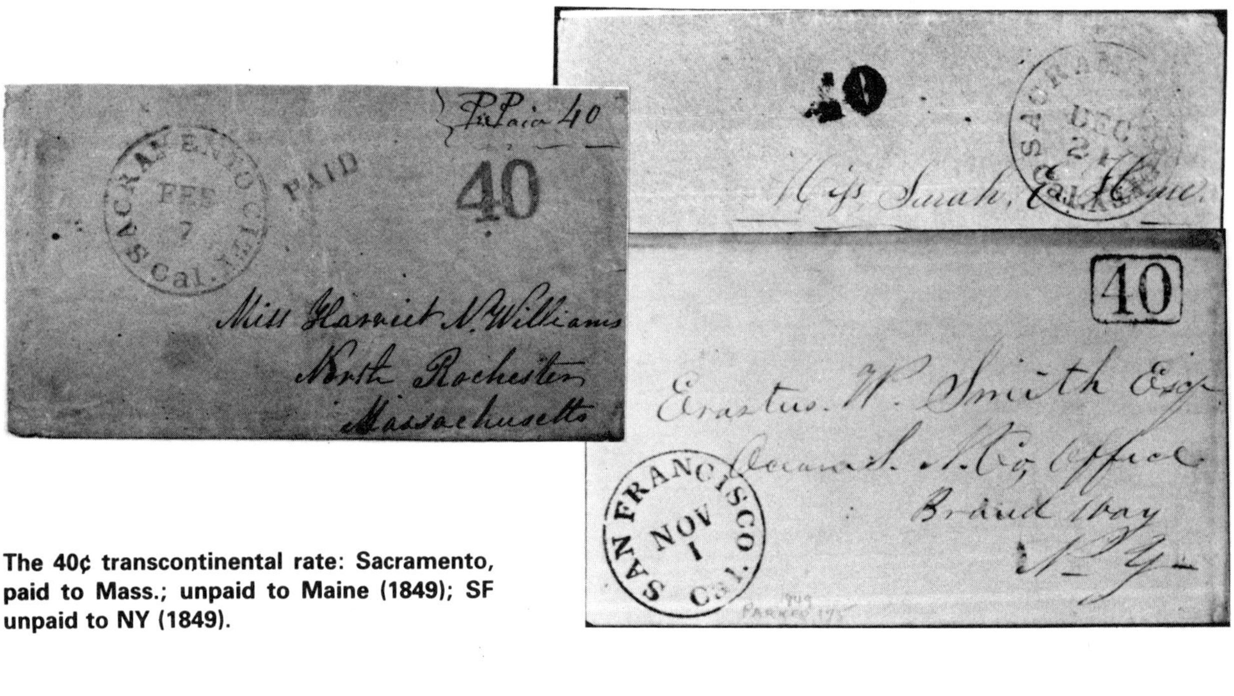

The 40¢ transcontinental rate: Sacramento, paid to Mass.; unpaid to Maine (1849); SF unpaid to NY (1849).

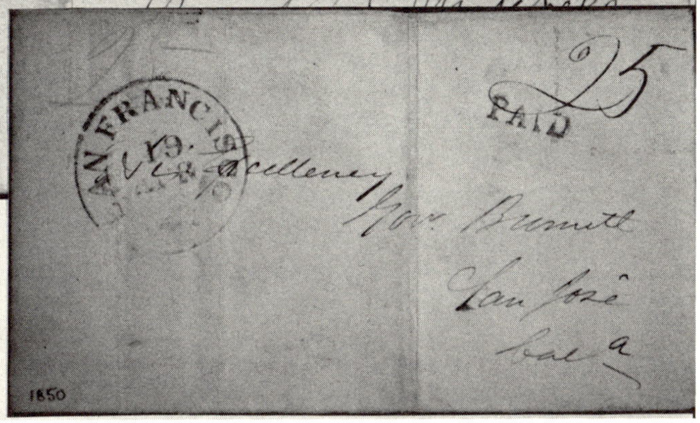

Examples of the 12½¢ intra-California rate: Sacramento to SF (1850), to San Jose (1851); Stockton to Sonora, oval handstamp (1851); Benicia to Sacramento (1850), all unpaid. Double rate SF to San Jose PAID 25 (1850).

by the addressee.

The postal Act of 3 March 1847 authorized the printing and use of the first U.S. postage stamps. These were issued on 1 July 1847 in denominations of 5¢ and 10¢. The same Act confronted the emerging problem of mail between the Eastern states and the West Coast. At the time this Act was passed, neither California nor the Southwest was part of the United States nor would they be until after the Treaty of Guadalupe Hidalgo ended the war with Mexico in February 1848. Consequently, the Act addressed itself primarily to the question of mail to Oregon, although it did also mention providing for mail at any other place on the Pacific Coast deemed expedient by the Postmaster General. It authorized the establishment of a post office

San Francisco drop letter (June 26, 1851) rated 2¢ due.

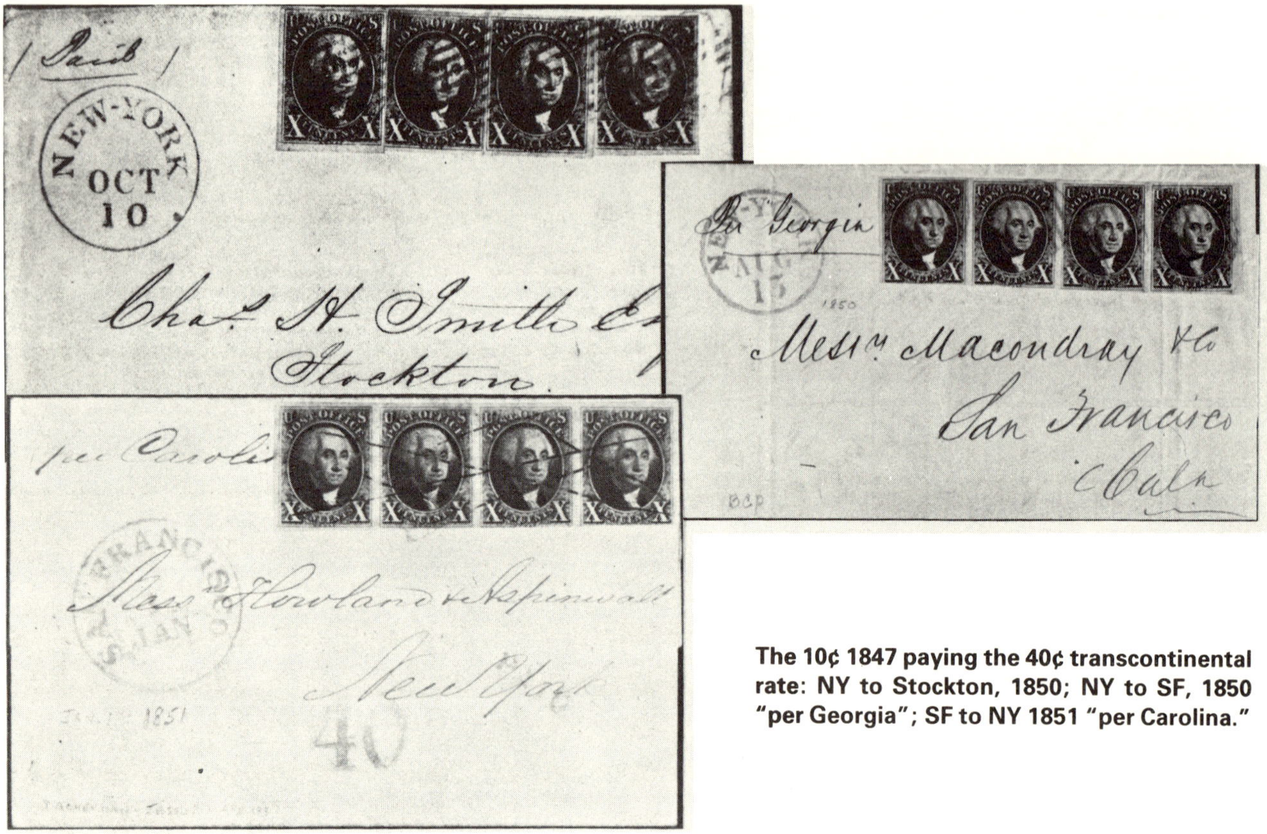

The 10¢ 1847 paying the 40¢ transcontinental rate: NY to Stockton, 1850; NY to SF, 1850 "per Georgia"; SF to NY 1851 "per Carolina."

at Astoria and set the rate of postage for single weight letters between the East Coast and Astoria or any other place on the Pacific Coast at 40¢.

Subsequent to the war with Mexico, the postal Act of 14 August 1848 authorized the establishment of post offices and the appointment of postmasters at San Diego, Monterey, San Francisco, and "other places on the Pacific Coast." It retained the 40¢ rate per half ounce between East and West, but decreed that mail between California towns and between California and Oregon would have a postage rate of 12½¢ per half ounce. Since there were no postage stamps in either of these denominations, and since correspondents were not required to use postage stamps on their letters, most mail to and from the West Coast was stampless and simply bore handstamped or manuscript notations to indicate the postage prepaid or due.

However, in a very few instances, multiples of the 10¢ 1847 stamp were used to make up the 40¢ rate. Since the Post Office Department did not supply the California post offices with postage stamps until late September 1851, after the 1847 issue had been demonetized, the stamps of the 1847 issue on eastbound covers had been hand-carried west by immigrants, sent in letters from the East, or purchased by argonauts at Panama City from the U.S. Post Office Agent there.

About 20 covers bearing 1847 stamps and

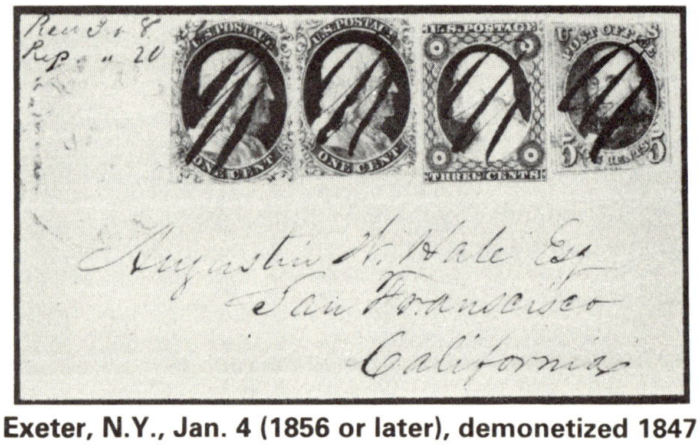

Exeter, N.Y., Jan. 4 (1856 or later), demonetized 1847 5¢ making up 10¢ rate to SF.

Strip of four of 10¢ 1847 cancelled by circled 80, carried by *Georgia* from NY to SF Jan. 28, 1850.

Four copies of 10¢ 1847 on folded letter from Phila. to SF Oct. 1, 1849.

Strip of four 10¢ 1847 from Baltimore to SF. Stamps cancelled by series of red 5s. Probably carried by *Cherokee* from NY July 8, 1850.

Pair of 5¢ 1847 tied by Tehama, Cal., May 17 postmark to envelope to Washington, D.C. Used long after demonetization, in 1855 or later, to pay 10¢ transcontinental rate.

Cover mailed Oct. 25, 1850, from NY to Culloma with pair of 5¢ 1847; also NY Oct. 26/40 postmark. At Culloma forwarded to SF at 12½¢. Stamps disregarded and total of 52½¢ collected from addressee.

associated with mails to or from California are recorded at the present time. The largest number (nearly half) show the 40¢ rate to California paid by four copies of the 10¢ stamp; in one case the double rate was paid by eight copies. The 5¢ stamp is quite rare on California mail. Only five examples are known; of these, three involve use after the 1847 issue was demonetized, in 1856 or later, the 5¢ stamp helping make up the 10¢ rate to or from the west coast.

Establishment of the Post Office in California

With the signing of the Treaty of Guadalupe Hidalgo in February 1848, California had become a possession of the United States and its inhabitants were American citizens. As such, they were entitled to the privileges and conveniences which the United States government provided for the general welfare of its constituents. This included a govern-

Remarkable cover from Austin, Tex., June 7, 1851, to Benicia with four copies of 1847 10¢ for rate to west coast. Before the letter reached California, rates had changed. Forwarded from Benicia Aug. 5, due 5¢.

ment-operated postal system. The United States, from its beginning, had accepted the obligation of postal service as part of the progress of civilization. By the start of 1849 there were about 17,000 post offices in the United States east of the Mississippi River. Yet there were no officially established offices in California.

On 7 October 1848 the Secretary of State informed William V. Van Voorhies in Washington that it was the intention of the President to send him to California to "make arrangements for the establishment of postal routes, for the transmission, receipt and conveyance of letters in California and Oregon . . . and the Congress is not unmindful of their interests . . . for California's inhabitants are entitled to all the blessings and benefits resulting from the best form of civil government ever established amongst men." Three weeks later, on 1 November 1848, Van Voorhies received an official letter from Postmaster General Dave Johnson notifying him of his appointment as a Special Postal Agent for California and Oregon. His instructions were such as to create the almost insurmountable problems that would bring the postal system into disrepute in California:

> . . . your first duty will be to proceed to the selection of suitable persons for postmasters at San Diego, San Pedro, Santa Barbara, San Luis Obispo, Monterey, and at such other points on the Pacific, at which United States steam mail packets shall touch . . . you will forward his bond . . . you will furnish him with proper blanks for post bills, quarterly returns . . . as no route into the interior of California has yet been created by Act of Congress, you will have to make the supply of each office situated in the interior conditioned upon the expense thereof being defrayed out of the net proceeds of such office. . . . At present no more can be contemplated than semi-monthly transportation by the cheapest mode of conveyance. . . . Horseback conveyance will vary from three to six dollars per mile per annum . . . the distance is to be counted but one way. . . . William Nelson, United States consul at Panama, will be the mail agent of the United States for the Pacific mail. You will promptly advise him . . . of every post office put into operation so that he may properly bag the mails for those places.

The letter contained further detailed instructions concerning mail bags and metal locks, forms and reports required, and a reiterated insistence upon economy of operation.

Van Voorhies arrived in San Francisco on 28 February 1849, on the maiden voyage of the steamer *California.* Two weeks later he wrote to the Postmaster General an accurate but despairing report on actual conditions in California which made the requirements of his office almost impossible to perform.

He stated that the *California* had not stopped in San Diego or Santa Barbara, therefore the mail for those places was still in his possession. He noted that postmasters would be difficult to hire, for men in public offices were quitting their jobs to take new ones at

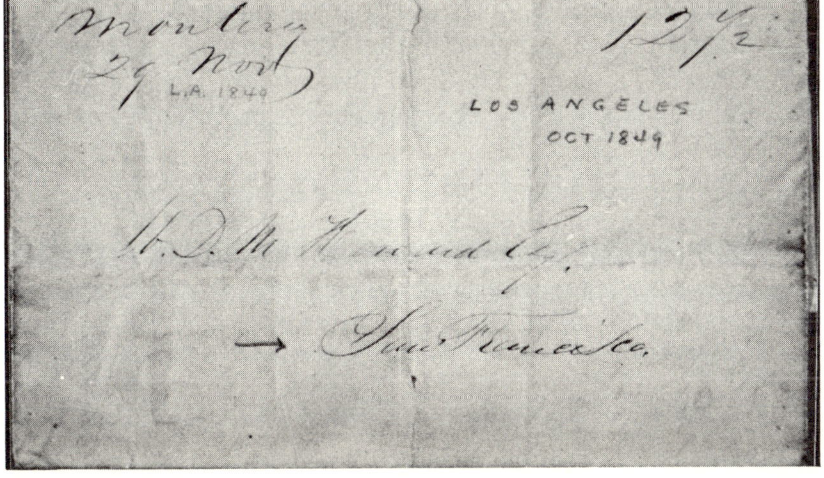

Ms. postmark of Monterey (1849). This is one of the offices Van Voorhies was instructed to open.

five times the compensation they could expect as a postmaster. "Eight to ten dollars per day are demanded and received for the most common services, and even these sums are refused during the mining season . . . teamsters may command from $200 to $400 per month . . . very ordinary horses are worth from $200 to $300. To contract with a Californian, therefore, to convey the mails, the department may well calculate a heavy *debit balance* over and above the proceeds of the post offices. . . . The compensation afforded postmasters in the States will be found wholly inadequate here, if the office is conducted separately from other business. . . ." He declared that merchants in some towns might be willing to act as postmasters only in order to attract customers to their stores. He enumerated the exorbitant costs of lodging, food, fuel, office rent, and furniture. He mentioned that he had persuaded a local merchant, Mr. C. L. Ross, to give him a small room in his store in order to open the mail bags from the *California,* and to make letters available for distribution. He suggested that all mail for California be sent to San Francisco since most of the recipients were there or in "parts adjacent . . . I venture to say that fully two-thirds of the citizens and residents of California are in the neighborhood of the mines." He reiterated that the emoluments of a postmaster were not sufficient "to induce the acceptance of the office by anybody unconnected with other business . . . unless some more increased compensation be in some way provided." Van Voorhies's reports were not received in Washington until 21 July 1849. By that time he no longer was the Postal Agent. A new Postmaster General had been appointed, and he had relieved Van Voorhies and had chosen a new Agent, R. T. P. Allen.

The Postmaster General's instructions to Colonel Allen were a repetition of those given earlier to Van Voorhies. Allen sailed from New York on 19 April on the *Falcon.* When he arrived in Panama City on 17 May he sent an interim report to the Postmaster General. He stated that the mail steamers were handling the mails improperly, that their primary interest was the lucrative passenger traffic, and that the Panamanian officials were unhappy with the payments made to them for the transmission of mail and express across the Isthmus.

The store in which the first San Francisco post office was established.

Allen arrived in San Francisco on 13 June 1849 and, ten days later, his second report reiterated many of the warnings implicit in the earlier reports of Van Voorhies. He explained again the high cost of living in California and the exorbitant wages demanded by workers. He stated once more that postage income could not equal postal expenditures. He estimated that the mail steamers would bring in at least 10,000 letters per month, a gross underestimation, and pleaded for authority to purchase a post office building. Allen also advised that he had, by letter correspondence, appointed three postmasters in Oregon.

Allen's third report, on 2 August, noted that he had been "out in the field," that the last mail steamer had brought in 18,000 letters, most of which were still in San Francisco, since almost no one was willing to be a postmaster in an outlying office in view of the paltry income.

On 29 August, Allen happily reported that he had established post offices in Benicia,

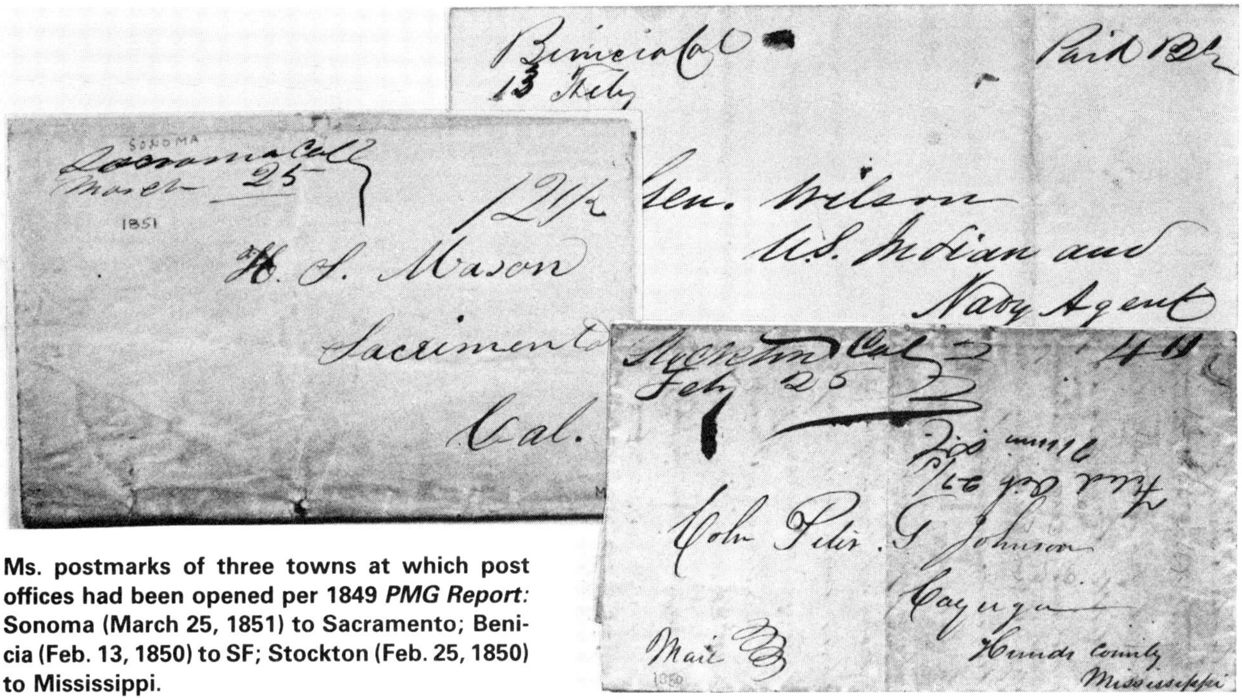

Ms. postmarks of three towns at which post offices had been opened per 1849 *PMG Report*: Sonoma (March 25, 1851) to Sacramento; Benicia (Feb. 13, 1850) to SF; Stockton (Feb. 25, 1850) to Mississippi.

Sacramento, Stockton, San Jose, Vernon, Culloma, and Sonoma, to be supplied weekly by water and by horseback. Then he unhappily added that his recommendations to improve the Panama transit had been ignored and the situation there had worsened, and that he still was having difficulty in finding postmasters and establishing mail routes because the pay was too small.

For his accuracy and honesty, Allen received a rebuke from the Postmaster General. In a letter of 10 December 1849, Washington reminded him that his failure to appoint postmasters in the coastal towns was a disobedience of orders, that his attempts to serve the public by making unofficial and temporary arrangements to distribute mail were illegal, that his authorizations for payments which exceeded postal income would not be honored, and that he had better get his "paperwork in triplicate" straightened out immediately.

From the viewpoint of the Postmaster General the rebuke was justified. Under the law Congress, not the Post Office Department, established postal routes and the only exception permitted was where a non-Congressionally established route generated revenues equalling its expense. No inland routes within California had been approved by Congress. The Postmaster General and the Post Office Department auditor were deservedly concerned when they discovered that the Special Agent was establishing post offices and postal routes at an expense that could not possibly be covered by the financial receipts of those routes. The emphasis upon setting up post offices in coastal towns was simply that these would be serviced by the ocean mail steamers at little expense to the Post Office Department, since the Navy Department was paying the cost of operating the steamers.

It almost seemed a comic opera of confusion, but no one in California was in the mood to laugh. Ten months had gone by and only nine post offices existed. Almost 100,000 people had arrived, each eager to hear from home, hungry for news from "civilization." Each mail steamer brought in thousands of letters, and they piled up in San Francisco, for it was not until 29 June 1850 that Sacramento became the second point of mail distribution to other California post offices. During 1850 only 13 more post offices were established, and the population had almost doubled again.

Crowds at San Francisco post office.

The only happy note was that the San Francisco post office finally had a building of its own.

Much has been written about the failure of the early California postal system to serve the needs of the ever increasing population in California. It is a story replete with emotional criticisms and complaints, too often accepted at face value by early postal historians. Yet, in fairness, it must be recognized that the early problems arose as much out of Congressional unwillingness to authorize the necessary money to improve the situation as it did from the bureaucratic slowness of the Post Office Department to meet an unexpected problem. Congress made the rules; the Post Office Department had to obey them.

Washington was more than 3,000 physical miles away from California, and was light-years away from understanding the unique postal problems of a California without roads or good transportation facilities, and which had experienced a totally unexpected and unprecedented population growth resulting from the Gold Rush. Existing, as Washington did, in an eastern environment in which wages and prices were relatively stable, and where post offices were served by a well-established network of postal routes on good roads connecting established towns, Congress was slow to comprehend the economic and transportation problems that were the hallmark of California's Gold Rush period. It was reluctant to authorize the deficiency funding which the Post Office Department needed to solve its western economic problems.

Despite the angry assertions of Californians, the Postmaster General did not ignore the reports from his special postal agents in the west. On 3 December 1849 the Postmaster General reported to Congress:

> Under contract with the Navy Department, steamships are now carrying mail on the Pacific between Panama and California once each month. By treaty the government of New Grenada is bound to transport the mail across the isthmus. . . . This service is being tardily and carelessly performed and the compensation under the treaty is probably inadequate. The amount required by the treaty to be paid to New Grenada for carrying that mail, being in no other way provided for by Congress, has been paid by this department up to this time. . . . Congress, on 14 August 1848, authorized the Postmaster General to appoint postmasters in California, and to appoint agents for making arrangements for the establishment of post offices and conveyance of mails in California and Oregon. . . all reasonable exertions are being

made to give the people there all the mail facilities the limited means legally applicable will afford. The laws regulating the post office duties and service are in many respects ill adapted to the circumstances and condition of that country. . . . The price of labor or personal service and the rent required for office rooms are such that no allowance now authorized by law can secure a post office in California. Nor can the mails be transported within the country for any compensation which the postage received there will furnish . . . provision must be made beyond the yield of postage there to meet the cost of even a very limited supply of mail accommodation in that country. This subject requires the early attention of Congress.

To reinforce his pleas that Congress should be more generous in its appropriations to meet these emergency conditions, in the same report the Postmaster General submitted statistics: "The California mails from New York alone in October numbered 13,362 letters and 15,571 newspapers. The return mails to New York fall but little short of this average number. Besides those from New York, separate mails are despatched to San Francisco from Charleston, Savannah and New Orleans." He then referred to the California Special Postal Agent's establishment of post offices in the interior at Benicia, Sonoma, Vernon, Sacramento, Culloma, Stockton, and San Jose, and his plans to immediately open post offices in Santa Barbara, Los Angeles, and San Diego. "He conjectures the service will cost from eighty to one hundred thousand dollars per annum. . . . Where prices are carried so extravagantly beyond all standards that elsewhere prevail, much difficulty must be experienced . . . as the law has subjected them to specific limitations. . . . The mail expenditures in California will exceed the post office revenues, and it is apparent that the action of Congress in modifying the existing laws in their application to Cal-

Covers with ms. postmarks from some of the additional offices established in 1850: San Diego to NY "Paid 40"; Napa to Benicia in 1853, unpaid double letter at 10¢; Nevada City and Placerville, both to Iowa due 40¢; Vernon (Nov. 1849) to Ohio, due 40.

ifornia service will be necessary."

In November 1850 the Postmaster General again reported to Congress: "The mail service in California and Oregon, and especially in the former, is still in an unsettled, crude and unadjusted state . . . not only very expensive, but somewhat uncertain and inefficient."

But Congress, long before, had determined that the postal service should be self-supporting, financed by its own generated revenues, and not subsidized as a public service. It did not regard mail service as similar to expenditures for national defense, where the public welfare justified deficit spending. To the pleas of the western pioneers that the postal system costs should be assumed by a nation eager to bind together a migratory people into a homogenous unit, it seemed to turn a deaf ear. It was not until 3 March 1851 that Congress authorized $66,000 for incidental expenses to postmasters in California in addition to the commissions allowed by law. Even this sum, granted only after lengthy debate, was woefully inadequate to meet the ever increasing needs of California's postal system. The result was continued poor internal postal service within the state. The problem was resolved only by local solutions which, during the critical years, kept communications open by utilizing private enterprises which substituted for a proper postal system, namely, the private expresses.

In his 1853 report to Congress the Postmaster General took cognizance again of the situation: "Our local mail service in California is strongly marked by two peculiarities: very high prices and great difficulty in giving the people the mail facilities they deserve." He noted that the cost of inland mail transportation in California was at least four times greater than in other parts of the Union, and that the sea and isthmian transportation costs far exceeded the postage revenue. Referring particularly to the expresses, the Postmaster General stated:

> And notwithstanding the government incurs a heavy unrefunded expenditure in supplying the citizen of that State with his letters, newspapers, and other mail matter, the citizen himself employs other facilities for the conveyance of his letters, and pays therefor at a rate from twenty to forty-fold greater than the government charges for similar service.
>
> The facilities referred to are the expresses. Government has not been able to keep pace in its organized mail arrangements with the movements of the miner. His settlement is suddenly made and rapidly extended long before the mail contractor and postmaster can be provided. The expressmen are at hand, and the wants of the miner are immediate. His orders are all sent by them to the central post office, to which, not knowing where his eventual location will be in the gold region, he has directed his letters to be addressed. The postmaster has favored the expressman in the prompt delivery of his letters, and has received in return a compensation much larger, per letter, than any commission ever allowed him by the government. The express charge is one dollar per letter, two dollars, and so on, rising in some instances to five dollars, according to distance. Once commenced, the habit of the miner of relying on the expresses is continued long after the post office and the mail-route have reached his neighborhood.

In 1860, in his *Annual Report* to the President, the Postmaster General noted that under the terms of a postal Act of 15 June 1860 Congress had limited the compensation (mail subsidy) for ocean service between the Atlantic and Pacific coasts to the amount of postage received on the mails conveyed. Since on 17 December 1859 the Post Office Department had been directed to send letter mail by overland routes, the steamer lines had simply refused to accept such a conveyance contract, and only the assurance of the President of the United States that he would urge Congress to modify the 15 June Act persuaded the steamer lines to temporarily continue their mail service. "I shall not regard the government as honorably acquitted of its obligation to the mail contractor without a full redemption of the pledge implied in the assurance which the President, from high considerations of the

Lines awaiting mail.

public interest, felt justified in giving." Referring again to the overland routes, the Postmaster General noted that such routes were established and maintained mainly for the advancement of certain national purposes not at all postal in their character. Noting also that such overland routes could not provide good mail service because of terrain and weather, mail subsidies would prove beneficial only to the carriers: "If their importunities and the importunities of their friends can not be withstood by the government, it would be far better that they and their descendants ... should be pensioned from the treasury, than that this unparalleled waste of the public money should be allowed."

Unfortunately, most Californians did not read the Postmaster General's reports, and they regarded the Post Office Department as a willing culprit for the unsatisfactory mail service they were receiving from the government. They still remembered as typical the scene described by Bayard Taylor years earlier: "Three steamers had come in with passengers but no mail. Finally on October 31, 1849 the *Panama* came in ... with 45,000 letters besides uncounted bushels of newspapers. ... The Post Office was a small frame building of one story, and not more than forty feet in length."

Taylor volunteered to help the mail clerks sort and prepare the mail for distribution: "The doors were instantly closed ... preparations were made for a long siege. The attack from without commenced ... knocks on the doors, taps on the windows ... beseeching calls ... they shouted, implored and threatened by turn ... offering large bribes." The clerks worked for 44 hours, then opened the windows for a partial distribution. The miners lined up on a "first come, first served" basis in lines extending for blocks, taking six hours for the man at the end of the line to reach a window — often to discover that he had no letter unless it was still in the unsorted bags. This went on for a week, men returned day after day to stand in line again, often paying $25 for a place in line nearer to a window. Then there were still thousands of letters for men who could not be in San Fran-

Elmira, N.Y., to Nevada City, June 11, 1851, 40¢ due. Rates changed before letter arrived. Forwarded to Sacramento: " Ford 5." At Sacramento "Adv'd" for 1¢. Total 46¢ due from addressee.

cisco to claim their mail.

And the advice of a letter writer in San Jose to a friend in San Francisco in 1855 was all too accurate: "Don't send me any more mail through the post office, for it takes a week to get here. Give it to Wells Fargo and I'll get it in nine hours."

Naturally Californians would not accept such conditions without voluble protest and vehement criticism. The *Alta California* newspaper on 23 November 1851 noted the low opinion of the Post Office Department held by the citizens who felt they were paying for a service not being rendered. Two years later, on 2 August 1853, the same newspaper quoted the feeling "we might as well be without mails and post offices, as for any good they do us." This cynical yet despairing attitude did not improve for a decade. Since much internal mail was being carried by expresses and stage lines, many citizens regarded it as unfair taxation to have to pay postage fees as well as express charges. There seemed to be little objection to paying express companies; the resentment was directed toward the Post Office Department in Washington.

Postal Acts from 1851

The postal Act of 3 March 1851 lowered the postage rate to 3¢ prepaid and 5¢ collect per ½ ounce for a distance of less than 3,000 miles, and to 6¢ prepaid and 10¢ collect for

Stampless covers showing the intrastate (under 3,000 miles) rates effective July 1, 1851: Chinese PAID 3 circle; Sonora unpaid with negative 5, and 5 in circle, both to Vallejo.

Stampless covers at 6¢ paid, 10¢ unpaid transcontinental rate: clockwise from lower left: Nevada City 3 PAID and PAID/3 to show 6¢; ms. "Paid 6" of Mok Hill; PAID/6 in circle from Georgetown; SF curved PAID/6; two varieties from Sacramento; first day of unpaid 10¢ rate from SF.

The 6¢ and 3¢ rates paid by stamp: double rates from SF to Washington, D.C., and Mass. paid by 1851 3¢ block of four and 12¢ respectively; Drytown to Marysville March 29, 1856, with 3¢ 1851.

over 3,000 miles, effective 30 June 1851. A drop rate of 1¢ was established. The Act also directed the issuance of a 3¢ stamp by the Postmaster General and "such other denominations as he may think expedient to facilitate the prepayment of the postage provided for in this Act." Under this authority the Postmaster General ordered the printing of 1¢, 3¢ and 12¢ stamps. In a notice dated 11 June 1851, he declared the use of the 1847 5¢ and 10¢ stamps to be invalid after June 30.

In an Act of 31 August 1852, Congress nullified the Act of 1845 which had expressly prohibited private expresses from carrying mailable matter over roads that had been designated as postal routes. The Postmaster General admitted awareness that this illegal conveyance by expresses, railroads, and ships was common, and noted that Congress had now legitimized the practice as long as such mail was in stamped envelopes. Therefore, the Postmaster General was authorized to provide all postmasters with "suitable letter envelopes with suitable postage stamps of such denominations as he may direct, printed or impressed thereon; such letters may be sent, conveyed and delivered otherwise than by post or mail. . . . As Congress has now provided convenient means for the lawful conveyance of letters both in and out of the mails, the law authorizing search for and seizure of all letters illegally carried will be rigidly enforced." In effect, this legalized California's already common practice, but also required its expresses to carry only prepaid mail. (Extant California covers reveal that the postage requirement was often ignored until unstamped letters were expressly forbidden on 1 January 1856.) Although stamped envelopes were authorized on 31 August 1852, they were not available until 1 July 1853. The government contract for their manufacture was held by George Nesbitt and Company from 1853 to 1870, thus collectors often refer to these covers as "Nesbitt entires." They were sold to the public for their postage value plus a small charge to cover the cost of manufacturing.

Important changes occurred with the Act of March 1855 (effective 1 April), which set the postal rate at 3¢ per ½ ounce for less than 3,000 miles and 10¢ for any greater distance such as to and from the West Coast. It prohibited collect mail after 1 April 1855 and required that all domestic mail be prepaid with affixed postage stamps or be placed in

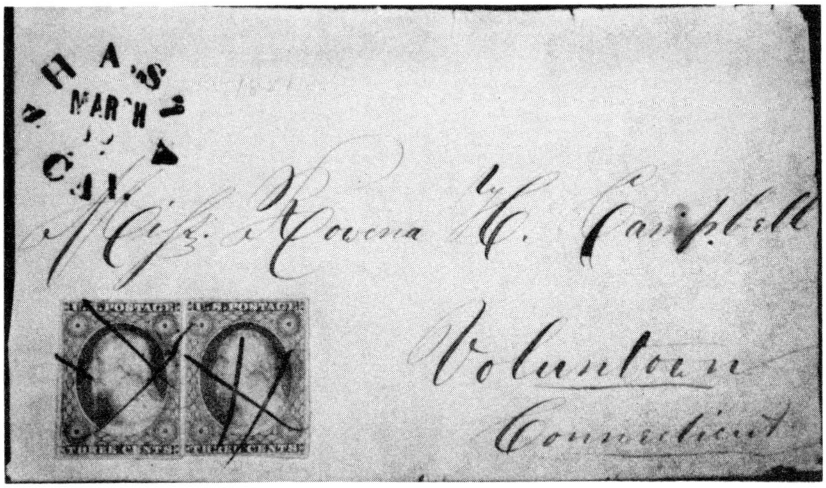

Shasta rimless circle on cover to Conn.

Marysville to Saratoga, N.Y., with segmented cork cancel.

Sonora to Mich., top margin pair cancelled by grid of squares.

The 6¢ transcontinental rate paid by a pair of 3¢ 1851.

Nesbitt entires: 3¢ Stockton to Sonora, "Fowd 3" to Columbia (1855 or later); Angels, 3¢ to Consumnes; Alpha, 10¢ to Phila., Aug. 2, 1858; Nevada City on 6¢ to Conn. (1855 or earlier).

stamped envelopes after 1 January 1856. Stampless covers after this date originated only in small post offices where postage stamps were temporarily unavailable or on mail to foreign destinations. More important to Californians, this Act again explicitly prohibited express companies and stage lines from carrying any mail unless the correct postage was affixed in United States stamps, or stamped envelopes were used.

This latter provision of the 1855 Act incensed California. The postal system in the state was still operating ineffectually and the express companies were delivering most of the internal mail. Witness an editorial which appeared in the *Alta California* on 17 April 1855:

> Mails and Expresses
>
> There is serious complaint made about the new Post Office arrangements, by which a postage of ten cents is required to be paid on each letter, whether carried by mail or by express. To us this appears to be one of the most outrageous impositions ever attempted by the Federal Government on California. By what right does the Postmaster require all express letters to be put in Government envelopes! Expresses may take letters from any foreign port to any port in the United States, and what right the government has to interfere to prevent them from taking letters from here to Panama, or from Aspinwall to New York, is more than we can perceive. It is clearly an imposition, and we ought not to submit to it. It is bad enough for us to be compelled to pay the highest postage, without being subjected to an illegal tax on matter with which the government has not, and ought not to have, anything to do. The mail service, anyway, is of doubtful use to California, and when superadded to that is a positive disability on individual enterprise which is not legal, and is made discriminative against California alone, it is time for a protest to be entered. If people are better served by expresses than by the government, it is surely unjust that they should be taxed for availing themselves of the best facilities. The injustice is too glaring to be submitted to without complaint and we trust that an earnest expression of the public sentiment in the matter may result in its speedy discontinuance.

The 3,000 mile distinction between postage rates caused some confusion among California postmasters. As long as the ocean mail

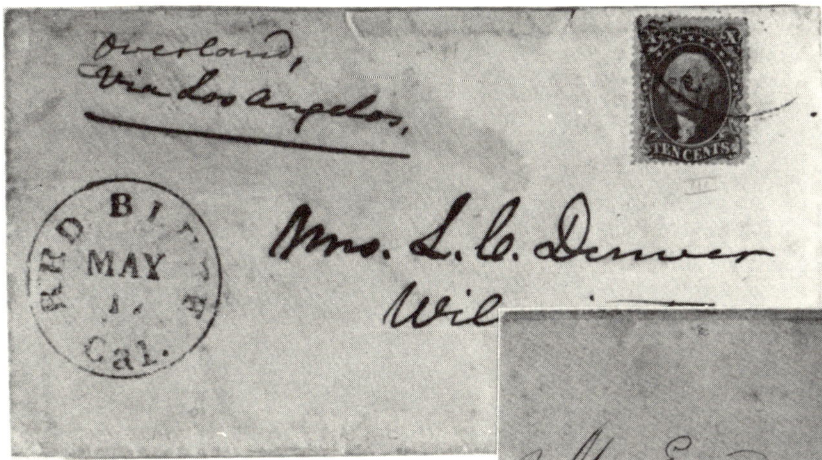

Red Bluff to Ohio, by overland mail.

Horse Town to Iowa.

Michigan Bluff on ladies' fancy envelope to Mass.

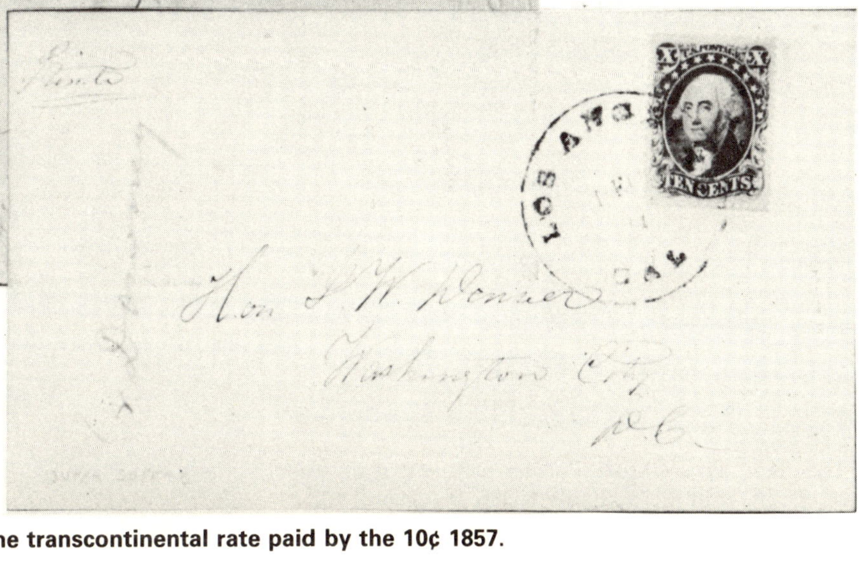

Los Angeles to Washington, D.C.

The transcontinental rate paid by the 10¢ 1857.

SF to Washington, May 16 1855. Entire 1851 issue amounting to 22¢; apparent overpay of 20¢ double rate.

route was used, with its eastern terminus in New York City and its western terminus in San Francisco, it was clear that this distance exceeded 3,000 miles, and the correct postage was 40¢ single rate until 1 July 1851, then 10¢ collect and 6¢ prepaid until 1 April 1855. But when the overland routes were established, no one seemed sure exactly which eastern towns were within the 3,000 mile zone and which beyond it.

The postal Act of February 1861 solved this problem. It stated that the 10¢ prepaid rate applied to "all letters conveyed in the mail from any point in the United States east of the Rocky Mountains to any state or territory on the Pacific, and from any state or territory on the Pacific to any point in the United States east of the Rocky Mountains." (This "over the mountains" rule was protested by Brigham Young in Salt Lake. The Mormons maintained, and were upheld, that while they were not east of the Rockies, neither were they on the Pacific, so mail to and from Salt Lake continued to be carried at the 3¢ rate.)

The next important action of the postal

Stampless covers from Eureka Bay and Auburn paid at 10¢ rate (presumably 1855); the 3¢ and 10¢ rates from Mendocino and El Dorado paid by 1857 issue stamps.

Stampless covers with the 10¢ transcontinental rate prepaid. Colusa PAID/10 in circle.

Handdrawn patriotic, PAID and circled 10 from Sonora.

Stockton 10 Paid postmark to Mass.

Mokelumne Hill PAID/10 to Illinois.

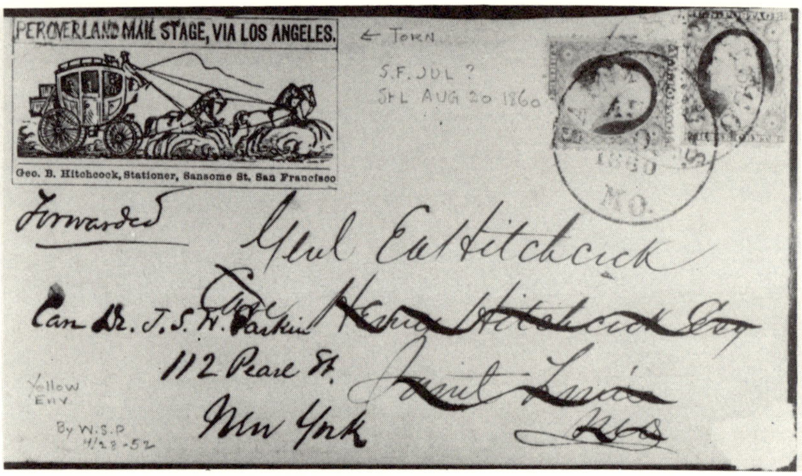

San Francisco to St. Louis, July 1860, 3¢ 1857 paying "over mountains" rate (under 3,000 miles). Forwarded from St. Louis Aug. 20, 1860, to NY with second 3¢ 1857. Total of 6¢ from coast to coast legally.

service was the invalidation and demonetization of the 1851-57 issue of postage stamps. On 20 April 1861 the hostilities that began the Civil War occurred; prior to this date nine southern states had seceded, beginning 20 Dec. 1860 — the secession of the others followed. Since postage stamps could be used as a form of currency as well as for postage purposes, and since large amounts of U.S. stamps were in the possession of postmasters in states which had joined the Confederacy and for which no value could be recovered, it was necessary to void the existing issue.

A contract was given to the National Bank Note Company of New York to print new stamps in a changed design, and to modify the design of the stamp upon stamped envelopes. Because of printing problems, the new stamps and envelopes were not available until the middle of August 1861. Upon receipt of the new postage, postmasters were directed to advertise that for a period of six days they would exchange the new stamps for the old ones, and after the exchange had taken place, use of the old stamps was invalid.

Unfortunately, the distribution of the new stamps was delayed in many cities, and existing covers reveal the confusion that ensued. It was impossible to get these new stamps immediately to the post offices in the far west, so the original plan permitted acceptance of letters mailed at post offices in California and other western states and territories until 1 November 1861. This date was later extended to 1 January 1862. In the meantime, however, irregular distribution

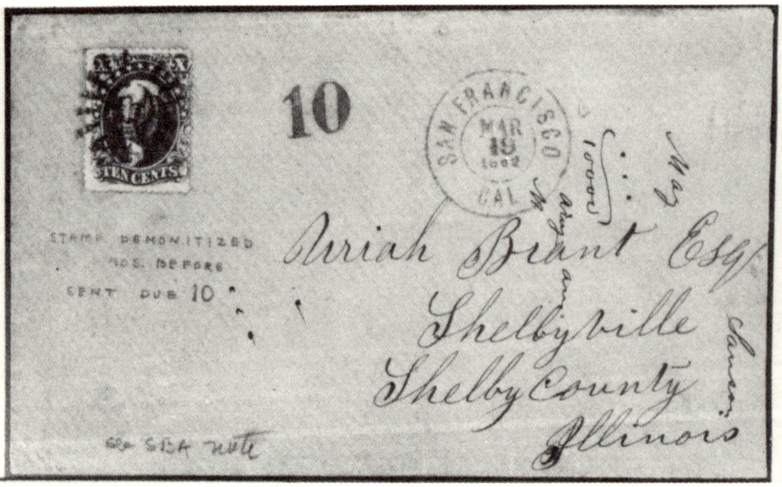

Unsuccessful attempt to use demonetized 10¢ 1857 from SF March 19, 1862. Rated 10 due.

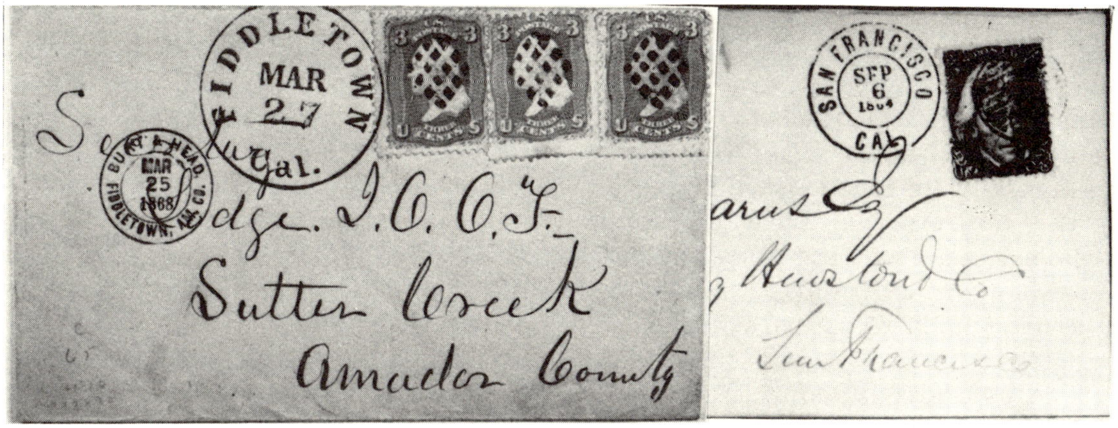

Triple rate from Fiddletown, 1868, with three 3¢ 1861; SF drop, 1864, paid by 2¢ Blackjack.

caused by short supplies of stamps had voided this part of the demonetization plan.

Replacement, which was performed at individual offices as the new stamps were received (notices being published allowing six days for the exchange) was largely accomplished by the first few months of 1862. Covers from California bearing old stamps are seldom encountered after that date.

The next important postal Act was that of 3 March 1863, effective 30 June. Again the postal rate was changed, this time to 3¢ per ½ ounce on all letters regardless of distance. The same Act required that the drop fee for intracity delivery must be prepaid by affixing 2¢ in stamps on a letter. Ship letters were charged double rates.

Despite its poor reputation the Post Office Department was sincerely trying to improve its service to the western citizens. By the end of 1852, 123 California post offices had been established and by 1869 California had 874 post offices. As early as 1859 the Postmaster General, in his annual *Report* to the President, stated that the expenditures for mail service to and in California, Oregon and the territories of Washington and Utah had amounted to $2,693,394 — an expenditure of $4.14 per person for the 650,000 souls there, since postal receipts from those areas totaled only $339,744. He compared this with the $12,272,000 spent to serve 30 million people east of the Rockies at a cost of 41¢ per person. He stated that as an act of simple justice this "gigantic service" of the west which had been established at the insistence of Congress should be charged against the public treasury, and that until a railroad had

Incoming ship letter (after July 1, 1863) delivered at port for double domestic: SHIP 4.

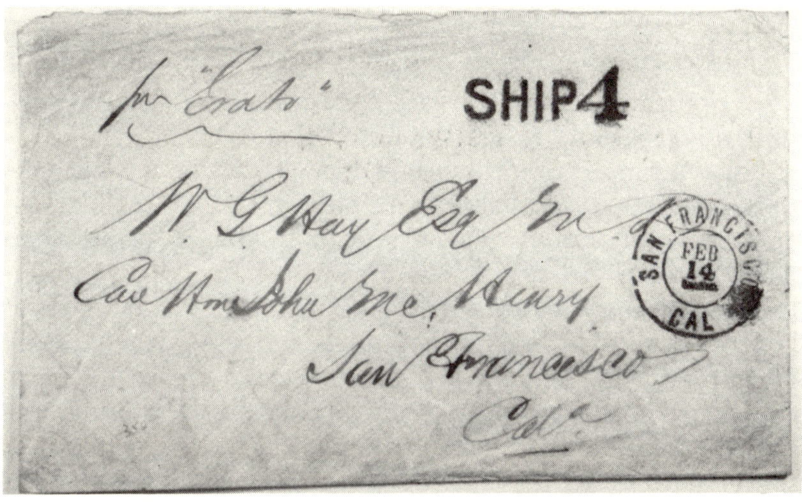

been constructed across the continent the conveyance of mail to the west would remain a problem. Though Congress was too frugal in its postal appropriations, the Post Office Department was well aware of its own inadequacies in serving the Pacific coast. Yet never during this entire period did Californians cease complaining about the postal service, for even when post offices existed annoying delays occurred in the reception, distribution, and dispatch of letters. The post offices continued to be inadequately manned and unable to cope with the volume of mail. By 1854 the mail-contract steamers were bringing in almost 80,000 letters per month. In 1858, 2,006,662 letters arrived in California — and 120,000 of them ended up in the Dead Letter Office in San Francisco. Mail from California to the East approximated 1,200,000 letters that same year. These statistics account only for letters carried by the mail-contract ships; many thousands more were carried in the letter bags of other steamers.

As the Postmaster General had informed Congress, it was these inadequacies of the government postal system which gave impetus to the development of the early expresses and stage lines as mail carriers. (These will be discussed in a subsequent chapter.) The inability of the post office to give prompt and adequate service to its patrons also led California to adopt an eastern enterprise known as the "locals."

San Francisco Locals

In our modern world we take for granted that the postman "on his appointed rounds" will deliver mail to our homes and offices. At the same time, if we wish, he will pick up our outgoing correspondence and take it to the post office for us. It was not always so! In California's early postal experience, a person trudged to the post office and inquired for his incoming letters. If he had written to someone, again he went to the post office, paid the postage or affixed the stamps he purchased there, and mailed his letter. If he wished to correspond with someone on the other side of town, it was back to the post office again to get the postage for "drop mail," and to leave the letter at the post office for the addressee to receive when *he* came to the post office to inquire for *his* mail. All of this took time, was inconvenient, and required a lot of walking, all too often in inclement weather. But that's the way it was — unless one subscribed to the services of a "local."

Locals were simply private firms that performed these services for the letter-writers. The advertisement of the western pioneer company, which appeared in the *Alta California* on 17 December 1853, is descriptive of the functions performed by all the locals:

> REED'S City Dispatch Post for the delivery of letters and parcels in all parts of the city. Office at Adams and Co., Montgomery Street. Boxes have been placed at convenient places for the reception of letters for city delivery. All letters entrusted to our care will be delivered with promptness and dispatch. Having secured box privileges of T. J. Henley, Esq., Postmaster, we are prepared to deliver Post Office letters to subscribers at the earliest moment after the arrival of the mails.
>
> REED & CO.

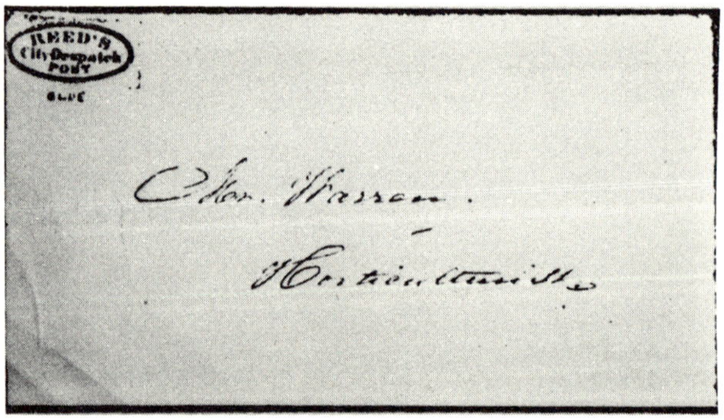

As this typical advertisement indicates, the local companies provided a threefold service for their subscribers:

(1) They made intracity deliveries. Although some advertised that they sold stamps for "drop mail" at their offices, the majority of their intracity-delivered letters have no stamps or postage charges indicated. The subscriber dropped his intracity letter in one of their street boxes, and they delivered it.

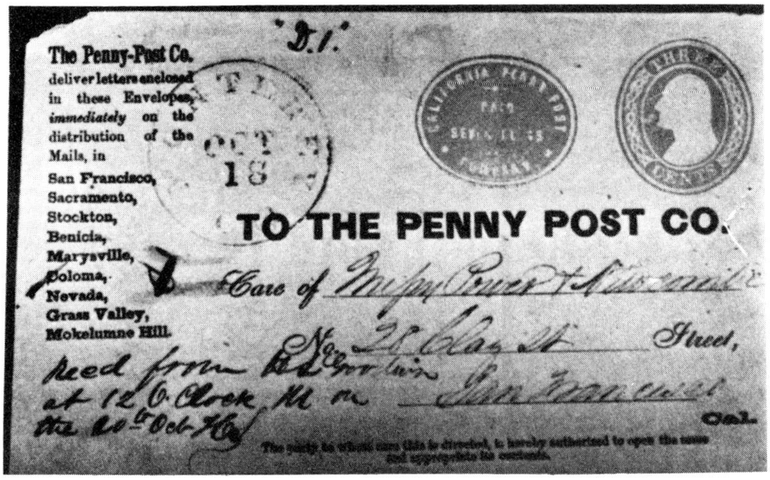

Oval vermilion California Penny Post and corner card ad on envelope postmarked Monterey.

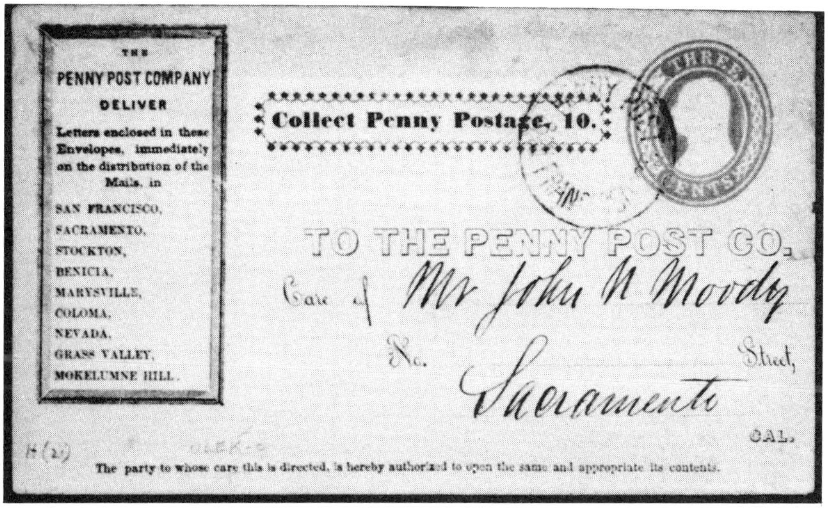

"Collect Penny Postage, 10" on envelope with boxed corner card ad. Double circle "Penny Post San Francisco" to Sacramento.

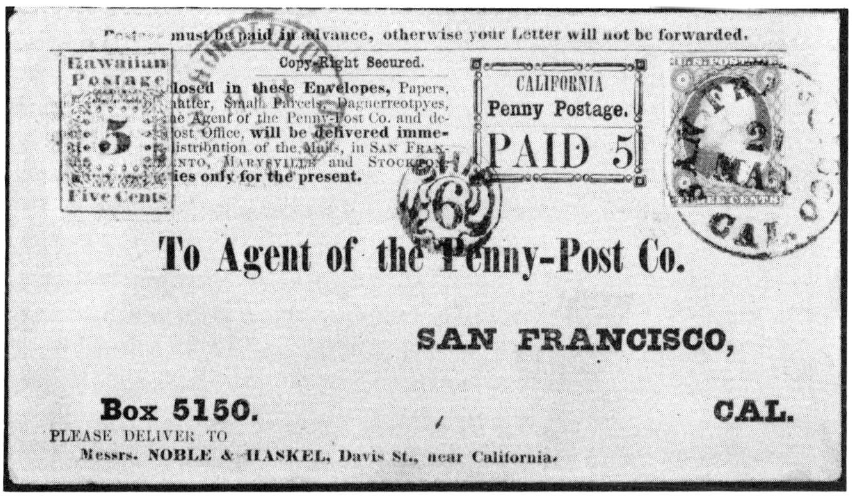

Penny Post envelope used from Hawaii to SF with Hawaii 5¢ Missionary, and SF fancy SHIP/-6; also 3¢ 1851.

There usually was a twice daily pick-up from these boxes.

(2) If the subscriber's letter was addressed to a destination outside San Francisco, the local would deliver the letter to the post office to enter the regular postal system. Some covers, with postage affixed, were delivered directly to the office of an express company or stage line for subsequent delivery. Wells Fargo seemed to have been a favorite of the locals.

(3) By arrangement with the postmaster, the local could rent a box at the post office and furnish the postal clerks with a copy of its subscription list. Consequently, when incoming steamers delivered their mail bags to the post office, the local could pick up all mail for those on its subscription list, and twice daily deliver letters and packages to home or office.

Penny Post

The largest of the locals was the so-called Penny Post, although its charges usually exceeded that figure. This company was started in June 1855 by H. L. Goodwin simply as a city delivery service. Like the other locals, its purpose was to deliver mail to the post office to be placed on outgoing steamers, and to obtain from the post office all incoming mail addressed to its clients and subsequently to make delivery to their homes and offices. Goodwin, however, was a businessman of vision and realized that other towns should appreciate this service also. Immediately he opened offices in Sacramento, Marysville, and Stockton. Within a few months, he added offices in Benicia, Coloma, Nevada City, Grass Valley, and Mokelumne Hill. In each of these towns the Penny Post branch office distributed all mail addressed to its subscribers.

To merchandize his service he not only purchased government envelopes, but had his own made. On all of these he printed franks and corner cards, rates, and advertising. He also used handstamps and adhesive labels.

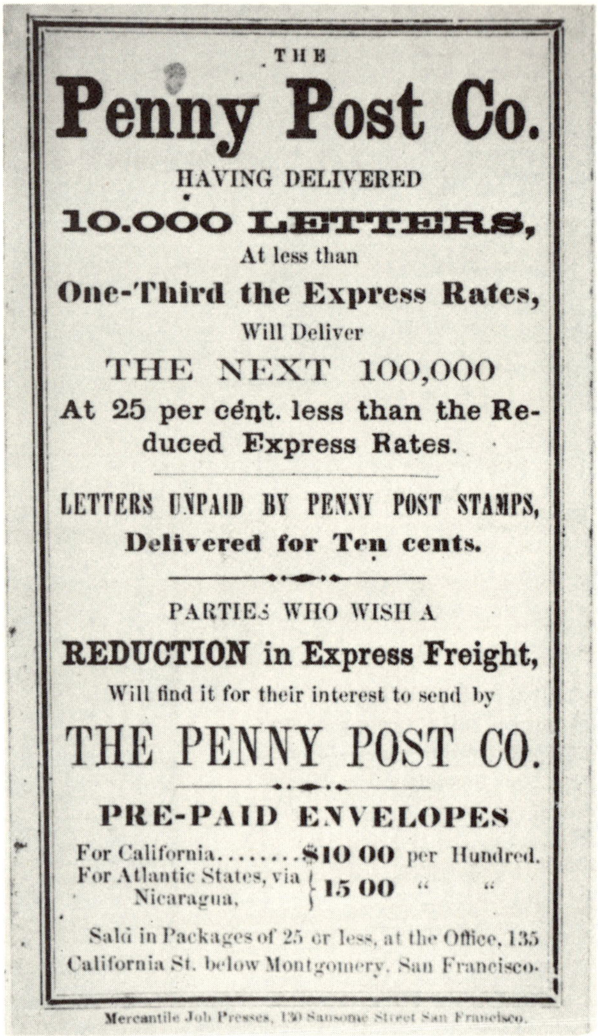

Penny Post ad card.

His envelopes were for sale in all of his offices; the customer chose the ones needed for his particular purpose, and the Penny Post guaranteed to do the rest. Despite all this activity, however, the Penny Post was short-lived, lasting only about one year. Active hostility of the San Francisco Postmaster was directly responsible for crippling his business. Nevertheless, the more than 20 varieties of his franked, labeled, and handstamped covers have made them a favorite of collectors and postal historians. All are of considerable rarity.

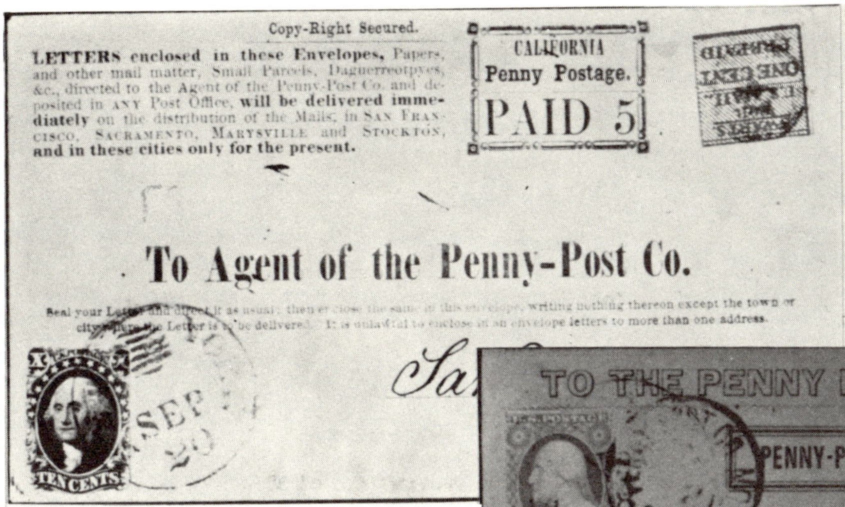

NY-SF with 10¢ 1857, Swarts 1¢, and PENNY POST PAID 5.

Stockton to SF, Penny Post PAID 5; four copies 3¢ 1851.

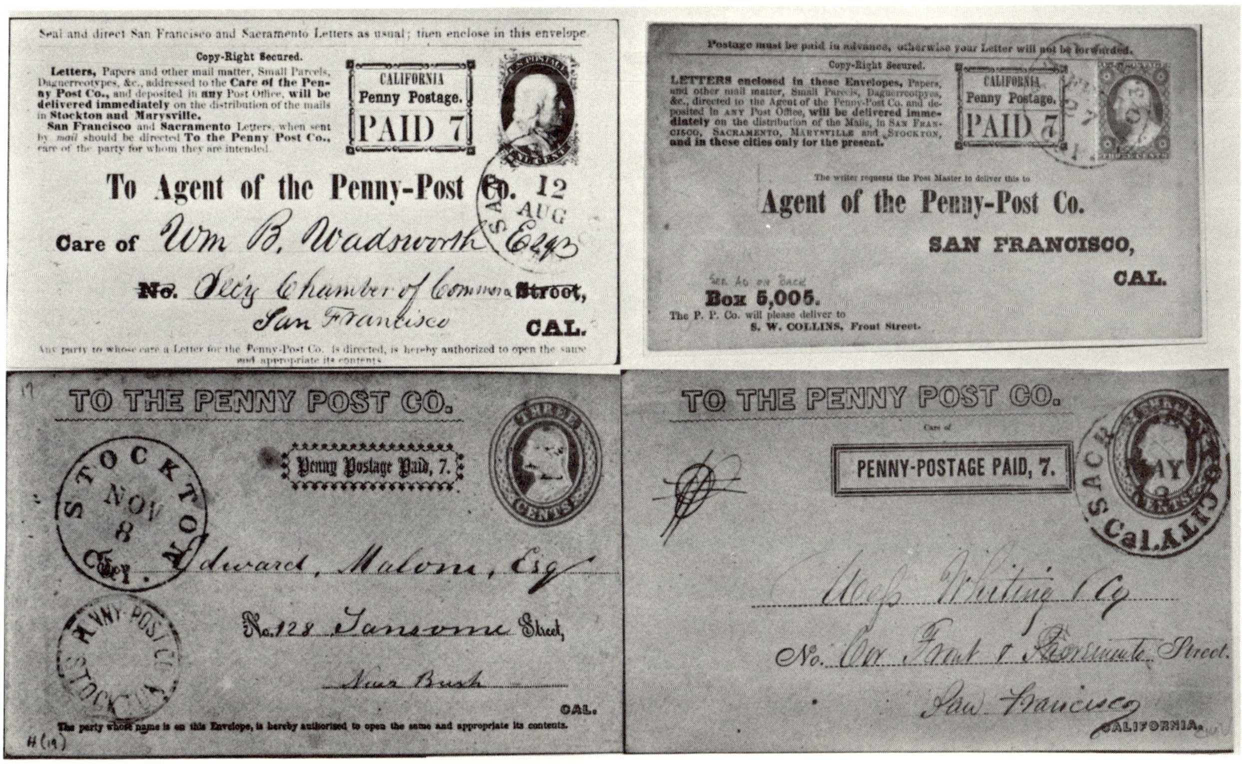

Penny Post Paid 7: different types with 1¢ and 3¢ stamps, 3¢ envelopes.

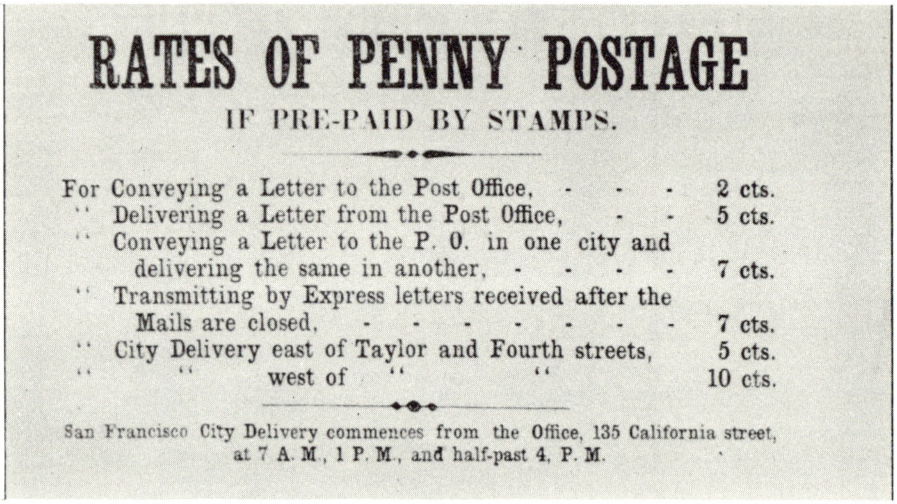

RATES OF PENNY POSTAGE

IF PRE-PAID BY STAMPS.

For Conveying a Letter to the Post Office, - - -	2 cts.
" Delivering a Letter from the Post Office, - -	5 cts.
" Conveying a Letter to the P. O. in one city and delivering the same in another, - - - -	7 cts.
" Transmitting by Express letters received after the Mails are closed, - - - - - - - -	7 cts.
" City Delivery east of Taylor and Fourth streets,	5 cts.
" " west of " "	10 cts.

San Francisco City Delivery commences from the Office, 135 California street, at 7 A. M., 1 P. M., and half-past 4, P. M.

Penny Post rates for various services.

Penny Post envelopes with large PAID 2, PAID 5, PAID 7, City Delivery, circular Penny Post San Francisco, PAID 5 label, and COLLECT PENNY-POSTAGE.

Merchants Exchange straightline on cover to NY by *Carolina* (Aug. 15, 1850) PAID 40; oval to Conn., PAID 40 and NY PAID (Oct. 1, 1850); oval to NY (Oct. 19, 1850) per *Columbus,* rated 40 due.

Merchants Exchange

The Merchants Exchange maintained letter bags on its premises and provided stamps and stationery simply as an adjunct or accommodation service for the convenience of its members. It also began its letter bag operation in 1849 when it started delivering mail to the San Francisco post office to meet the departure times of the mail steamers. Before stamps were available, the agent of the Exchange would collect in cash the amount due for postage on a letter, and then deliver the money with the letter bag. When the non-subsidized ships began competing with the Pacific Mail steamers, the Exchange prepared letter bags for these vessels also.

The Merchants Exchange was a social and business clubhouse where the merchants and businessmen of San Francisco could meet to confer, read newspapers, write letters, and eat and drink together. Current mercantile information and news were available and the sailing dates and arrivals of vessels were posted daily. By November 1849 the Merchants Exchange had its own building on Washington Street, and one became a member by subscription. Out-of-town businessmen visiting San Francisco could become short-term members and have access to all the facilities offered by the Exchange, including the use of its letter bags. All letters placed in the bags were handstamped, either with a black or red oval or a black or red block-letter straightline.

Known Locals

Advertising by handstamps, printed franks, and adhesive labels, the locals became popular and appreciated, as is evidenced by the number of them which are known to have existed:

California City Letter Express (1863-66)
The California Letter Express Co. (1864-66)
California Penny Post (1855-56)
Carnes City Letter Express (1855-56)
Comstock's City Letter & Parcel Delivery
Earley's San Francisco Letter Express (1861)
G. & H. City Express (1864-65)
Hoogs & Madison Dispatch Post (early 1860s)
Wm. E. Loomis City & Package Express (1867-71)
Merchants Exchange (1849-54)
Penny Express Co.
Penny Post (1855-56)
Private Post Office (1864)
Public Letter Office (1864)
Reed's City Dispatch Post (1854-55)
Robinson & Co. San Francisco Express (1862-64)
San Francisco City Dispatch Post (1851)
San Francisco City Letter Express
San Francisco Letter Express (1865)
Swift's City Express (1856-59)

The existence of locals in the eastern United States long antedated their appearance in California. In larger cities, such as Boston, New York, Baltimore, Philadelphia and Cleveland, private firms had for decades been providing home delivery service for a fixed fee. The Post Office Department, resenting this intrusion of private enterprise into government business and also acknowledging a responsibility to alleviate inconvenience to postal patrons, began efforts to provide better service. By authority of the Act of 2 July 1836 providing for carriers, on 1 August 1842 the New York postmaster purchased the City Dispatch Post and appointed government carriers.

In succeeding years the Post Office Department continued its efforts to control local delivery and eliminate competition from the locals by improving service and reducing rates. These efforts failed, for the private locals increased their deliveries as frequently as five times daily and provided service to suburban areas outside the city limits.

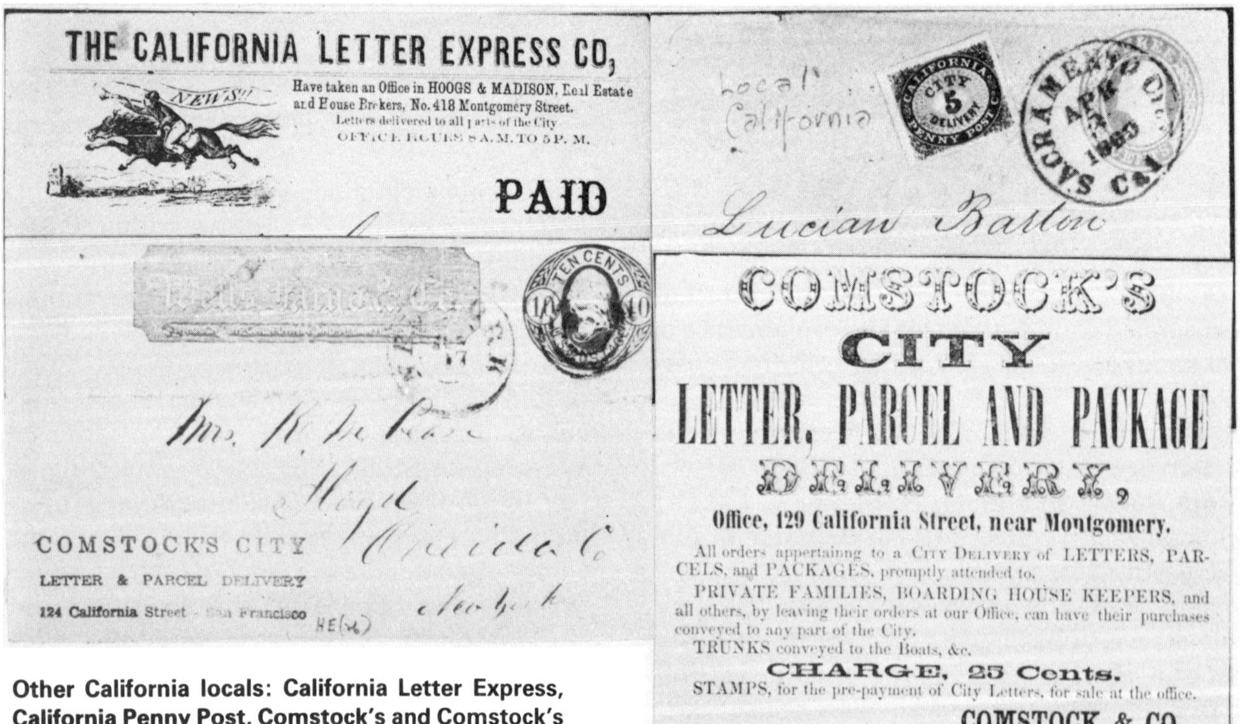

Other California locals: California Letter Express, California Penny Post, Comstock's and Comstock's ad.

More California locals: Earley, City G & H, Hoogs & Madison, Public Letter Office, Robinson, San Francisco Letter Express.

The next effort of the government was contained in the Act of 3 March 1851, which authorized the Postmaster General to establish post routes within cities and towns having presidential post offices, to set up mail deposit boxes at strategic locations, and to provide for the employment of mail carriers who would pick up mail at the boxes for delivery to homes and offices. The local postmaster could set a fee of either 1¢ or 2¢ per letter, to be paid by the receiver or sender.

A further Act of the same date provided for letter carriers for post offices in California and Oregon, allowing the postmasters of those areas to set a delivery fee not exceeding 5¢ per letter, 2¢ per newspaper, and 2¢ per ounce for other mailable material. However, since the first Act of 3 March 1851 stated that the total compensation paid to letter carriers could not exceed the postal receipts from this activity at any post office, the special provisions for California and Oregon were of little value to small towns. Equally affecting the larger cities was the fact that these authorized mail routes manned by the employed and bonded letter carriers served only the downtown areas and left the suburban areas unserviced. In California the problem was accentuated by the fact that as late as 1859 more than one half of the post offices reported net receipts of less than $100 per year.

An Act of Congress on 27 July 1854 again addressed the problem: "The Postmaster General is authorized to empower the special agents of the Post Office Department in the State of California and in the Territories of Oregon and Washington to appoint letter carriers for the delivery of letters from any post office in the said State and Territories, and to allow any letter carrier, so appointed, to demand and receive a sum for all letters, newspapers or other mailable matter not more than may be recommended by the postmaster for whose office such letter carrier may be appointed: *Provided,* That not more than twenty-five cents shall be charged for any letter, newspaper or ounce of other mailable matter." However, no evidence exists that this Act was implemented in California.

On 11 October 1857 the San Francisco postmaster announced that his office was prepared to deliver letters to the residences and places of business of all persons who would leave their names and addresses at his office. The continued existence of the private locals in that city indicates that they still outserviced and underpriced any effort to establish a post office carrier service.

On 3 March 1863 (effective 1 July) all mail carrier fees were abolished and free city delivery by salaried mail carriers was authorized — but only in 49 cities, all east of the Mississippi River. Once again, California was a neglected stepchild. It was not until 3 January 1887 that free delivery service was extended to the whole nation — but even then in a restricted form. It was directed only for cities with a population of 50,000 or more, and permitted for towns which generated postal receipts in excess of $10,000 per year. Since the majority of California post offices could not meet those requirements, free home delivery of mail did not become common in California until the early 20th century.

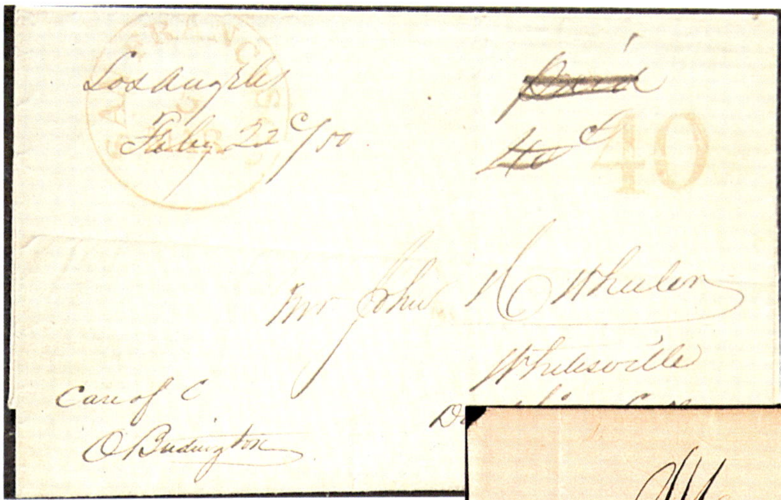

Los Angeles Feby 22/50 to Florida, rated 40 at SF. Earliest known postmarked cover from LA.

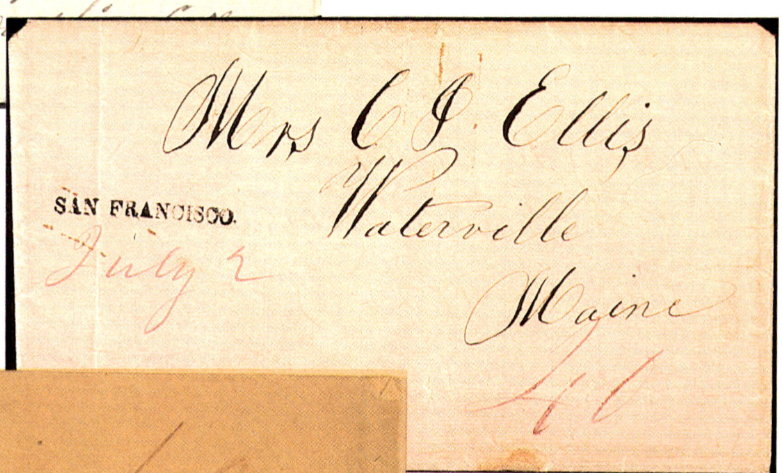

Straightline SF July 2 on cover to Maine, rated 40.

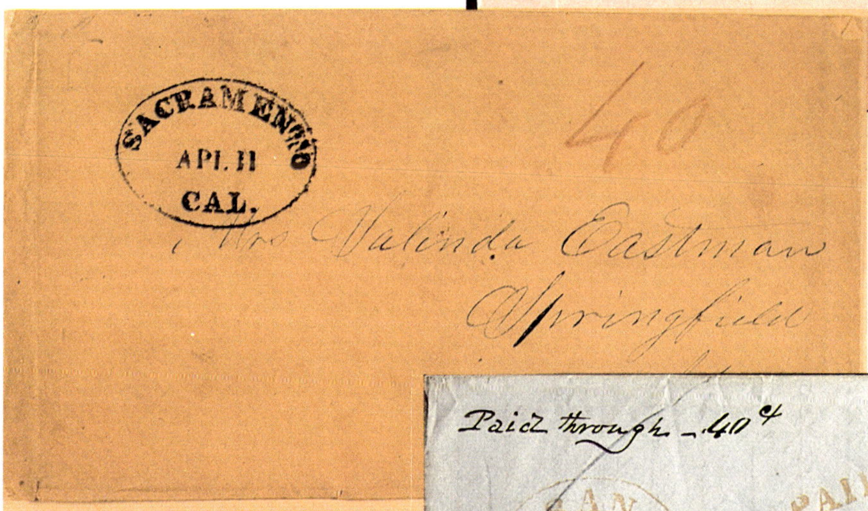

Oval Sacramento Apl 11 to Massachusetts, due 40.

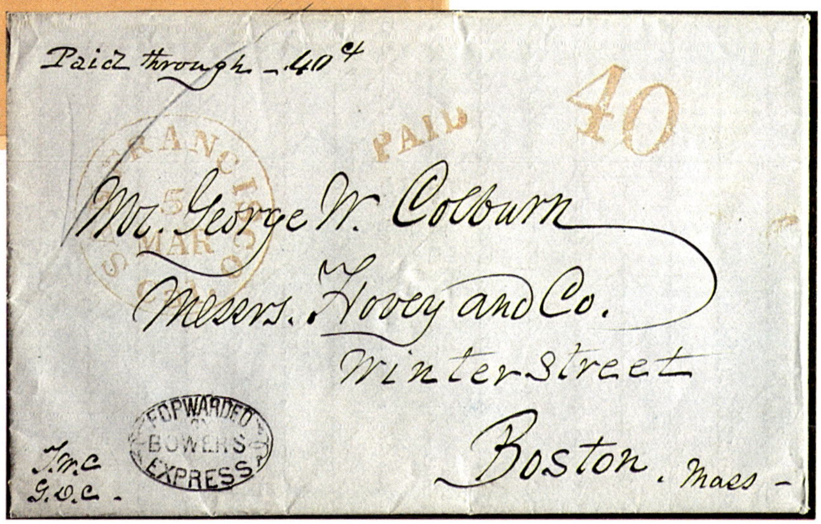

SF to Boston PAID 40. Also Bowers Express forwarding oval.

NY to SF; carried by *Philadelphia* Feb. 7, 1850. A double letter paid by eight copies of the 1847 10¢.

Sacramento to SF Feb. 26, 1850. Intrastate postage rated 12½¢ due by handstamp.

Oval handstamp of San Francisco Letter Express on local letter carried entirely outside mails.

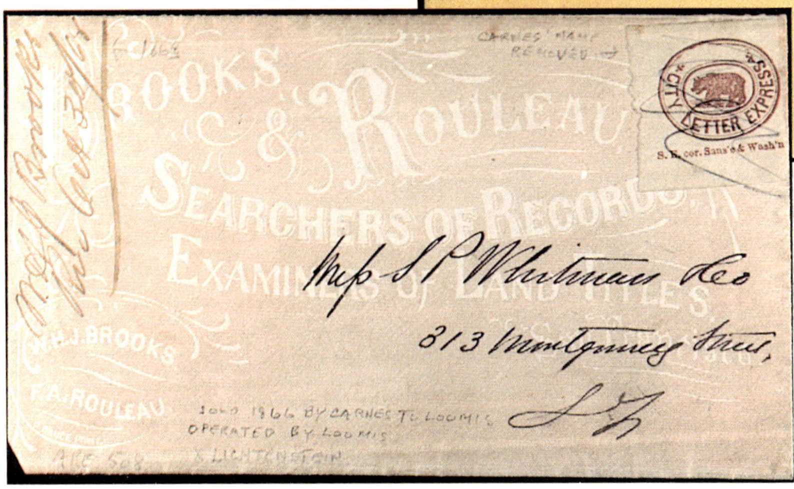

City Letter Express adhesive ("Carnes" removed after Loomis bought firm) on 1868 allover advertising cover.

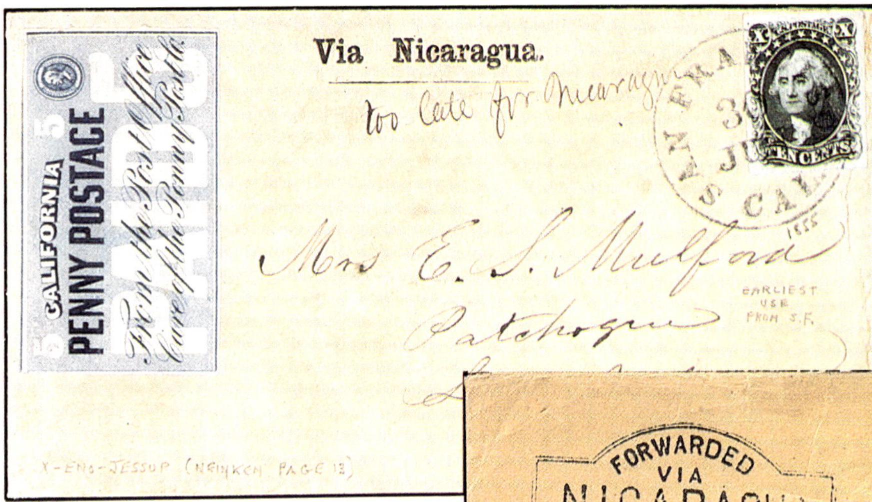

California Penny Post adhesive on cover from SF to Long Island in 1855. Intended to go "Via Nicaragua" but marked "too late."

Noisy Carriers FORWARDED VIA NICARAGUA to Del. with pair of 3¢ 1851 paying rate.

Noisy Carriers fancy box on 6¢ envelope to Phila. Four copies of 1¢ to make up 10¢ rate, 1855.

SF 30 SEP (1850), boxed 30 (collect) on incoming cover at Panama rate.

Sonora Nov 28 (1853) to Mich., with pair of 3¢ 1851. Carried on last voyage of *Winfield Scott* which was lost in fog on Anacapa I. Dec. 2 on the way to Panama. The mails, passengers, and cargo were saved.

SF to France 16 Mar 1852 PAID 26 for carriage in Brit. open mail by Am. packet.

Sacramento to Chile in 1861; three copies of 10¢ 1861 + 1¢ on 3¢ envelope making up 34¢ single rate via Panama.

SF to Lombardy, Italy, in Bremen/ Hamburg mail, 1862. Double letter paid by 46¢ in 1861 issue stamps.

SF to Vienna, Austria, by PCM. Rate paid by 1861 30¢ on all-over advertising envelope.

Hawaii to Switzerland by PCM, with three 10¢ 1861 and two each 1¢ and 3¢ issues for 38¢. Hawaiian 5¢ postage paid in cash. The PCM rate was 33¢, thus 35¢ U.S. (33¢ + 2¢ ship) would have been enough.

From Hawaii, with Hawaii 5¢ (Boston engraved) and U.S. 12¢ 1851 paying rate to Mass.

Hawaii 10 Dec 1867 to R.I. with 10¢ 1861 and oval Hawaiian Steam Service. Hawaiian postage paid in cash.

Ms. "Don Pedro's Bar" and straightline OVERLAND to Mass., paid by 1857 issue stamps.

Triple rate cover by overland mail from Benicia to Mass., 1861. Four horse coach.

Blue MARYSVILLE PAID BY STAMPS on overland envelope to Baltimore with 10¢ 1857.

St. Louis to Marysville, 1860, at 3¢ rate. Six horse coach on envelope published in Marysville.

California Pony Express and St. Joe running pony in red on 10¢ Nesbitt from St. Joe to SF, 1860. Ms. "Pd $5."

Virginia City pony express cover with 25¢ red stamp.

California Pony Express New York, and St. Joe running pony from NY to SF.

Oval SF running pony and $1 stamp on patriotic cover to Germany in PCM, 1861. Atchison, Kan., postmark ties 30¢ 1860 paying rate to Germany.

Double circle WF SAC MESS. on 10¢ Nesbitt to Phila. by *Golden Age,* 1860.

Oval WF NEVADA to SF with 3¢ 1851. Red TOO LATE.

WF STOCKN MESSENGER on paste-up to Mariposa composed of two 3¢ star die envelopes. Also 3¢ 1857.

Rhodes & Whitney's Express Weaverville oval on 3¢ envelope with elaborate frank of Wines & Co.'s Express.

Chapter 4

Early California Steamship Mail

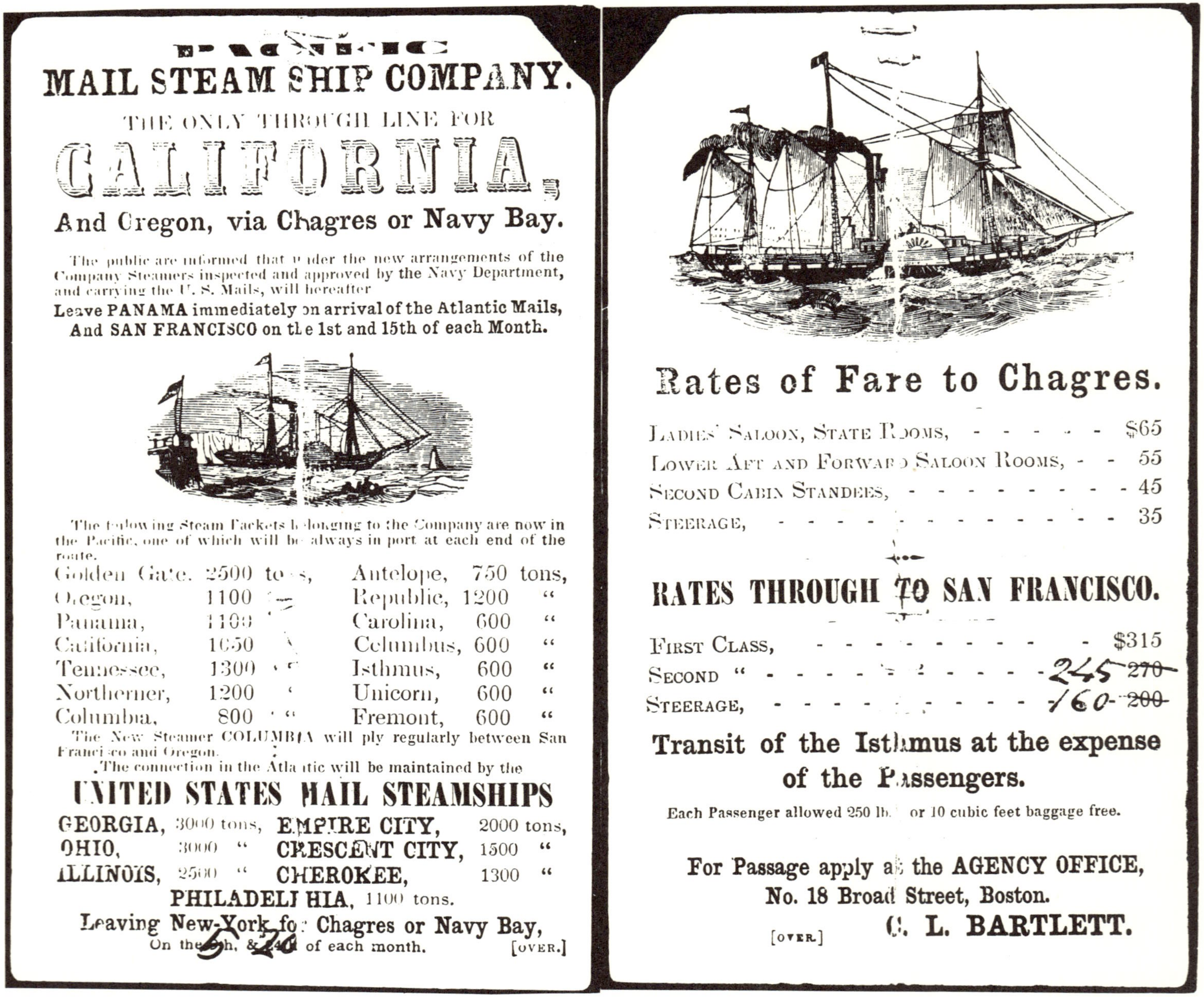

PACIFIC

MAIL STEAM SHIP COMPANY.

THE ONLY THROUGH LINE FOR

CALIFORNIA,

And Oregon, via Chagres or Navy Bay.

The public are informed that under the new arrangements of the Company Steamers inspected and approved by the Navy Department, and carrying the U. S. Mails, will hereafter

Leave PANAMA immediately on arrival of the Atlantic Mails,
And SAN FRANCISCO on the 1st and 15th of each Month.

The following Steam Packets belonging to the Company are now in the Pacific, one of which will be always in port at each end of the route.

Golden Gate,	2500 tons,	Antelope,	750	tons,
Oregon,	1100 "	Republic,	1200	"
Panama,	1100 "	Carolina,	600	"
California,	1050 "	Columbus,	600	"
Tennessee,	1300 "	Isthmus,	600	"
Northerner,	1200 "	Unicorn,	600	"
Columbia,	800 "	Fremont,	600	"

The New Steamer COLUMBIA will ply regularly between San Francisco and Oregon.

The connection in the Atlantic will be maintained by the

UNITED STATES MAIL STEAMSHIPS

GEORGIA,	3000 tons,	EMPIRE CITY,	2000 tons,
OHIO,	3000 "	CRESCENT CITY,	1500 "
ILLINOIS,	2500 "	CHEROKEE,	1300 "

PHILADELPHIA, 1100 tons.

Leaving New-York for Chagres or Navy Bay,
On the 5th, & 20th of each month. [OVER.]

Rates of Fare to Chagres.

LADIES' SALOON, STATE ROOMS, - - - - - -	$65
LOWER AFT AND FORWARD SALOON ROOMS, - -	55
SECOND CABIN STANDEES, - - - - - - - - -	45
STEERAGE, - - - - - - - - - - - - - - -	35

RATES THROUGH TO SAN FRANCISCO.

FIRST CLASS, - - - - - - - - - -	$315
SECOND " - - - - - - - - - -	245 ~~270~~
STEERAGE, - - - - - - - - - -	160 ~~200~~

Transit of the Isthmus at the expense of the Passengers.

Each Passenger allowed 250 lbs. or 10 cubic feet baggage free.

For Passage apply at the AGENCY OFFICE,
No. 18 Broad Street, Boston.

[OVER.] C. L. BARTLETT.

Front and back of sailing card.

Early California Steamship Mail

Frequent reference has been made to ships carrying mail between California and the eastern United States. From the days of the first American pioneers in the far west, most letter correspondence between the Pacific Coast and "back home" had been entrusted to officers of passing ships. When California, Oregon, and Washington became American possessions with a consequent growth of American population, the flow of mail became increasingly important. Obviously the most rapid method of transporting postal material was by sea. Thus it was that the newly acquired mail route across the Isthmus of Panama gave birth to a significant chapter of California postal history: California steamship mail.

The Panama Route

For almost 300 years the narrow isthmus connecting North and South America had been a center of interest to foreign powers. The Spanish conquistadores and subsequent

The Steamship *Brother Jonathan*.

The harbor at Chagres, on the Caribbean side of Panama.

colonists who established Spain's vast empire in the western hemisphere had used the isthmus as a transportation route for communication, supplies, and treasure. When the Hispanic-American wars of independence from Spain during the early 1800s were successful, the use of the isthmus declined but did not completely cease. From 1826 onward there were American proposals to establish a communication system across the isthmus, primarily to serve whaling and sealing ships and merchants on the Pacific coast, but also to be used by diplomatic and naval personnel on inter-station transfers. The British were also interested in the area. In 1840 Britain chartered the Pacific Steam Navigation Company to ply the Pacific coast of Hispanic-America with two steamers and gave the company a mail subsidy. In 1846 the British Royal Mail Steam Packet Company, which had served the West Indies since 1842, extended a line (mules and canoes) across the Isthmus of Panama to connect with the ships of the Pacific Steam Navigation Company.

While the British pioneered in this 19th century isthmian shipping, the United States also was increasingly recognizing the importance of the Panama route for contact with Americans on the Pacific coast. By 1843 hundreds of American emigrants had followed the Oregon Trail into the Willamette Valley. The town of Astoria at the mouth of the Columbia River had become an important shipping port. An 1846 treaty with Britain had established the 49th parallel as the international boundary to the Pacific, and Oregon and the area later to be Washington then were American possessions. Rapid transportation of merchandise and mail became a significant concern. On 12 December 1846 a treaty was signed with the Republic of New Grenada (Colombia), in which Panama was a state, giving the United States the right of way and the right of transit across the isthmus. In the same year the United States declared war on Mexico and in February 1848 California became a United States possession, adding more American citizens with whom there was a need for more rapid communication.

Beginning with a 3 March 1845 Act of Congress dealing with foreign mail service, there were attempts to open a mail line across Panama to serve the north Pacific. The Postmaster General began soliciting bids from individuals and companies to provide the service. Various bids were received but foundered upon disputes over the frequency of service, the types of vessels to be used, the money amounts of the subsidies, and the Navy's desire to supervise ship construction. After almost two years of Congressional debate and bills, finally on 3 March 1847 Con-

Different steamships illustrated on "Via Panama" covers to N.H., Conn., and Oswego, NY, all from SF, and Timbuctoo to Mich. Last two California printers.

gress passed a law directing the establishment of a postal route from Atlantic ports, across the Isthmus, and to Oregon with stops at Monterey and San Francisco, the through service to be subsidized in the amount of $100,000. Simultaneously, on the same day, Congress authorized the Secretary of the Navy to provide the steamers which would carry the mails and to advertise for bids for their construction, the costs to be amortized over a ten year period by annual subsidies from the Navy Department.

Although oceangoing steamships had just emerged from their infancy, and such steamers still bore masts for sails in case of engine breakdowns, the Navy knew that sailing warships were becoming obsolete. This Act would permit the Navy to supervise the construction of eight new steamers readily convertible to Navy warships in an emergency. The bids were to call for the construction of five ships, each of 1,500 tons, to steam from New York to Chagres, and three ships, two of 1,000 tons and one of 600 tons, to serve in the Pacific from Panama (City) to San Francisco and Astoria.

Both the Navy Department and the Post Office Department were to find this arrangement objectionable. The Navy Department, while pleased to have potential warships available, was distressed by the limitation of one naval officer of the rank of lieutenant and four midshipmen as watch officers on each ship which was otherwise

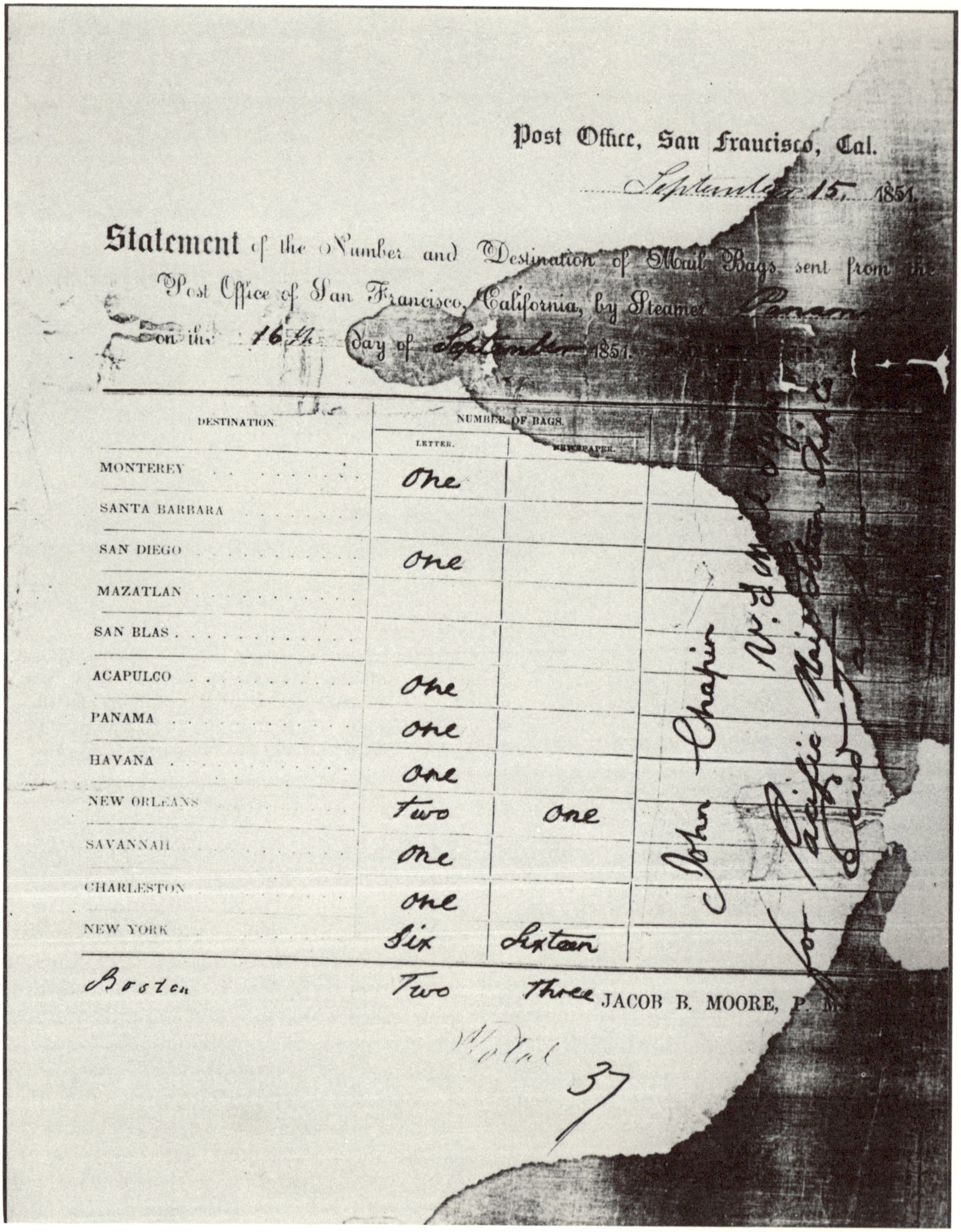

Post Office, San Francisco, Cal.

September 15 1851.

Statement of the Number and Destination of Mail Bags sent from the Post Office of San Francisco, California, by Steamer Panama on the 16th day of September 1851.

DESTINATION	NUMBER OF BAGS. LETTER.	NEWSPAPER.
MONTEREY	One	
SANTA BARBARA		
SAN DIEGO	One	
MAZATLAN		
SAN BLAS		
ACAPULCO	One	
PANAMA	one	
HAVANA	One	
NEW ORLEANS	two	One
SAVANNAH	One	
CHARLESTON	one	
NEW YORK	Six	Sixteen
Boston	Two	three

JACOB B. MOORE, P. M.

Total 37

John Chapin

Post Office statement of mail carried by the *Panama* from SF on 16 September 1851.

manned by civilians over whom the Navy had little control. When it became obvious that the transportation of mail and emigrants would be the primary function, the Navy resented paying the bills while the private contractors and the Post Office Department substantially had the right to direct the operation of the vessels. The Post Office Department disliked the arrangement because it had no real control over the contractors while at the same time it was required by law to send the mails over this route on these vessels.

The Act authorizing the Navy to establish a steamship line to Oregon specified that the contract from New York to Chagres be given to Albert G. Sloo, who accepted on 20 April 1847. Open bidding would, however, be received for the Pacific line. The contract called for the construction of five steamers to be commanded by a naval officer and four midshipmen, and stated that "the Secretary of the Navy shall exercise control over said steamships and shall have the right at any time to take them for the service of the United States." A further provision was that each ship would carry free of charge a postal agent to supervise the transportation of mail. Mr. Sloo, however, had neither the capability nor the desire to construct and operate the shipping and mail line, so he subsequently assigned his contract to the Law, Roberts and McIlvaine syndicate, a group of wealthy men already in the transportation business. This syndicate was incorporated by the New York Legislature on 23 March 1848 as the United States Mail Steamship Company. Their assurance that they could begin operation with two ships by 1 October 1848 was accepted, and they were granted a mail subsidy of $290,000 per year.

The competitive bidding for the Pacific line began on 4 May 1847. Seven months later, on 16 November, a contract was awarded to Arnold Harris, a Tennessee politician who immediately assigned the contract to the Aspinwall, Howland and Chauncey syndicate, an experienced group of shipping operators. On 12 April 1848 they were incorporated by the New York Legislature as the Pacific Mail Steamship Company. They were granted a ten-year mail subsidy of $199,000 per year, beginning 1 October 1848. The original contract had called for the Pacific line to provide for three steamships to run on a monthly schedule from Panama (City) to Astoria, Oregon, with intermediate stops. This requirement was modified by the Secretary of the Navy on 19 June 1848 to allow the steamships to proceed only to San Francisco, then to provide service to Oregon by sailing ships, and the Oregon port could be the mouth of the Kalumet River in lieu of Astoria. Congress, on 3 August 1848, specified that the Pacific steamers would "stop, deliver and take mails at San Diego, Monterey and San Francisco."

Then came the news of the gold discovery

Cover carried by the *Crescent City* of the Howard Line. First steamer to clear for Panama from NY.

Covers carried by PMSS steamers *Oregon* (both 1850), *California*, all at 40¢ unpaid; and by *Panama*, a first day of the 6¢ paid/10¢ unpaid rate.

in California, and the Gold Rush of 1849 began, with every sailing line on the east coast besieged by clamoring hundreds seeking passage to the gold fields. The first steamer to clear for the Isthmus was the *Crescent City* of the Howard Line, which had no mail contract but plenty of passengers willing to pay exorbitant rates. To meet the emergency situation the Atlantic mail contractor, the United States Mail Steamship Company, received permission to begin service with three small steamers of about 600 tons each, the *Falcon, Isthmus* and the *Orus.* Other Atlantic seaboard steamers also loaded passengers and headed for Panama.

The United States Mail Steamship Company The Pacific Mail Steamship Company

In the meantime the Pacific Mail Steamship Company had built the *California* (1,050 tons), the *Oregon* (1,100 tons), and the *Panama* (1,100 tons). Because these ships had to steam around Cape Horn, they had started before the Gold Rush got under way. The *California* with accommodations for about 250 passengers had left New York on 6 October 1848, with instructions to stop in San Diego and Monterey upon reaching California. After rounding the Horn she stopped in Peru and picked up 80 Peruvians who had heard of the gold discovery. When she reached Panama on 17 January 1849 and found over 700 argonauts waiting, she crowded aboard 365 passengers and steamed on north, finally arriving in San Francisco on 28 February 1849. Her crew promptly deserted and headed for the gold hills. One of the passengers on this maiden voyage was William Van Voorhies, the Postal Agent with orders to establish a postal system in California and to start the flow of mail. Thus it is an historical irony that when Van Voorhies closed a mail bag containing his report to the Postmaster General and letters from local residents for eastern shipment from San Francisco on 15 March 1849, there was no American ship to carry it, so he dispatched this mail by the Peruvian bark

Callao to Panama.

On 23 February 1849 the *Oregon,* having rounded Cape Horn, docked in Panama to find approximately 1,200 impatient Americans fighting for accommodations. When the *Oregon* arrived in San Francisco on 1 April, she wisely anchored under the guns of the U.S. Warship *Ohio* to prevent her crew from deserting. So it was that when the *Oregon* left San Francisco on 12 April 1849 as the first Pacific Mail Steam Ship Company vessel to return to Panama, she carried the first *contract mail* eastward. The steamer *Panama* arrived in Panama in May 1849 and encountered over 2,000 angry gold-seekers demanding passage. She anchored in San Francisco on 4 June 1849. All three steamers had overloaded in Panama, yet they still left more than 2,000 people stranded there, either to wait for the steamers to return or to find their own way to California.

The Pacific Mail Steamship Company, nevertheless, was now in place and ready to begin its contract to provide monthly mail

Covers by US Mail SS Co. steamers *Ohio,* PAID 40, 1850; *Illinois,* PAID 6; *Cherokee,* 40 unpaid, 1851; *Georgia,* 80¢ unpaid, a double letter from Liverpool; new PMSS *Tennessee,* 1851, and *Philadelphia,* both PAID 40.

service between San Francisco and Panama. Yet when the *Oregon* came in from Panama on 18 September and the *California* on 10 October, there was no mail aboard. Plenty of passengers but no letter bags!

Since the original treaty with New Grenada had given the United States the right of transit across Panama, in October 1848 the Pacific Mail Steamship Company's offer to transport mail across the Isthmus for $2,900 per year had been accepted. However, in 1849 the New Grenada government assumed the task. Their fee was based on the weight of mail carried and was nominal, for no one anticipated the Gold Rush. Suddenly the deluge of people and mail descended, 50 times more than expected. Unfortunately the New Grenada government had farmed out the transportation to private Panamanian firms, and these firms demanded substantial increases in payment. When the Postmaster General could not increase payment because of his appropriation limitations, the Panamanian transporters often refused to move the mail, allowing it to accumulate, and sometimes even dumping the bags into the jungle to rot.

The Californians knew nothing of this argument, but they did know that they wanted their mail. They bitterly blamed the steamship line. Fiercely accusatory letters and editorials filled the *Alta California* newspaper. Fortunately, a new agreement with Panama raised the fee, appointed a postal agent in Panama to supervise the isthmus transportation, and mail began arriving again in California. In fact, such a flood of it arrived that Californians had a new object of anger: the inability of the California postal service to distribute the mail it received.

Both mail steamship lines tried to improve their service. On the Atlantic run, the United States Mail Steamship Company placed in service the *Ohio* (2,400 tons) on 20 September 1849, the *Cherokee* (1,200 tons) on 13 December 1849, the *Georgia* (2,700 tons) on 28 January 1850, the *Philadelphia* (1,000 tons) in March 1851, and the *Illinois* (2,100 tons) on 26 August 1851. In the Pacific the Pacific Mail Steamship Company chartered (later purchased) the British steamer *Unicorn* (650 tons) to begin service on 1 December 1849. Between April 1850 and November 1851 five other ships were added to the Pacific line: the *Tennessee* (1,300 tons), the *Carolina* (600 tons), the *Columbia* (800 tons), the *Fremont* (600 tons), and the *Golden Gate* (2,000 tons). Upon arrival in San Francisco on 7 May 1850 the small *Carolina* began running to Oregon to complete the Pacific Mail Steamship Company's contractual obligations.

On 27 October 1849 the privately-owned *Senator* (750 tons) had arrived in San Francisco from around Cape Horn. For some years she engaged in river traffic between San Francisco and Sacramento. Later she plied the coastal waters touching the ports from San Francisco to San Diego. Five months

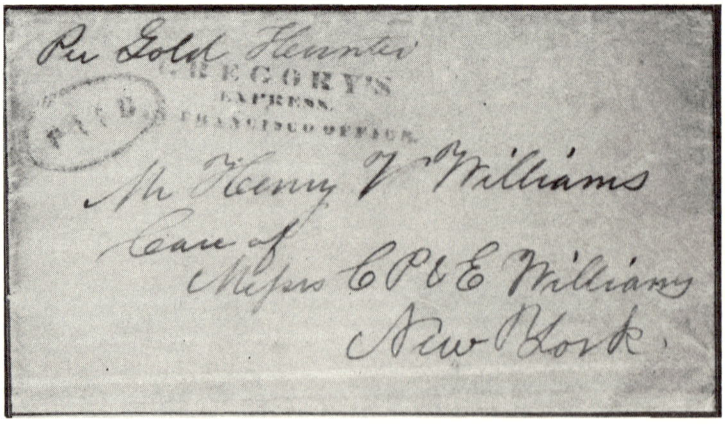

By *Gold Hunter* and Gregory's Express to NY, outside mails.

later, on 29 April 1850, another privately-owned smaller steamer, the *Gold Hunter* (436 tons), arrived to engage in the river trade. Shortly thereafter she was running north to Oregon, next south to Mazatlan and San Blas, Mexico. Neither of these ships had a mail subsidy, but both carried letter bags "outside the mails," and thus were in competition with the subsidized line.

Law's Line of Pacific Steamers

The mail contracts were held by the two steamer lines, the United States Mail Steamship Company and the Pacific Mail Steamship Company. As traffic grew, the mail subsidies were increased, and the fees for carrying gold, freight, and passengers generated huge profits. George Law, president of the Atlantic company, became greedy. He decided to compete also on the Pacific side. In May 1850 advertisements appeared in the *Alta California* for Law's Line of Pacific Steamers. On 4 May 1850 the *Isthmus* (600 tons) arrived in San Francisco, followed in June by the *Columbus* (500 tons), in August by the *Republic* (820 tons), and on 25 October by the *Antelope* (650 tons). Each ship immediately began operating between San Francisco and Panama, carrying passengers and treasure — and mail!

Although by government contract the Pacific Mail Steamship Company had the exclusive right to carry all mail, Law offered the San Francisco postmaster the opportunity to use his ships as mail carriers, and thus increase the frequency of mail delivery. Knowing the mood of the mail-hungry Californians, on 14 May 1850 the postmaster agreed to Law's offer, on the condition that senders must indicate on their envelopes a wish to use Law's steamers and that such mail carry the name of the ship used. The agreement was a breach of contract, but it pleased the public. However, this private decision of the San Francisco postmaster was terminated after five months, when the Postmaster General decreed that all mail sent through the San Francisco post office must go only by contract

Covers by Law Line steamships *Isthmus* and *Columbus* showing ms. endorsements and straightline handstamps. All 1850.

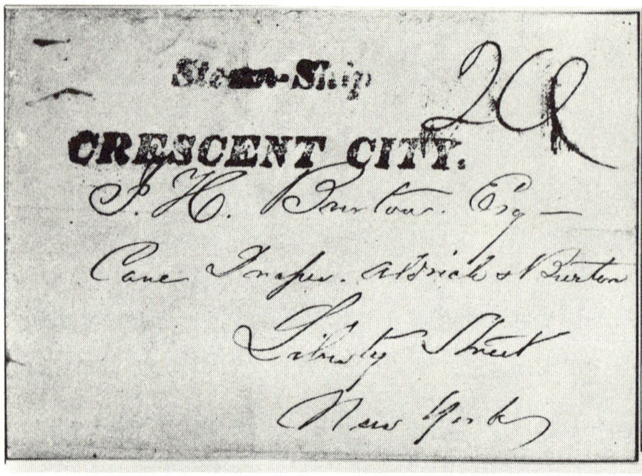

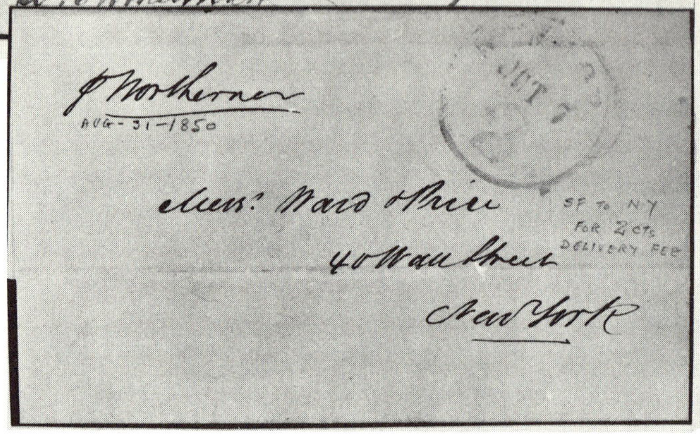

Handstamps of *Crescent City*, 1849 (20¢ rate to or from Chagres), and *Empire City*, 1849, both steamships on NY to Chagres run for Howard Line; and endorsement *Northerner*, 1850, Howard vessel on SF to Panama route.

mail steamers.

The Pacific Mail Steamship Company, of course, had protested the private decision of the local postmaster to expedite the mail service. The Post Office Department followed the law when on 10 October 1850 it forbade the San Francisco postmaster to use any ship not part of the mail-subsidy monopoly as a mail carrier. At the same time it tried to solve the problem by directing the Pacific Mail Steamship Company to increase its sailings to twice each month, on the 1st and the 15th, a service which the line had already voluntarily initiated on 15 July. It further ordered that each sailing should be accompanied by a government mail agent. It promised to increase the mail subsidy and did so in 1855 and after to $348,250 per annum.

There was no way, however, that the Post Office Department could enforce its edict that only the subsidized lines were authorized to carry mail. It could make it applicable only to letters actually posted at the post office. The edict could be evaded simply by not using the post office and by taking letters directly to ships willing to accept them. The attempt of the Post Office Department to maintain a complete monopoly was a failure.

Howard & Son's Empire Line

In the meantime the very lucrative business of transporting passengers, freight, express, and gold brought another competitor into the western ocean: the Howard & Son's Empire Line. Already operating on the New York to Chagres route since December 1848, first with the *Crescent City* (1,300 tons) and later the *Empire City* (1,700 tons), Howard dispatched three steamers to the Pacific. Successively in June, August, and September 1850 the *Sarah Sands* (1,500 tons), the *Northerner* (1,200 tons), and the *New Orleans* (800 tons) began runs between San Francisco and Panama. These were followed in 1851 by the *Commodore Stockton* (450 tons) and the *Monumental City* (750 tons). Each carried unofficial mail bags. To attract customers for his other services, Howard advertised that his line would carry mail "free" if the letters bore the correct postage.

With so many steamers operating, rate-cutting began. The Pacific Mail Steamship Company, because of its mail subsidy and the tremendous profits generated in 1849 and early 1850, was in the strongest position. As suddenly as he had entered the scene, Howard retrenched. He sold the *Sarah Sands* and the *Northerner* to Pacific Mail Steamship Company to become part of their Pacific fleet. Howard did not make a clean break, however. He retained the *New Orleans* and added the *Union* (600 tons) to compete on the Panama-San Francisco run. Since the *Union* sank after two months of sailing, and the *New Orleans* was sold to an Australian line in December 1852, Howard's competition was negligible. The Pacific Mail Steamship Company now decided to give George Law a taste of his own medicine. It purchased Howard's *Crescent City* and *Empire City* to compete in the Atlantic under the Pacific Mail flag as a rival of the United States Steamship Company.

With so much competition, all the steamer lines had been caught in the ruinous rate-slashing which had cut passenger fares in half and reduced the gold-carrying commission from 5 percent to 1½ percent. The two original subsidized lines quickly came to agreement. Law sold his Pacific-operating steamers, the *Isthmus* (later renamed the *Southerner*), *Republic, Antelope,* and *Columbus,* to Pacific Mail Steamship Company. They, in turn, sold Law the *Crescent City* and the *Empire City*. The status quo of 1849 now was restored, and each of the lines had an enlarged fleet for the runs from New York to Chagres and from Panama to San Francisco. Business peace did not last long, however, for a new and serious disturbing element arose.

The Nicaragua Route

There had long been proponents of a different route across Central America, *i.e.,* across the Isthmus of Nicaragua. Although

Map of route across Nicaragua.

the Nicaraguan route was 170 miles long, compared with the 70 miles of the Panama crossing, all but 12 miles of it could be covered by boats. Twelve miles inland from the Pacific Ocean lay Lake Nicaragua, 96 miles long and 39 miles wide. From the lake's southeast corner, the San Juan River ran into the Caribbean Sea.

Located on a higher plateau, Nicaragua had a more healthful climate than Panama. Furthermore, it shortened the journey from New York to San Francisco by about 400 miles, and from New Orleans to San Francisco by 525 miles. As early as 1849 the great shipping magnate of New York, Commodore Cornelius Vanderbilt, had been granted a contract by the Nicaraguan government to survey and build a cross-isthmus canal. Though he made no canal-building efforts, Vanderbilt transferred his contract to his newly organized company, the Accessory Transit Company, thus assuring himself of

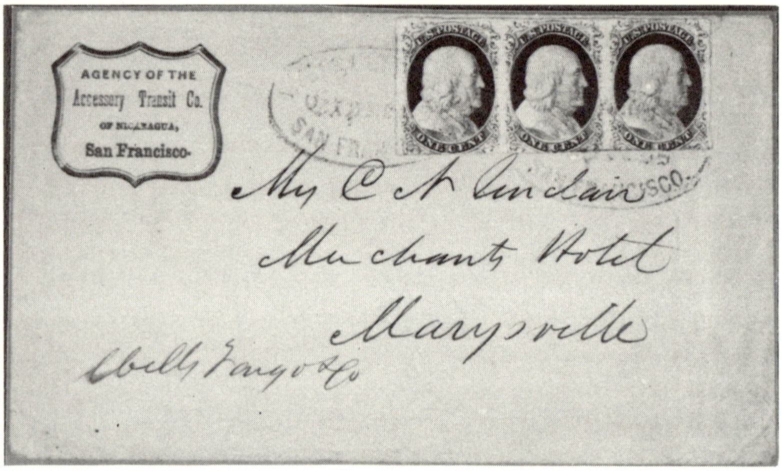

Cover showing corner card of Accessory Transit Co., SF office.

control of the overland passage across Nicaragua. He listed its stock on the New York Stock Exchange, watched it rise from $3 to $50 per share in a few years, and reaped millions in profits.

Since he already controlled the land route across Nicaragua, Vanderbilt decided to challenge the two steamship monopolies. He knew that sailing ships and small steamers already were calling at Realejo on the Pacific side of Nicaragua as well as at San Juan del Norte harbor where the San Juan River emptied into the Caribbean. He was aware that some argonauts earlier had chosen this Nicaragua route on their way to the gold fields. He had observed the advertisements of numerous sailing ships appearing in eastern newspapers which touted the advantages of going to California via Nicaragua.

Vanderbilt's Independent Line

On 2 July 1851 the Vanderbilt steamer *Pacific* (1,000 tons) arrived in San Francisco. On the same day the *Alta California* carried advertisements for Vanderbilt's Independent Line. The advertisements asserted that the *Pacific* would run to San Juan del Sur, that coaches and lake and river steamers would carry passengers across Nicaragua to San Juan del Norte, and that the new steamer *Prometheus* (1,200 tons) would complete the Atlantic journey to New York. It was the beginning of a new shipping war, for on 11 July 1851 another Vanderbilt steamer, the *Independence* (600 tons), was in San Francisco ready for the run to San Juan del Sur. A second new steamer, the *Northern Light* (1,750 tons) was scheduled to make the journey on the Atlantic side of Nicaragua. Vanderbilt seems to have been the first to use what Ernest A. Wiltsee has called "propaganda" on the letters carried by his ships. Each letter bore a handstamp "Via Nicaragua in Advance of the Mails" or "Via Nicaragua

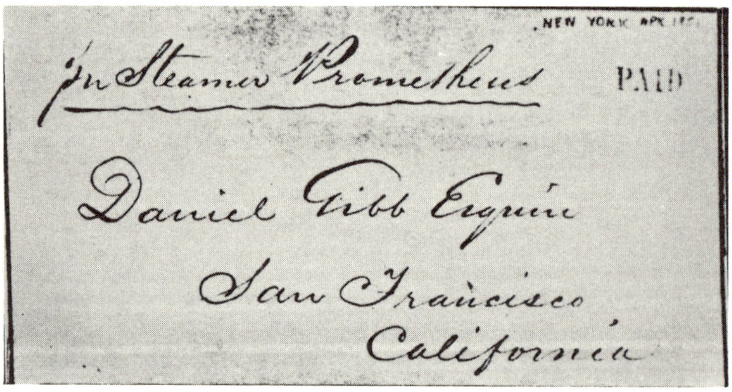

Cover carried from NY to SF by Vanderbilt *Prometheus* in April 1851.

Covers with "Via Nicaragua" slogans: straightline on circular to Boston; boxed and STEAM/SHIP to Maine; oval to Virginia; straightline to Ohio; boxed, and ms. on 6¢ envelopes; boxed on double with 12¢ stamp. All 1853-54.

Ahead of the Mails."

The Pacific Mail Steamship Company responded by diverting the *Constitution* (530 tons) from a coastal port service and putting her on the Panama run. Vanderbilt replied by purchasing the *Gold Hunter* and the *Monumental City* for his Pacific fleet. Pacific Mail Steamship Company answered by again diverting another of her coastal steamers, the *Frémont,* to the Panama service and in December 1851 added a new steamer, the *Golden Gate,* to its fleet. Vanderbilt immediately retaliated by placing the *North America* (1,500 tons) and the *Independence* (600 tons) on his San Francisco to Nicaragua route, and the *Daniel Webster* (1,050 tons) and *Brother Jonathan* (1,300 tons) to serve the Caribbean port. This shipping war and rate-cutting were so expensive that for a short time both sides retreated. Vanderbilt leased the *Gold Hunter*

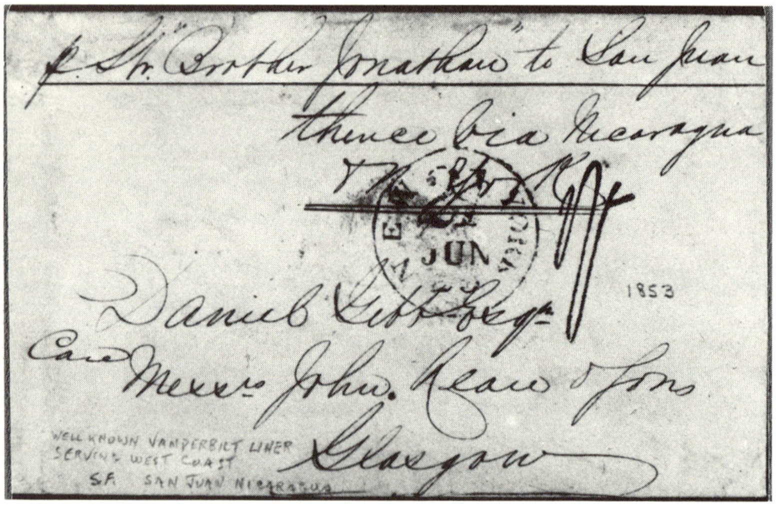

A cover carried by Vanderbilt *Brother Jonathan* in 1853 to San Juan and via Nicaragua to NY outside mails. Entered mails at NY unpaid to Scotland. Rated 1/- due at Liverpool.

and the *Monumental City* to the Empire Line. Pacific Mail Steamship Company sold the *Frémont* to the South Pacific Steamship Company, which then on 28 November 1851 began the first steamer service between San Francisco and Hawaii. When, on 27 February 1852 Vanderbilt's Pacific fleet flagship, the *North America,* was lost near Acapulco, he secured the *New Orleans* and the *Monumental City* from the Empire Line to fill the vacancy.

New York & San Francisco Line (New York & California Line)

Then another competing line appeared. On 29 April 1852 a new and luxurious steamer, the *Winfield Scott* (1,300 tons), owned by a new company known by interchangeable names (the New York & San Francisco Line from April 1852 to March 1853 or the New York & California Line for four months, March to July 1853) arrived in San Francisco to enter the Panama run and connect with the *United States* (1,200 tons) at Chagres. Vanderbilt was not intimidated and he immediately brought around the Horn two more ships, the new *S.S. Lewis* (1,100 tons) and a rebuilt *Brother Jonathan* (2,000 tons). In reply, in the fall of 1852 the New York & San Francisco Line brought to the west coast the *Cortes* (1,100 tons) to connect with an Atlantic steamer, the *Uncle Sam* (1,400 tons), at the new port of Aspinwall in Panama. Vanderbilt placed new steamers on the New York to San Juan route: the *Northern Light* in May 1852 and the *Star of the West* (1,200 tons) in

Oval slogan marking of Vanderbilt *Sierra Nevada*.

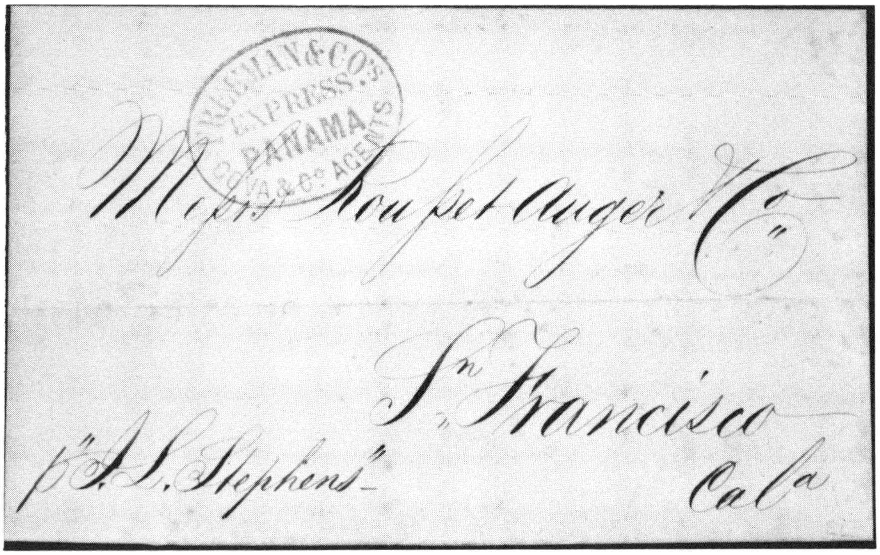

Carried by PMSS *J.L. Stephens;* forwarding handstamp of Freeman's Panama office.

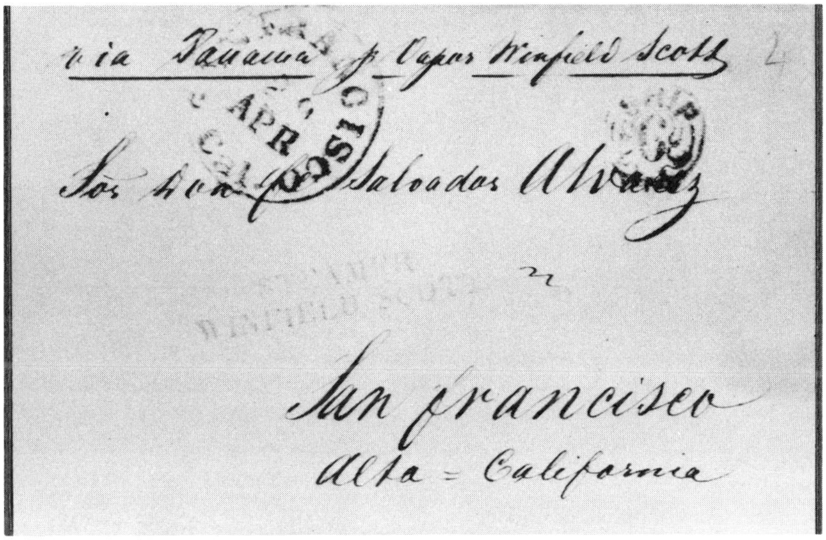

Above: WINFIELD SCOTT handstamp on cover rated as ship letter at SF. Right: same marking on cover postmarked Mokelumne Hill and addressed to Illinois, PAID 6. Carried on last voyage of *Winfield Scott* lost in fog on Anacapa I. Dec. 2 en route to Panama. The mails, passengers, and cargo were saved.

October 1852. Also he purchased the *Sierra Nevada* (1,250 tons) and on 24 March 1853 placed her into Pacific runs. In September 1852 the Vanderbilt Line had established a record time of 25 days from New York to San Francisco. From then on his advertisements read "through ahead of any other line" and asserted that all of his ships would carry letter bags for mail and papers.

The Pacific Mail Steamship Company also put its new steamer, the *John L. Stephens* (2,500 tons), on the Panama route in April 1853. Soon after, the New York & California Line, as the New York & San Francisco Line was now known, gave up and went out of business, selling the *Cortes* to the Nicaragua Steamship Company (formerly the Accessory Transit Company), and the *Winfield Scott* to the Pacific Mail Steamship Company.

When the *Winfield Scott* began as a New York & California Line steamer, someone, perhaps the Captain or the Purser, was so proud of his ship that he had a handstamp made (STEAMER WINFIELD SCOTT) and used it on all letters which the ship carried. When the Pacific Mail Steamship Company

SF newsletter to Phila. carried by *Sonora;* 10¢ rate paid by 1855 stamp.

SF newsletter rated as circular, paid by 1¢ 1851. Endorsed to *California.*

Newsletter with 10¢ 1855 stamp carried by *Golden Age* June 20, 1857. Struck DUE 10 as rate to Panama was 20¢.

bought the steamer, the handstamp continued to be used until the ship was lost on Anacapa Island in the Santa Barbara Channel on 2 December 1853. This is the only individual ship's name handstamp known to be used on the Pacific by the Pacific Mail Steamship Company prior to 1869. However, they did have a printed corner card for a two-page semi-monthly San Francisco Newsletter which they carried eastbound.

One other handstamp marking used between 13 November 1850 and 29 June 1852 by route agents aboard ships of the Pacific Mail Steamship Company must be noted. It is the

Covers with PAN. & SAN FRAN. S.S. marking, two at 30¢ Panama rate, one paid by 1847 stamps; last 40¢ transcontinental rate due.

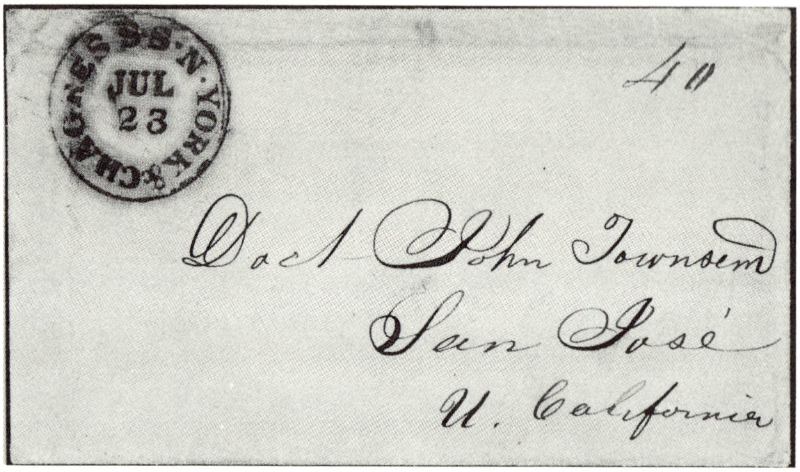

N. YORK & CHAGRES S.S. marking on cover to San Jose. Used on NY-Chagres portion of route.

dated circular handstamp "PAN & SAN FRAN. S. S." (*i.e.,* Panama and San Francisco Steam Ship). Both the United States Mail Steamship Company and the Pacific Mail Steamship Company, as part of their mail subsidy contracts, had agreed to carry at their own expense a government agent appointed by the Postmaster General to have charge of and to safeguard the mail being transported. Mail bound west from New York and east from San Francisco was placed in mail bags which were locked and sealed, to be opened only by the destination postmaster for subsequent distribution. However, mail was also picked up at ports of call, such as Savannah, Charleston, and Havana by the Atlantic ships, and at Panama, Mazatlan, Acapulco, San Diego, etc., on the Pacific side. In addition, ships' passengers whiled away the tedium of the voyages by writing letters. Since the regular mail bags were closed, these loose letters were placed in an open mail bag for delivery to the destination post office. The "PAN & SAN FRAN. S. S." handstamp may have been applied to this loose mail by the government mail agent aboard ship, thus being a route agent marking. Research has not yet revealed a final answer to the question, nor is it known why this handstamp was used for less than two years. Fewer than 20 covers are extant bearing this marking. Compounding the problem is the fact that due to illness and interim delays between a postal agent's quitting his job and the appointment of a new agent, some voyages of mail contract steamers were made without a postal agent aboard.

A very recent discovery of a single cover bearing a similar dated handstamp but reading "N. YORK & CHAGRES S. S." and addressed to San Jose, California, during the 40¢ rate period, supports the belief that these handstamps were route agent markings.

Three other possible route agent markings, all manuscript notations, appear on letters carried by the Pacific line. Although accepted as authentic route agent markings by some authorities, they are considered by others to have been written by a ship's officer or postal clerk. A cover to New Orleans in 1851 has the notation "U.S.M.S., Panama Sept 17." The second has the manuscript cancellation "P M Steamer/Oregon" on the stamps, while the third is marked "Pacific S. Ship, Way, San Francisco, Feb. 2" (1850).

The Nicaragua Steamship Company

On 10 April 1853 a startling advertisement appeared in the *Alta California.* It was an announcement signed by Cornelius K. Garrison that the Nicaragua Steamship Company steamer, the *Brother Jonathan,* would sail on 15 April carrying passengers, treasure, and a mail bag. But the *Brother Jonathan* was a Vanderbilt Independent Line ship and Garrison for years had been Vanderbilt's

Nicaragua Line oval slogan used on cover from SF to Vermont.

General Agent! And what, exactly, was this new company? It was the result of one of Vanderbilt's few mistakes. In late 1852 Vanderbilt had decided to take a lengthy European vacation on his planned new 1,860-ton yacht, the *North Star,* which was launched in the spring of 1853. Before leaving, in a paper transaction he sold his ships to his own company, the Accessory Transit Company of Nicaragua, for $2,700,000 and named Charles Morgan and Cornelius Garrison as General Agents to serve as "Trustees" of that company. While Vanderbilt was gone, Garrison, also a millionaire, quietly bought controlling interest in the stock of the Accessory Transit Company. On Vanderbilt's return, he was confronted with a *fait accompli:* he no longer owned either the steamers or the vital right of way across Nicaragua. He bowed to the inevitable, accepted a multimillion dollar offer for a release of all claims to his Nicaraguan Atlantic and Pacific fleets, and Vanderbilt's Independent Line ceased to exist.

The new line, the Nicaragua Steamship Company, continued advertising "through ahead of the mails", meaning of course ahead of the U.S. mails carried under contract by the Pacific Mail Steamship Company and the U.S. Mail Steamship Company. It continued the Vanderbilt practice of placing a handstamp marking on all letters: "Via Nicaragua in Advance of the Mails." It also stated that all of its steamers would carry mail bags with free transportation of all letters properly paid by stamp. On 15 July 1853 its newspaper advertisement listed the names of its ships: on the Pacific side the *Sierra Nevada, Cortes, Pacific,* and *Brother Jonathan:* on the Atlantic side the *Northern Light, Prometheus, Daniel Webster,* and *Star of the West.*

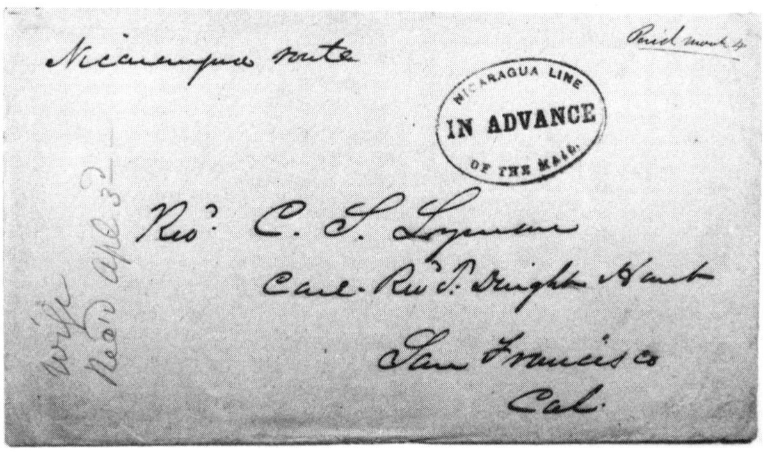

Nicaragua Line oval on cover to SF from the east.

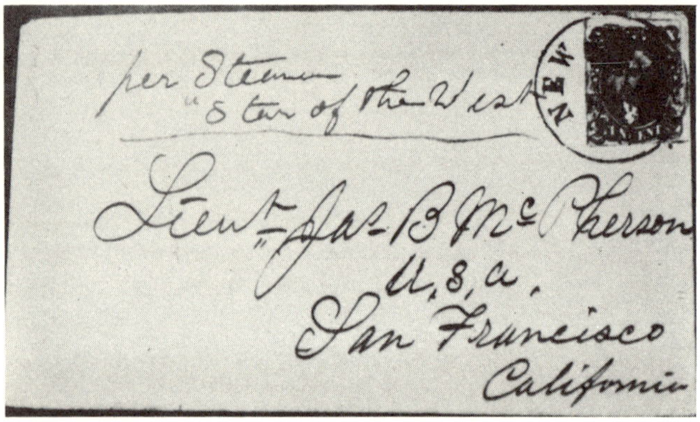

Cover from NY to SF carried by Nicaragua Line's *Star of the West.*

There are covers in existence with a printed corner card showing a side-wheeler steamer with the inscription "New York & California Steam Ship Line/for San Francisco/via Nicaragua/C. Morgan, Agent, etc." These are probably envelopes used by the Nicaragua Steamship Company of which Morgan was the New York agent and Garrison was the San Francisco agent. Advertisements in May 1855, soliciting business for the New York & California Steamship Line, indicate the Accessory Transit Company of Nicaragua as the proprietor.

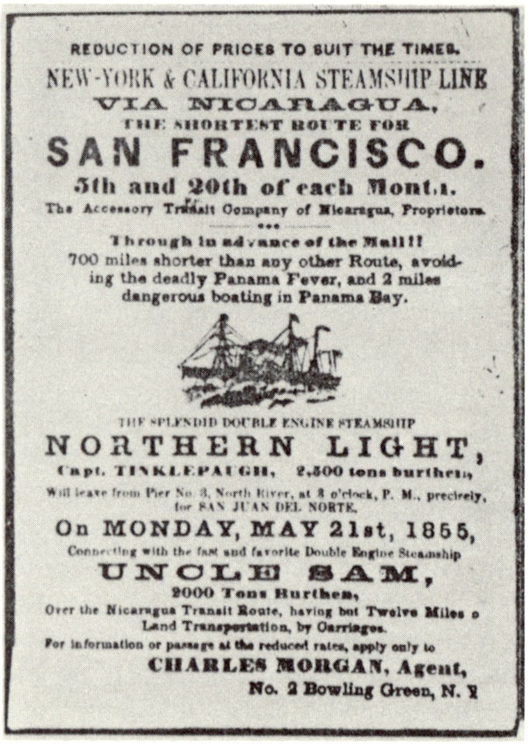

REDUCTION OF PRICES TO SUIT THE TIMES.

NEW-YORK & CALIFORNIA STEAMSHIP LINE

VIA NICARAGUA,

THE SHORTEST ROUTE FOR

SAN FRANCISCO.

5th and 20th of each Month.

The Accessory Transit Company of Nicaragua, Proprietors.

Through in advance of the Mail!!

700 miles shorter than any other Route, avoiding the deadly Panama Fever, and 2 miles dangerous boating in Panama Bay.

THE SPLENDID DOUBLE ENGINE STEAMSHIP

NORTHERN LIGHT,

Capt. TINKLEPAUGH, 2,500 tons burthen,

Will leave from Pier No. 3, North River, at 3 o'clock, P. M., precisely, for SAN JUAN DEL NORTE,

On MONDAY, MAY 21st, 1855,

Connecting with the fast and favorite Double Engine Steamship

UNCLE SAM,

2000 Tons Burthen,

Over the Nicaragua Transit Route, having but Twelve Miles o Land Transportation, by Carriages.

For information or passage at the reduced rates, apply only to

CHARLES MORGAN, Agent,

No. 2 Bowling Green, N. Y

Advertisement of the New York & California Steamship Line.

Independent Opposition Line

Vanderbilt was not one to enjoy defeat. Although he had taken his millions, now he returned for more. He promptly organized a new steamship company, the Independent Opposition Line, and moved its operation to the Panama route. Thus, on 20 September 1853 there arrived in San Francisco as a Vanderbilt steamer the *Uncle Sam*, followed in May 1854 by the *Yankee Blade* (1,800 tons). His yacht the *North Star* was converted into a steamer for the Panama to New York route. Vanderbilt was back in business again. He renewed the practice he had begun earlier of using handstamps on letters as adver-

Corner card of New York & California SS Line, NY office.

Oval slogan handstamps — Yankee Blade and North Star, Uncle Sam and North Star — of Vanderbilt Independent Line.

ments of his line. Now his markings read "Independent Line Ahead of the Mails, UNCLE SAM [or YANKEE BLADE] and NORTH STAR via Panama."

To meet the increased competition, the Pacific Mail Steamship Company added the *Golden Age* (2,200 tons) and the *Sonora* (1,600 tons) to its fleet. Vanderbilt promptly lowered rates to a level that would bankrupt all of the companies, knowing that someone would have to buy him out again. He was correct. At a huge profit he sold the *North Star* to the U.S. Mail Line, and the *Yankee Blade* and the *Uncle Sam* to the Nicaraguan Steamship Company. While he had exacted his pound of flesh, now he wanted the blood to go with it. There was still revenge to be obtained for what he regarded as "the theft" from him of the Nicaraguan Accessory Transit Company. By January 1856 Vanderbilt had purchased sufficient stock to regain control of the Transit Company.

At this time a new element entered to frustrate Vanderbilt: the filibustering activities of William Walker. Walker was an American military adventurer with no wars to fight — so he created his own wars. He had come to California in 1850, began gathering a band of American mercenaries, and in 1852 invaded Baja California and proclaimed the annexation of the Mexican State of Sonora. Mexican armies soon drove him back across the border and he retreated to San Diego to regroup.

Next, he carefully analyzed the chaotic civil-war-stricken status of Central America. On 5 May 1854 with his mercenaries he invaded Nicaragua which was torn by a civil war, pacified the country, defeated an invading army from Costa Rica, and in June 1856 was elected president of Nicaragua. Other Central American states refused to accept this "Yankee mercenary," and invaded Nicaragua from all sides. British warships blockaded Nicaraguan ports to prevent supplies from reaching Walker. The United States, embarrassed by the whole affair, sent the warship *St. Marys* to demand that Walker surrender. On 1 May 1857 Walker capitulated, and with 463 Americans was deported to San Francisco. Three years later Walker made a final effort. In 1860 with 100 men he invaded Honduras and quickly captured the city of Trujillo. Promptly the British warship *Icarus* placed Trujillo under its guns and forced Walker to surrender. The British commander delivered Walker to the Hondurans, and on 12 September 1860 William Walker was executed by a firing squad.

This account of William Walker's filibustering is included because of its effect on steamer transportation of the mails. Walker's actions were strongly motivated by the desire to gain control of the Accessory Transit Company and its attendant potential fortune for himself and his associates. Walker was a friend of Cornelius Garrison and Charles Morgan but despised Cornelius Vanderbilt, a sentiment which was heartily returned. When Walker was invading Nicaragua, he had seized the Accessory Transit Company's lake and river steamers. When in 1856 Vanderbilt regained stock control of the Transit Company, Walker claimed that the Transit Company had defrauded Nicaragua of millions of dollars, so he canceled its contract, seized its assets and closed the route. He waited until Garrison could organize a new company, The New Nicaragua Steamship Company, and could place the *Sierra Nevada* (1,250 tons) and the *Orizaba* (1,450 tons) on the Pacific side and the *Cahawba* (1,650 tons), the *Texas* (1,200 tons) and the *Tennessee* on the Atlantic, then he reopened the route for Garrison's use, which lasted only from April 1856 until March 1857. The completion of the Panama Railway in January 1855 made use of the Nicaraguan route unprofitable. There is little doubt that Vanderbilt used his vast financial resources to aid the British in their blockade of Nicaraguan ports, and his considerable political influence to persuade the American government to intervene and remove Walker from Nicaragua.

The Peoples Line

In October 1860, M. O. Roberts organized The Peoples Line with the *Moses Taylor* (1,400 tons) and the *America* (2,000 tons). Although these steamers operated in the Pacific waters from 1862 until 1868, while their owner changed the company's name from The Peoples Line to the Central American Transit Company and then to the North American Steamship Company, running sometimes via Panama and sometimes via Nicaragua, they would be of little interest to postal historians except for the rarity of the handstamped covers of these two ships. During their years of operation, they carried letter bags for only a few months on the Pacific.

While all this was going on, the Pacific Mail Steamship Company with its four large steamers, the *John L. Stephens, Golden Gate,*

Cover from William Walker in Nicaragua in 1856.

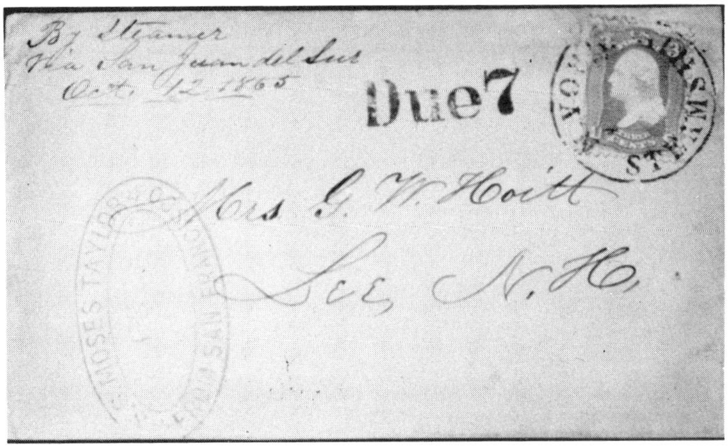

Cover with double oval cachet "SS MOSES TAYLOR FROM SAN FRANCISCO." Short paid and Due 7.

Golden Age, and *Sonora,* was carrying an average of 1,100 passengers on each of its San Francisco-Panama runs. The Panama Railway had been completed in January 1855, the Isthmus could be crossed in five hours, and the Panama passage had lost its terrors. Then in 1859 the original ten-year mail contract of the U.S. Mail Steamship Company expired. Although the contract was first awarded to D.H. Johnson, in the end Vanderbilt was the successful bidder for the new mail contract. By stock purchase and cooperative agreements he also had control of the Pacific Mail Steamship Company by 1861. Vanderbilt in association with the Pacific Mail Steamship Company now had complete control of the mail subsidy steamers on the Panama route, until September 1865 when Vanderbilt sold his stock and retired.

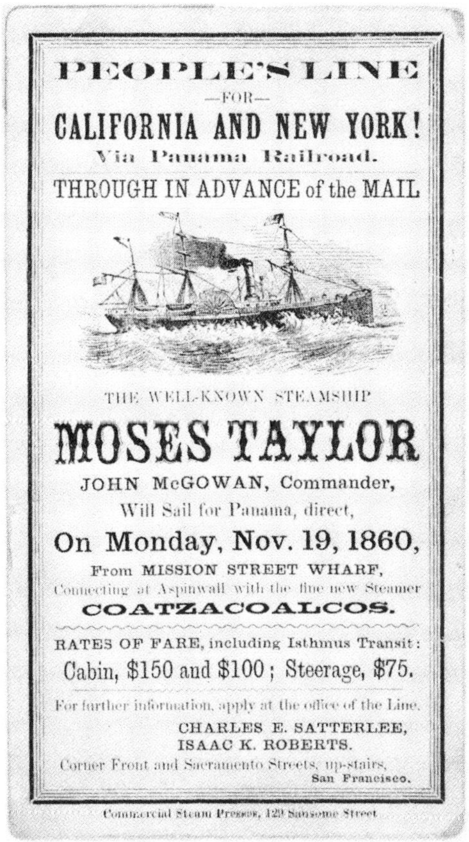

Advertisement for *Moses Taylor* sailing from NY on Nov. 19, 1860. This sailing was later cancelled.

Via Tehuantepec

There is one short chapter in the story of isthmian mail which must be noted for the postal historian. It is the account of the dramatic failure of mail transmission via Tehuantepec. The Gadsden Treaty of 1853 with Mexico had given the United States the right to convey sealed letter bags across Mexico without interference by Mexican officials. Before the treaty was in effect, Postmaster General Hubbard had signed a conditional contract with Messrs. Carmick and Ramsey to grade a road from the Pacific Ocean at Ventosa to the Coatzacoalcos River at Suchil, a distance of 125 miles. From Suchil to Minatitlan, 95 miles, a river steamer would provide passage to the Gulf of Mexico. Carmick and Ramsey spent over $300,000, then a new administration in Washington refused to pay

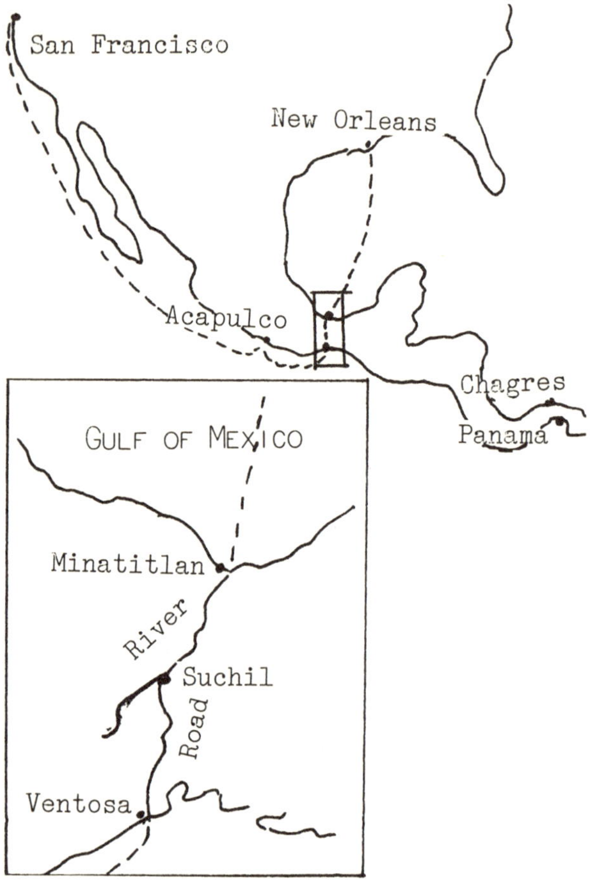

Map of Tehuantepec route.

them since the contract was conditional on appropriation of the funds by Congress.

This was an omen of the fate of the "via Tehuantepec mail." In June 1858 a contract was signed with the Louisiana Tehuantepec Company to establish a mail route from New Orleans to San Francisco for a subsidy of $286,000 per year, twice monthly service by ship to Minatitlan, then via the 1853 route of Suchil to the beaches of Ventosa, then ship service to Acapulco and to San Francisco by Pacific Mail Steamship Company steamers. The Louisiana Tehuantepec Company provided a 17 inch draft steamer for the river passage and 20 Concord coaches for the 125 miles of land travel.

It was a well-intentioned effort. As the Postmaster General reported to Congress on 24 May 1858, the Tehuantepec route provided six days closer connection with California and lay within the Gulf of Mexico "which is destined to be hereafter an American sea, over the whole of which the United States must exercise control . . . therefore the transit across Tehuantepec in time of war would be more easily protected from falling into the hands of an enemy." Unfortunately, almost

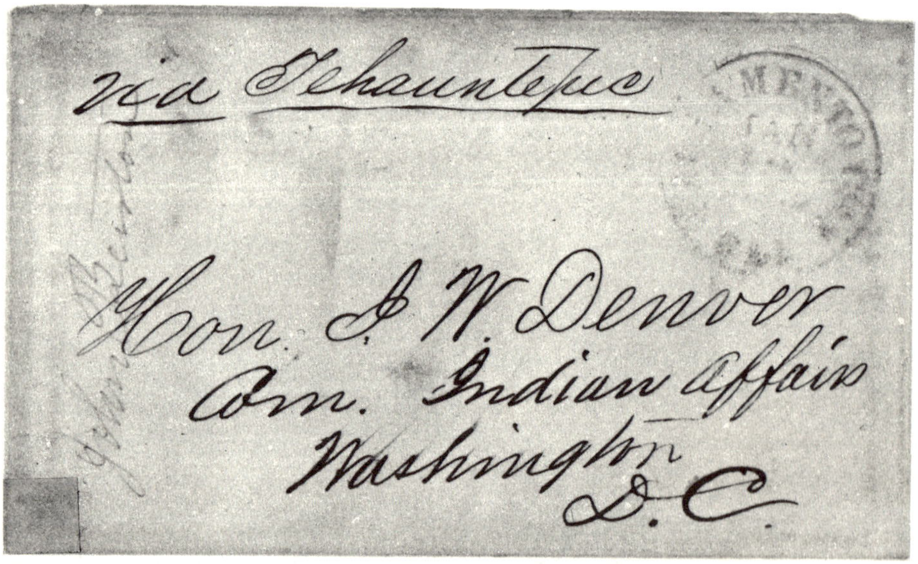

Cover postmarked Sacramento Jan. 19, 1859, and endorsed "via Tehuantepec." Addressed to J. W. Denver, Commissioner of Indian Affairs, Washington. No postage charged.

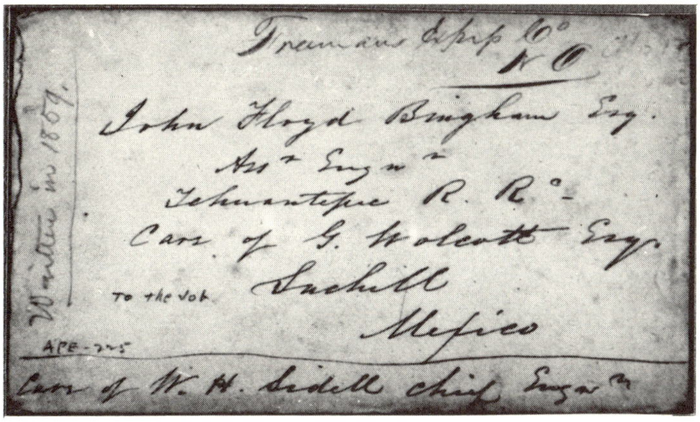

Letter carried by Freeman Express to Suchill on the Tehuantepec route.

no one used the route. It had been established ten years too late. It operated only from 1 October 1858 to 27 September 1859. New York steamer lines did not connect with it, and its primary value was to New Orleans and the Mississippi River area. Not more than 1,000 passengers, and only 34,598 letters, were carried in its coaches. Only Freeman & Co.'s California express used the route extensively. The mail subsidy had cost the government $286,000 and the total postal receipts were $5,277. When the annual contract expired, everyone was glad to let it die and the route ceased to exist. The postmaster in San Francisco had aided in its demise by decreeing that no letters would be placed in the Tehuantepec mail bags unless the senders wrote "via Tehuantepec" on the envelopes.

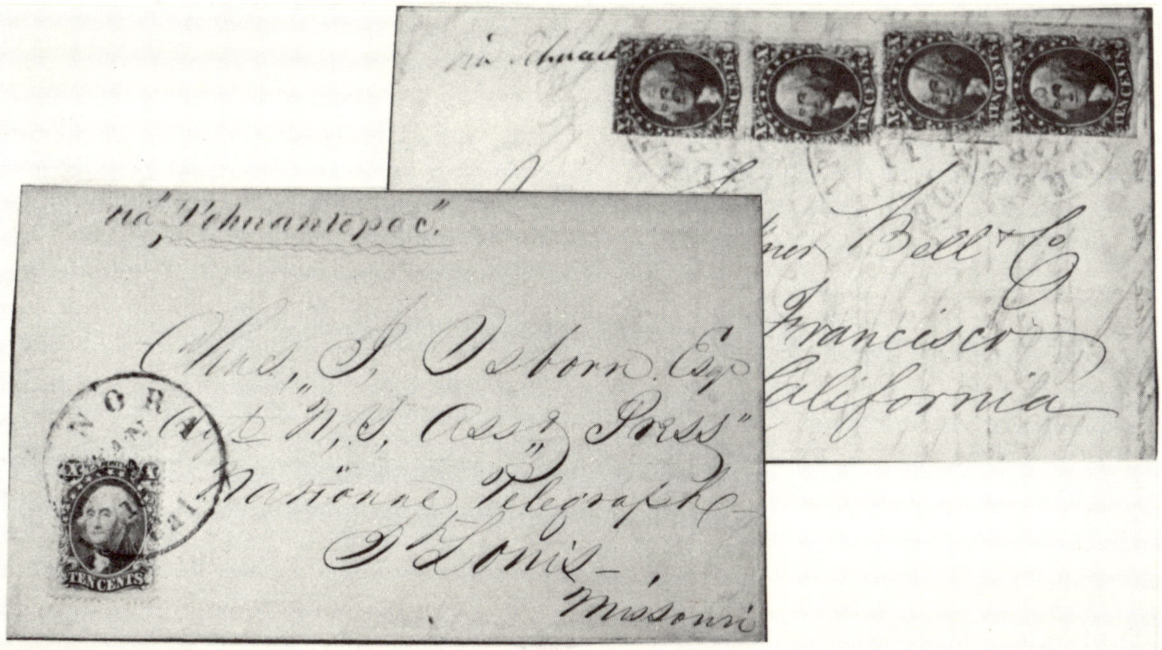

Covers sent "via Tehuantepec:" quadruple letter from New Orleans to SF; another west to east from Sonora to St. Louis.

Sailing routes between the east and west coasts.

The Last Decade

The last decade in the account of the Gold Rush steamers had come. It had long been asserted that a good overland route across America would be superior to dependence on lengthy sea voyages. In 1858 came the opening of the Butterfield route. In 1861 the transcontinental telegraph line was completed to give immediate access to important news. Yet the last decade of the steamers was a period of great activity, although of lessened competitive rivalry. The *Ariel, Champion, Costa Rica, New York, Northern Light, North Star, Ocean Queen, Atlantic, Baltic, Henry Chauncey, Alaska, Arizona, Rising Star, Ericsson, Santiago de Cuba, San Francisco, Nebraska, Nevada, Arago,* and the *Fulton* steamed the route from New York to Chagres or Aspinwall. On the Pacific side, the previously mentioned *California, Cortes, Golden Age, Golden Gate, John L. Stephens, Orizaba, Oregonian, Sonora,* and *Uncle Sam,* and also the *Hermann* (1,750 tons), *St. Louis* (1,600 tons), *Washington* (1,650 tons), *Colorado* (3,700 tons), *Golden City* (3,600 tons), *Sacramento* (2,650 tons), *Montana* (2,700 tons), and a new *Constitution* (3,300 tons) ran from Panama northward. Some vessels, at one time or another, operated on both coasts.

Passenger traffic was large in both directions, for steamers were more comfortable than stage lines and could accommodate the thousands of travelers who moved east and west. While the overland stages carried increasingly large amounts of letter mail, all bulky printed matter continued to go by sea, as did surprisingly large numbers of letters, partly because mail by steamer went from coast to coast in 18 to 22 days as opposed to 23 to 25 days by stagecoach to St. Louis. The vast amounts of freight and merchandise needed in the west could be transported only by steamer. During the Civil War almost all the gold and silver bullion from the western mines, needed to support the Union, was shipped by sea.

The story of the steamers thus far has dealt primarily with passage to and from the ports of the Isthmus. To complete that picture, attention must be given to the vital link of the transit across the Isthmus of Panama. As mentioned in an earlier chapter, the Atlantic steamers at first terminated at the small village of Chagres, a shallow harbor where passengers were taken ashore by lighter for a fee of $2 per person. The traveler then had to make his own arrangements for the 70-mile journey across the Isthmus. Hollowed-out log canoes, about 25 feet long and three feet wide, holding two to four passengers and three or four native boatmen to pole the canoe, then

The *Golden Gate* was destroyed by fire off Manzanillo, Mexico, on 27 July 1862, with heavy losses.

Covers carried by steamships *Goliah, Southerner, Cortes, Sonora* and *St. Louis.* "Sonora" notation is probably route agent marking.

struggled up the Chagres River about 45 miles to the town of Gorgona or five miles beyond to Cruces, at a usual cost of $50 per boatload. It was usually about a four day river trip. From Gorgona or Cruces the traveler hired mules to carry him down the rough trails to the city of Panama on the Pacific, at a charge of anywhere from $15 to $30 for this one-day travel.

The difficulties of this isthmian passage were somewhat eased when the Pacific Mail line contracted with local expresses and forwarding companies to safeguard and supervise the transportation of people, baggage, freight, and treasure for a uniform tariff between ports. The Atlantic line also placed shipping agents in Chagres and the new port of Aspinwall, and both lines cooperated in the improvement of the mule road from Cruces to the Pacific.

Even before the beginning of the Gold Rush, groups of shipping financiers had proposed the building of a trans-isthmian railway. After three years of political maneuvering in Congress, William Aspinwall, Henry Chauncey and John L. Stephens were authorized to negotiate with the Panamanian government. On 29 May 1850 the contract was ratified by the Grenadian Congress and railroad construction began about 1 December 1850. It was a builder's nightmare, through mountain passes, rough terrain and jungle, across rivers that rose and fell 40 feet in 24 hours, and over deep swamps. Nevertheless, on 28 January 1855 the first trains rolled from the Atlantic to the Pacific. The dreaded Panama passage had been conquered, and by December 1855 two trains were running daily in each direction with a schedule of four hours from port to port.

The transportation of mail across the Isthmus was the obligation of the government of Panama under an old postal convention of 6 March 1844. The agreed charge was 30¢ per

Train landing at Aspinwall.

pound for the first 100 pounds and 12¢ for any additional poundage. The United States consul at Panama (City) was appointed as the mail agent. Because of the poor service by the native government the postal contract was abrogated by mutual agreement. A new contract was authorized with the Panama Railroad Company, service to begin on 1 January

Panama Railroad train at San Pablo.

1852 at a rate of 22¢ a pound for letter mail. During the three ensuing years of rail construction, the company also successfully transported the mail, and upon completion of the railroad retained its mail contract in an annual subsidy of $100,000.

But on 10 May 1869 the transcontinental railroads of the Central Pacific and the Union Pacific were joined, and the isthmian steamers ceased to be the chief link between Pacific America and the rest of the nation. During two decades approximately 450,000 persons had been transported by these steamers from Atlantic ports to California, and over $750 million in gold and silver treasure had been carried from San Francisco to the east coast. For the postal historian only postal peculiarities associated with the steamer activities remain to be noted.

The Bisects of 1853

A postal curiosity of California sea mail during this period was the use of bisect stamps. As noted earlier, between 1 July 1851 and 1 April 1855 the postage rate between California and the eastern states was 6¢ prepaid. Since there were no 6¢ stamps, a sender usually would use two 3¢ stamps on a cover. However, prior to 10 November 1853 there was no postal law or regulation which forbade a sender or postmaster from cutting stamps in half, and this had sometimes occurred with the 1847 10¢ issue which was bisected to provide 5¢ postage. The same procedure was used with the 1851 12¢, creating two 6¢ stamps. A listing of 12¢ bisects appeared in the *Chronicle* of the U.S. Philatelic Classics Society in August 1972; it shows that five of seven known Sonora bisect covers were mailed in 1851 and 1852 and the bisected stamps accepted as valid postage.

During July, August, and September 1853 the supply of 3¢ stamps became exhausted in San Francisco and subsequently in some towns which received their stamp supplies from that city postmaster. During this time the San Francisco postmaster bisected many 12¢ stamps. At least 61 San Francisco postmarked covers with bisects are known from this period, as are one from Rough and Ready, two from Sonora, one from Shasta, and four from Sacramento. Many of these covers were carried as ship letters via Nicaragua. Seventeen were unrecognized by receiving post-

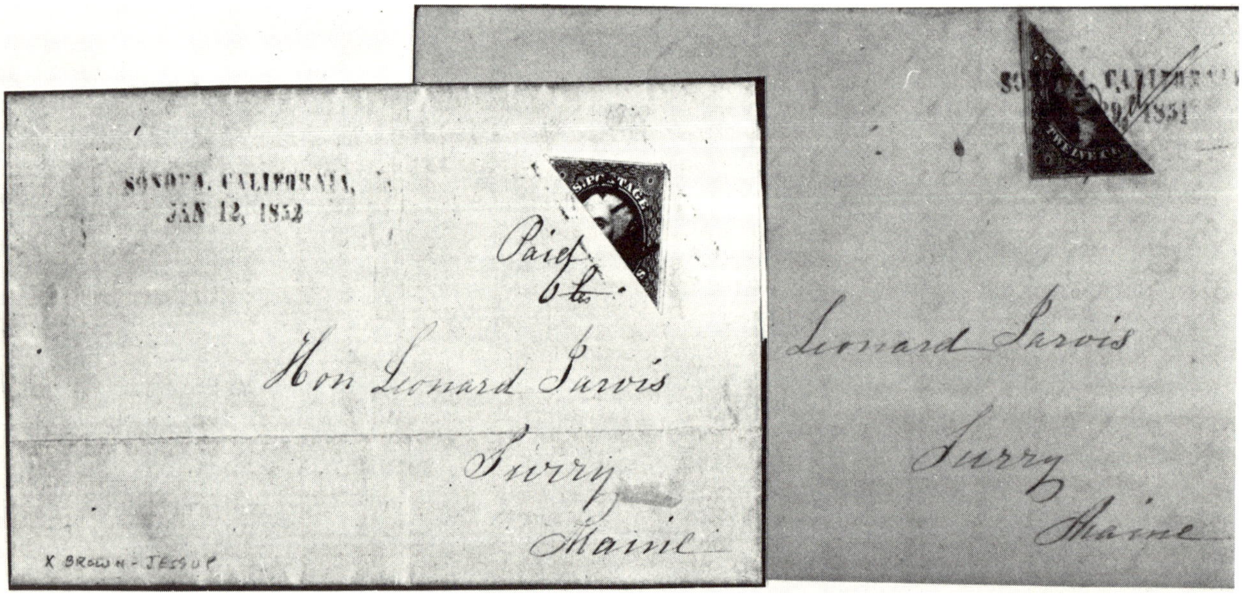

Lower left and upper right diagonal bisects of the 12¢ with Sonora straightline postmarks, Dec. 29, 1851, and Jan. 12, 1852.

SF to Boston 16 AUG (1853), upper right diagonal of 1851 12¢ paying transcontinental rate.

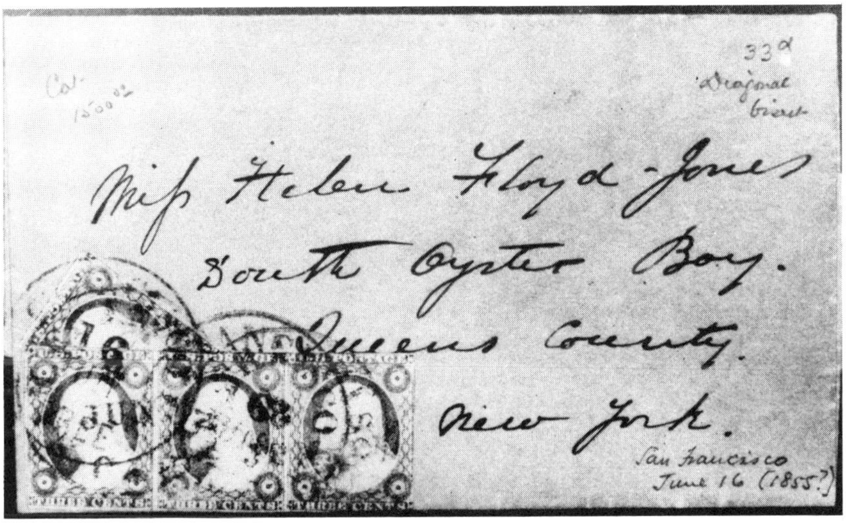

Three copies of the 3¢ 1851 plus a bisected 3¢ to make up 10¢ rate from SF to Oyster Bay, probably 1855.

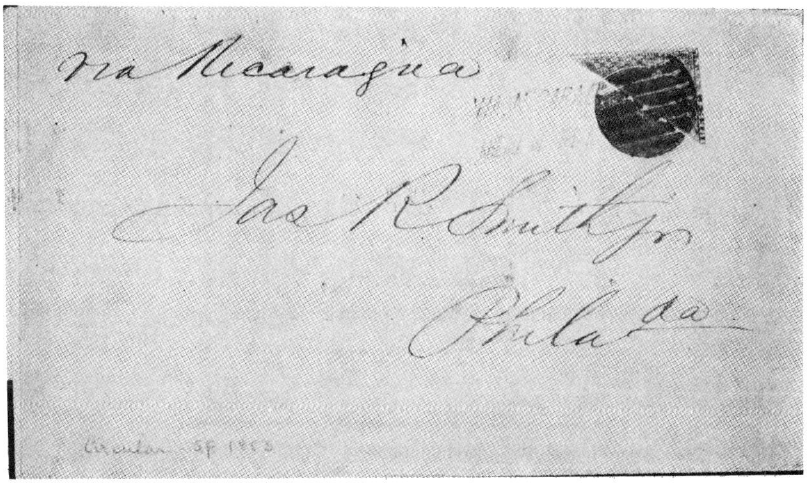

A circular from SF to Phila. in 1853, with a bisected 3¢ used to pay the 1¢ circular rate. Stamp tied by VIA NICARAGUA handstamp.

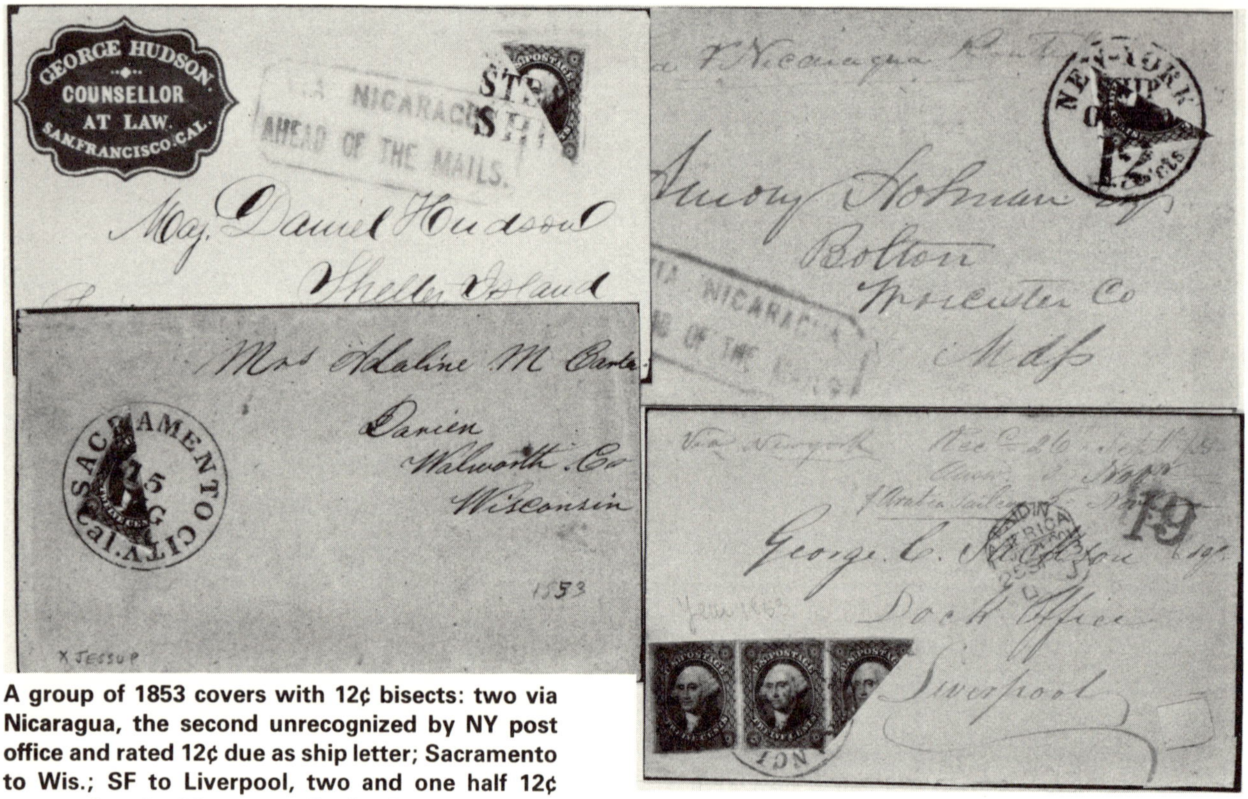

A group of 1853 covers with 12¢ bisects: two via Nicaragua, the second unrecognized by NY post office and rated 12¢ due as ship letter; Sacramento to Wis.; SF to Liverpool, two and one half 12¢ overpaying the 29¢ rate to England.

masters and postage was collected from the addressee, evidently as the result of inquiries made to the Postmaster General, for he resolved the confusion in a 10 November 1853 Circular which stated: ". . . neither does the law authorize the use of parts of postage stamps in prepayment of postage." A few California covers with bisects are known after this date, but were usually not recognized as valid by the east coast post offices.

Letter Bag Operators

There have been many references in this chapter to the service provided by steamship lines in carrying mail even though they had no mail subsidies. This practice produced a unique group of entrepreneurs in San Francisco. They were the "letter bag operators," individuals who, for a small fee, would deliver a citizen's mail to any steamer on its sailing date. These operators would determine the date, then display it at their offices and advertise it in the local newspapers. Some of these operators requested that mail be brought to their offices, others constructed mail boxes and scattered them throughout the town, and ran a pick-up service.

Prior to the Act of 1855 which required that all mail carry postage stamps beginning January 1, 1856, a sender could leave his letters with the letter bag operator either prepaid or to be paid by the recipient. Very early, however, some of the letter bag operators established two requirements: (1) the name of the steamer to be utilized must be written on the envelope; (2) the correct postage must be affixed. Extant covers reveal that neither requirement was enforced very rigidly. These requirements, however, allowed the operators to deliver their letter bags either directly to the non-contract ships or to the post office for the subsidized Pacific Mail Steamship Company's steamers.

The letter bag operators were a phenomenon peculiar to San Francisco. They were

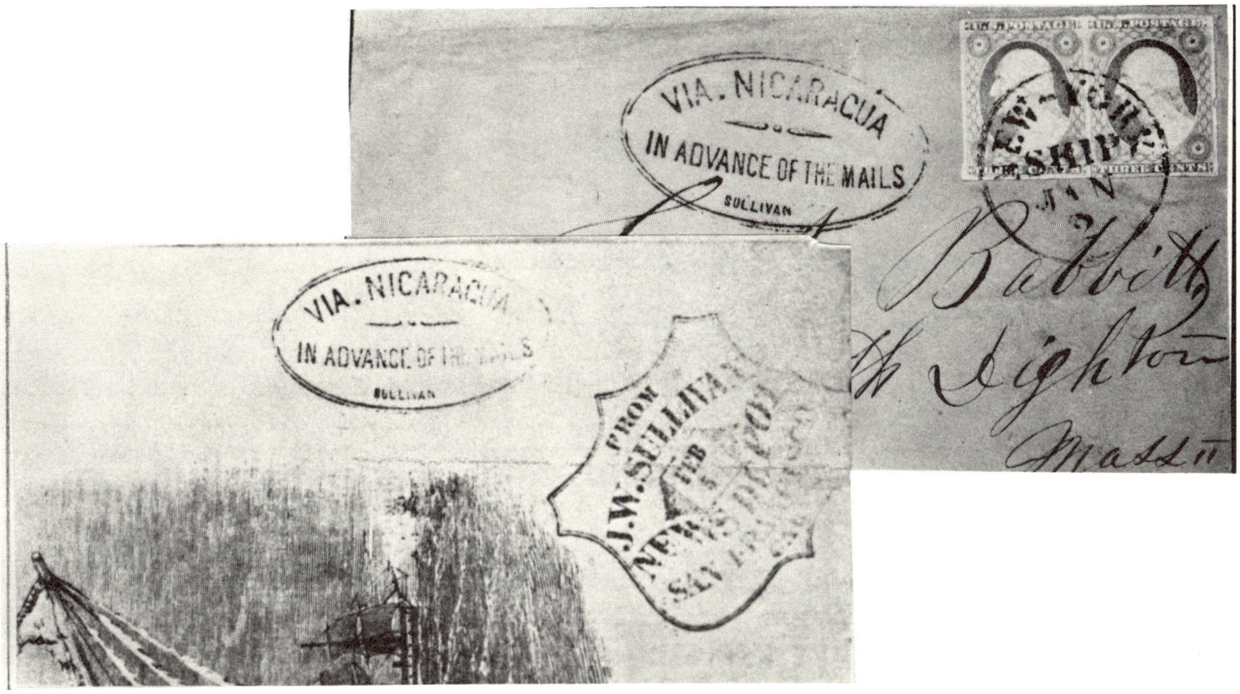

Sullivan via Nicaragua, and shield handstamps.

successful because of the mail-hungry Californians who, anticipating prompt replies, were eager for their correspondence to reach "home" as quickly as possible. It was worth the small fee charged to know that their letters would leave San Francisco on the first ship bound east. This service had begun when Law's Opposition Line of steamers in 1850 advertised that it would carry all mail brought to its office. As other competitive lines began serving San Francisco, they too offered the same service, often stating that their ships would not charge the usual carriage fee. It was this which gave birth to the letter bag operators. Among the more prominent of these operators were J. W. Sullivan, G. L. Leland, Still and Company, and Charles Kimball who as "The Noisy Carrier" became the largest of the operators. These men placed their handstamp markings on letters to advertise their service.

John W. (Jerry) Sullivan, who specialized in delivering newspapers all over California, also found it profitable to provide letter bags

Boxed Via Nicaragua/Leland marking.

Circular marking of Still's N.Y. Bookstore.

for mail sent on the Nicaragua route. For this reason he adopted the slogan of Vanderbilt's lines, "Via Nicaragua, in Advance of the Mail" and identified his company by placing his name at the bottom of the oval handstamp.

George L. Leland, later a partner in the Leland and McCombe Express Company, copied Sullivan's practice of using the Nicaraguan slogan, only he chose to sign a rectangular handstamp using the second slogan format, "Via Nicaragua Ahead of the Mails".

Still and Company was a large book store in New York which opened a branch in San Francisco. The San Francisco store, selling books, newspapers, etc., developed a large clientele. As a service to his customers, Still installed a letter bag in his store.

The Noisy Carriers, with Charles P. Kimball as proprietor, deserves special consideration because of the magnitude of its operation and the variety of its handstamps. Charles Kimball had arrived in San Francisco in July 1849. Three months later he had organized Kimball & Company's Express to carry mail and parcels between San Francisco and Sacramento by boat. It was not a financial success. In February 1850 he established a city delivery service in San Francisco, which also was a financial failure. Next he began a business of picking up newspapers from arriv-

Long Wharf, San Francisco, where Noisy Carrier's Publishing Hall was located.

Noisy Carrier markings: negative box; Independent Line; arc forwarder's; oval "per Golden Gate"; fancy scroll; tiny straightline; boxed "77 Long Wharf."

ing steamers and selling them by walking the streets of San Francisco, shouting their names and prices in rhyming verse, clever but noisy. In September 1850 he published a *City Directory* of businesses in San Francisco. This too was not a financial success.

Finally he found his *forte;* he opened a book, newspaper, and stationery store and publishing house at 77 Long Wharf. From this location he also operated a letter bag business to every steamer that sailed. Some of his mail went through the post office and traveled the Panama route, but his primary connection was with the Vanderbilt Lines and their successors using the Nicaragua route. He initiated the use of propaganda handstamps, which soon were copied by his competitors. He printed franks and illustrations of several

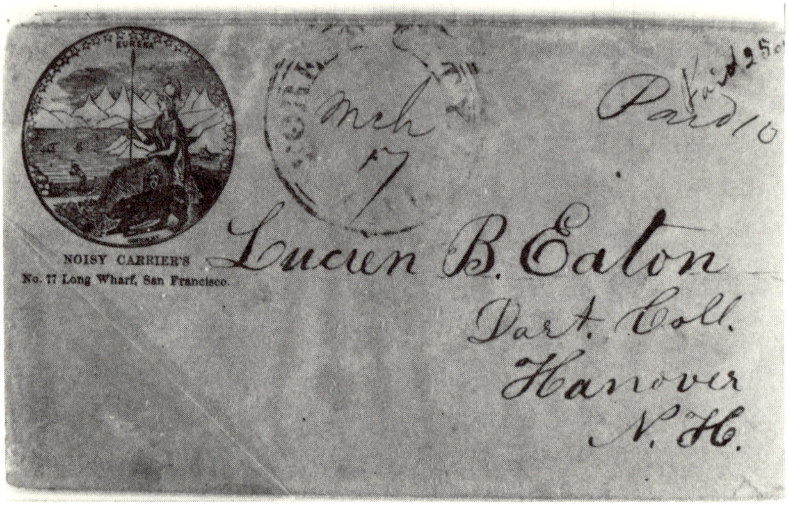

Noisy Carrier corner card on cover from Forest City.

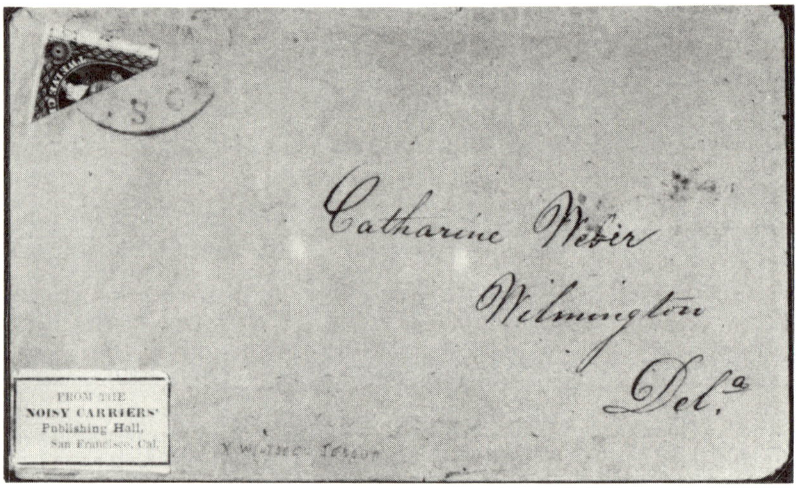

Noisy Carrier label on cover with 12¢ bisect from SF to Delaware.

NC RR/POST (Noisy Carrier River Route) on cover from SF to Sacramento.

kinds on envelopes as well as illustrated letter sheets and sold them. (Noisy Carrier corner card envelopes are classified often as "miners envelopes." Examples are in the chapter on Illustrated Mail.) After a shaky beginning, Kimball became one of San Francisco's more successful businessmen until he retired in 1860. Postal historians value highly his handstamps and printed franks.

Forwarders

Closely akin to the letter bag operators were the "forwarders," except that they functioned in larger geographical areas and served more remote customers. Unlike the letter bag operators, who were concerned with placing California's mail on outbound steamers to the east coast, the forwarders were firms in Cuba, Panama, Mexico, Peru, Hawaii, Australia, the British Isles, the Orient, and California, which would accept mail from those locales and guarantee to place it on a ship bound for the letter's destination.

Among those that functioned as forwarders were some of California's larger express companies, such as Freeman & Co., Berford & Co., Wines & Co., Pacific Union, Wand & Davis, Gregory & Co., Adams & Co., Pacific Express & Co., A. M. Hinckley & Co., and Wells Fargo. These express companies had company agents in foreign countries who would facilitate the delivery of mail in and out of those nations.

The majority of forwarders, however, were commercial firms within foreign countries where there were no U.S. postal agents, who, as an extra service, would accept mail and forward it on. Sometimes a forwarder would utilize the services of still another forwarder until finally one of them would be in a position to place a letter into the hands of a ship's mail agent. These forwarders were business firms with an official position to maintain, therefore they were trustworthy and efficient in serving their clientele. They relieved the individual correspondent of the problem of finding a ship and making his own arrangements.

Obviously these forwarders must have charged a fee for their services to the sender, but the amounts are not known, and letters do not bear any markings relevant to such charges. From the early 1800s most forwarders used a handstamp to mark the letters for which they had accepted responsibility. Among the most prized by collectors are those few covers which were handstamped by Thomas Larkin when he was the U.S. Consul in Monterey, California, before the Mexican War.

Larkin forwarding oval on letter dated May 23, 1845, from SF to Boston.

San Francisco Private Forwarders

Since San Francisco was the hub of California's universe, numerous forwarders were located there. As mentioned earlier, many of the established express companies with San Francisco offices forwarded mail all over the world. Also, individual firms performed this function, placing their markings on letters:

Agard, Foulkes & Co.
Aldrich Merrill & Co. (1865)
Alsop & Co. (1857)
Barron, Forbes & Co.
Charles Brooks & Co. (1862-64)
J. Le Cacdina & Galley
W. T. Coleman & Co. (1856-57)
Gildmeester de Fremery & Co.
Godeffroy, Sillem & Co.
Hawes San Francisco & New York Express (1854-57)
Hellman Bros. (1860s-1870s)
Henry Hentsch (1860)
Heynemann Peck & Co.
Hussey, Bond & Hale
J. Lambert
Henry Lund (1860s)
Macondray & Co. (1850) (red oval)
John B. Newton & Co.
Noisy Carriers (1850-53)
Palmer & Co. Express (1850-51)
Pascal & Dubedat & Co.
G. B. Post (1852-56)
Charles H. Shaw & Co. (1856)
George N. Shaw & Co. (1852)
J. W. Sullivan (1849-53)
Surroner & Co.
Sweeney & Baugh (1849)
A. Vignier
Wand & Davis Co. (1864-65)
Rob't Wells & Co. (1850)
Dickson de Wolf

G. B. Post forwarding handstamps: box on cover from Honolulu Oct. 16, 1854, to Mass. PER "RESTLESS" and also oval Via Nicaragua; oval on 1852 cover to Honolulu "Per Baltimore."

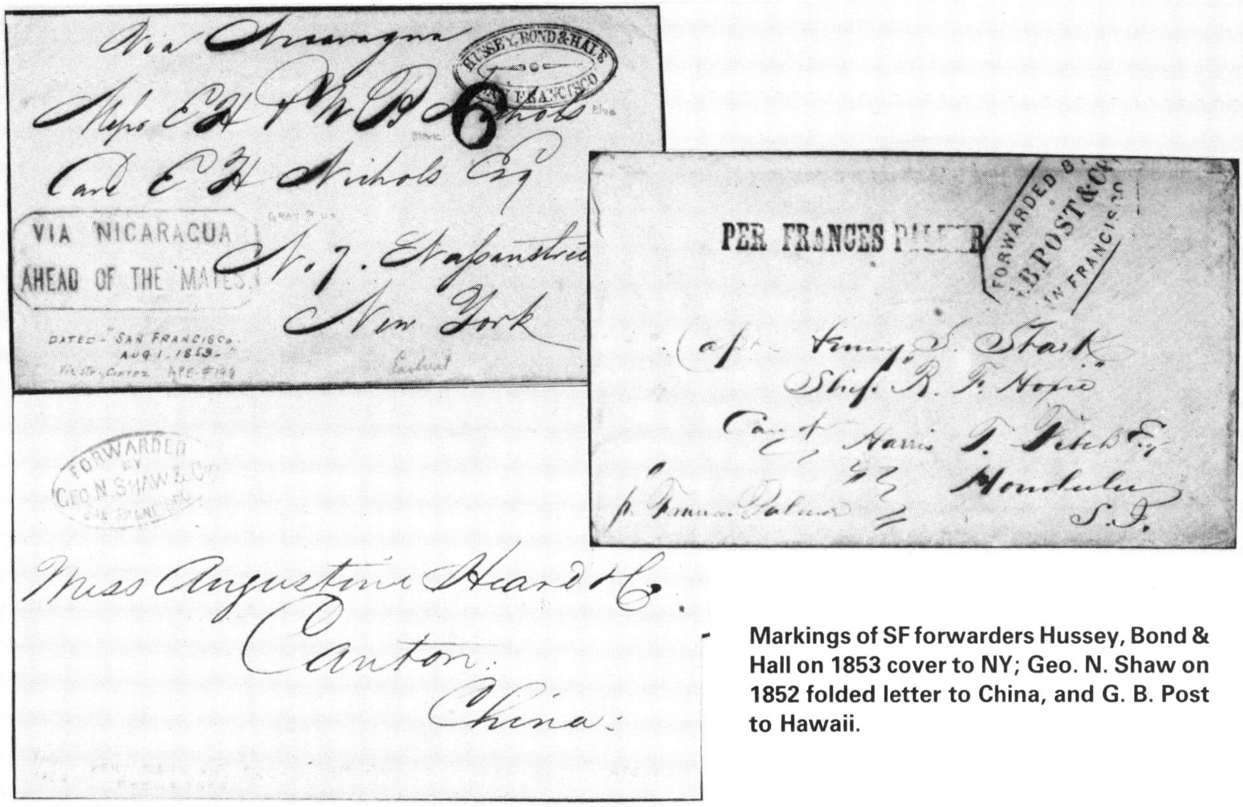

Markings of SF forwarders Hussey, Bond & Hall on 1853 cover to NY; Geo. N. Shaw on 1852 folded letter to China, and G. B. Post to Hawaii.

Since our primary interest is in California's postal history, the subsequent lists of forwarders are those known to have serviced California mail.

Hawaii Forwarders

The following forwarders are known to have had offices in Hawaii, and to have handstamped letters:

Dudley C. Bates (1861)
C. Brewer & Co. (1844)
Bush & Co. (1850)
Coady & Co.
A. P. Everett (1852)
H. T. Fitch (1855)
G. D. Gilman (1851)
Gregory's Honolulu Express (1850s)
Hackfeld & Co.
Makee Anton & Co. (1850)
Miller & Co. (1851)
Mitchell & Co. (1852)
Mitchell & Hutchins (1851)
Porter & Ogden (1850-52)
Russels & Co. (1865)

Oval cachet of Starkey Tanun on cover from Honolulu via SF to Oswego, N.Y., in 1850.

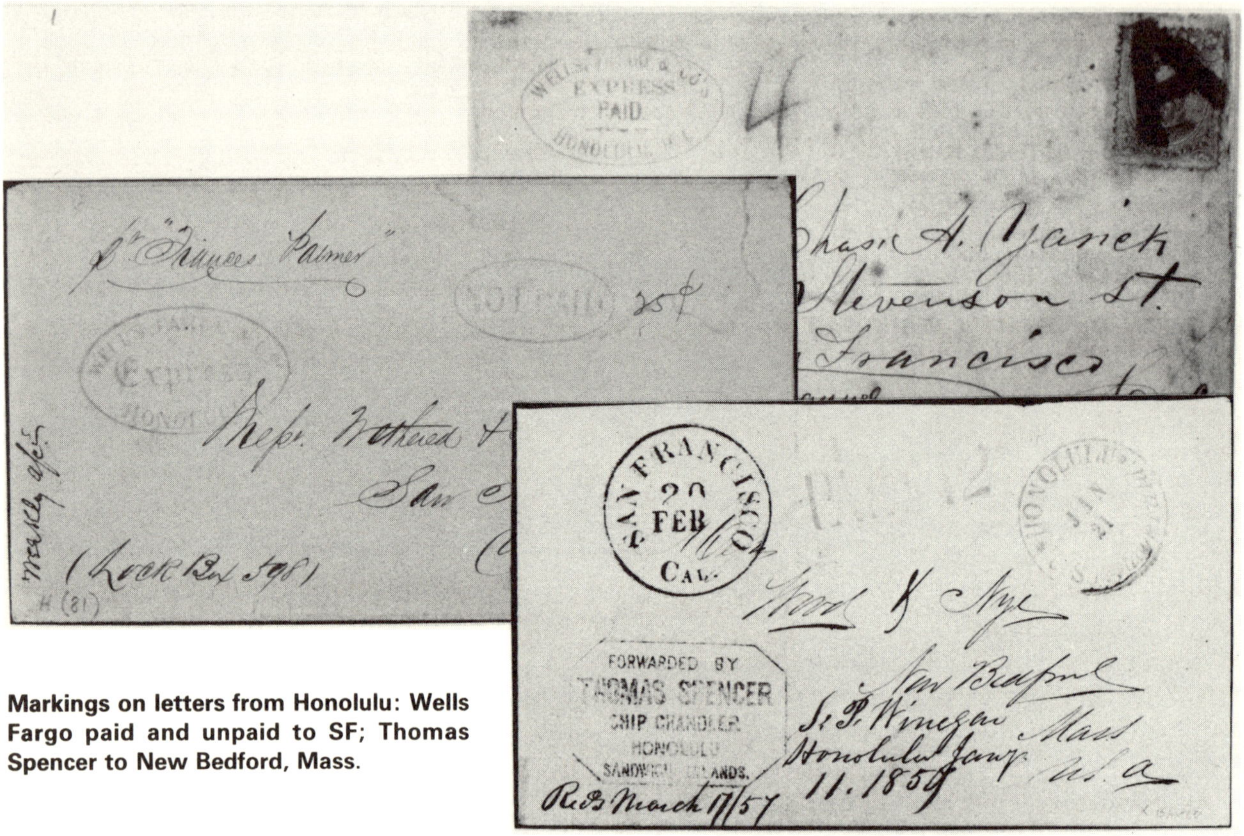

Markings on letters from Honolulu: Wells Fargo paid and unpaid to SF; Thomas Spencer to New Bedford, Mass.

Thomas Spencer (1850-57)
Starkey Tanun & Co. (1849-50)
U.S. Consulate
Waldo & Co. (1846)
Wells, Fargo & Co.

Mexico and Philippines Forwarders

In Mazatlan, Mexico, as early as 1843 William Scarborough & Co. was handstamping mail as a forwarder. Mott Talbot & Co., Bayley, Traub & Co., Leftman y Gutriel, and P. Fort & Co. were also Mazatlan forwarders. Forwarding mail through Manila was Russel & Sturgis. This is only a partial listing of forwarders from these countries, but represents the forwarders' handstamps most commonly found on California mail.

Panama Forwarders

Panama had many forwarders of U.S. mail. It was served by agents of several California expresses: Adams & Co., Berford & Co., Freeman & Co., A. M. Hinckley & Co., Pacific Express Co., The U.S. and California Express Co., and Wells, Fargo & Co. The United States Consulate was a forwarder in 1861 and 1862. But Panama also had its private forwarders, who used handstamps or manuscript notations:

E. & H. Abrahams (1864)
M. de Alcazar (1866-69)
J. B. F. Arrivet (1851)
N. Brandon & Co. (1869)
Frederick W. Byrne (1844-46)
Cabrero, Hourquet & Cia. (1859-62)
Campbell, Jones & Co. (1851)
Jose Antonio Cespedes
Corwine, Bro. & Co. (1850)
Cova & Co. (1851-55)
T. R. Cowan (1853-62)
Cross & Co.
J. B. Donalisio (1855-60)
Druce, Eckart & Co.
L. G. Duckworth (1861)
L. Ehlers y Ca. (1867)
I. B. Feraud (1834-35)

George Fisher
Geo. W. Fletcher (1854-55)
C. A. Fox (1847)
C. J. Fox (1858-60)
Garrison Fretz & Co.
F. Gogorza y Ca. (1860-61)
J. Hawes & Co. (1854-56)
A. N. Henriquez (1866)
Hurtado y Hermanos (1854-60)
Jimenez Hermanos (1858-61)
Mosquera Hurtado y Ca. (1850-51)
P. M. Leay (1868-69)
W. P. Maal (1868-76)
Pedro M. Merino (1860-64)
A. Morrell (1857)
Wm. Nelson (1841-56)
Obarrio, Planas & Perez (1866)
Justo Paredes
Samuel Piza & Co. (1866)
Victor Plise
Horquet Poylo & Ca. (1866-72)
D. A. Robinson Jr. (1861)
de Salba & Co.
Ramon Leon Sanchez
E. & T. Serruys & Co. (1850-52)
Smith & Lewis (1849-53)
Earnest Zachrison (1844)
Zachrison, Nelson & Co. (1849-51)

Cova & Co., Panama, oval on 1856 folded letter to SF.

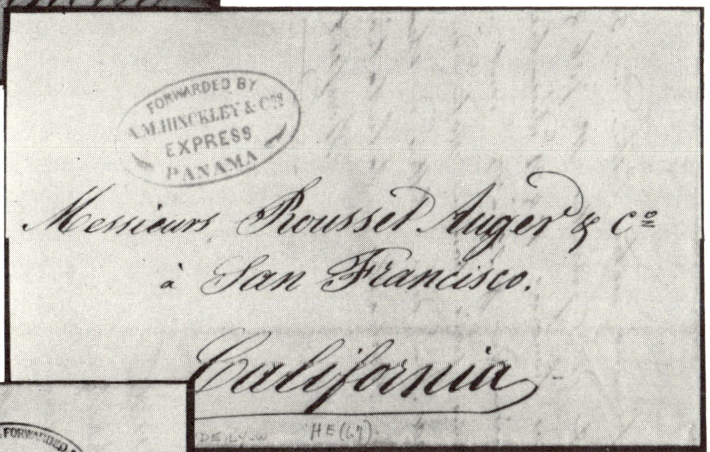

A.M. Hinckley & Co., Panama, oval on 1855 cover to SF.

Geo. W. Fletcher, Aspinwall, Ramon Leon Sanchez and Druce, Eckart & Co., Carthagena, ovals on cover to NY.

TransOceanic Mail

Hawaiian Mail

A primary source of ship mail traffic began very early between the United States and the Hawaiian Islands. Following the discovery of this isolated island chain 2,100 miles west of the California coast in January 1778 by the English sea captain James Cook, their latitude and longitude had become part of navigators' maps and charts. Their ports became important places to provision and water. In 1790 the first two American ships had anchored in Hawaiian waters. During the next 30 years the many ships engaging in the lucrative seal and sea otter fur trade and sandalwood commerce with the Orient were using the Islands as ports of call. When American whaling ships began plying the Pacific in 1819, more than 80 American vessels utilized Honolulu as a supply center. In 1820 the American missionaries arrived, devoted to both religion and education.

When the famed "hide and tallow" trade developed with California in the next two decades, the New England shipmasters rounding Cape Horn often found that prevailing winds and ocean currents carried them out to the Islands. By 1840 American merchants were firmly established in Hawaii, most of the important administrative positions in the native government were held by Americans and Britons. While California's pueblos were sleepy little villages, Honolulu was an important town with over 600 foreign residents, and publishing a government newspaper, *The Polynesian.* During the early California Gold Rush of 1849 and 1850, the stores in Honolulu supplied much of the merchandise for the gold fields. Hawaiians, nicknamed "Kanakas" in western slang, were so numerous in the mining camps that 13 California mining communities bore the names Kanaka Bar, Camp, Flat, etc. Since California and Hawaii were only about 20 sailing days apart, their postal history became intertwined.

At first the rather sparse mail between California and Hawaii followed normal ship letter practices. The United States' acquisition of California and the subsequent Gold Rush changed all this. The volume of mail increased considerably and much of it was handled through San Francisco, as this routing was quicker and more reliable than the previous long journeys by sailing vessel around the Horn or via the Orient. When the Hawaiian monarchy established an official postal system in late 1850 mail from Hawaii began to carry both Hawaiian and U.S. markings. In addition a few of California's early express companies and forwarders provided independent service.

Folded letter from NY to SF forwarded by G. B. Post to Honolulu, 1853. Honolulu to NY carried by Gregory's Express and handed to Boyds at NY for delivery.

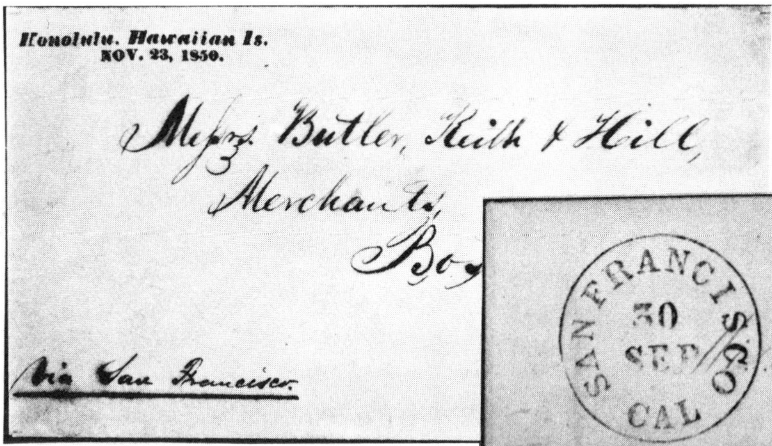

Straightline Honolulu postmark on cover to Boston.

Postage paid in cash in Hawaii: PAID/8/SHIP: U.S. 2¢ ship fee and 6¢ transcontinental.

The Hawaiian postal decree of 20 December 1850 announced a rate of 10¢ per ½ ounce letter to or from San Francisco, U.S. charges being extra. Following the reduction in U.S. rates effective 1 July 1851, Hawaiian charges were reduced on 13 September 1851 to 5¢ per ½ ounce on outgoing letters. This combined with a U.S. ship letter fee of 2¢ plus 6¢ postage if prepaid, or 10¢ unpaid, on letters to the Atlantic coast. To the despair of collectors many of these charges were paid in cash and no Hawaiian indicative markings appear on the covers.

The first Hawaiian postal marking was a straightline postmark used in 1850-51. The first Hawaiian stamps, the famed "Missionary" issue, appeared in late 1851: 2¢ for newspapers and printed matter, 5¢ Hawaiian internal postage (to port of exit), and 13¢ to cover 5¢ Hawaiian internal, 2¢ U.S. ship letter fee and 6¢ postage for over 3,000 miles. Some postmasters in the eastern United States might not recognize the 13¢ stamp because it was inscribed "Hawaiian Postage" (or so many correspondents in Hawaii feared), so a revised version with the legend "H.I. & U.S.

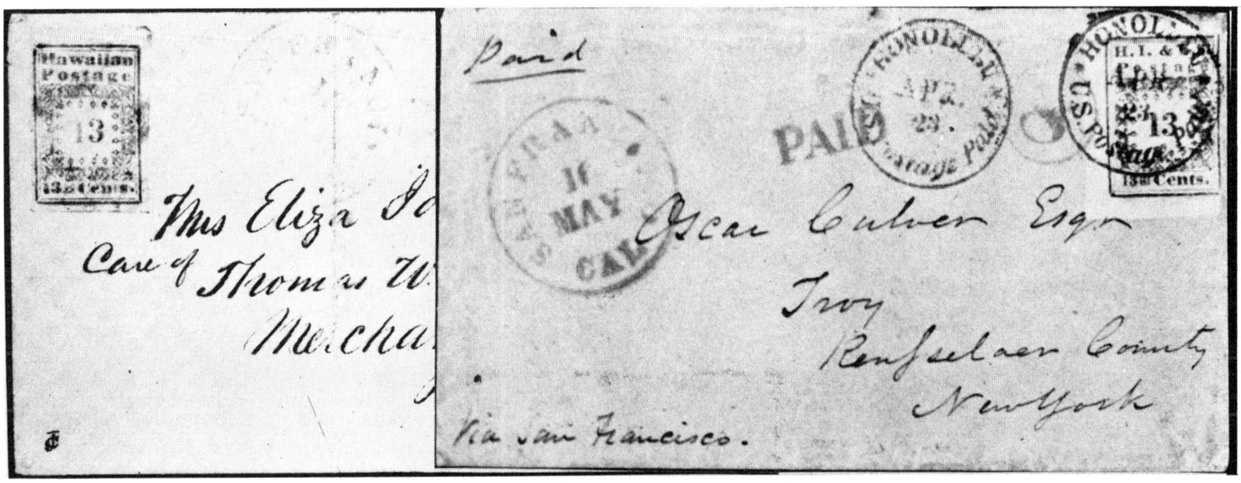

Honolulu to the U.S.: 13¢ Missionaries (first and second issue) paying entire postage.

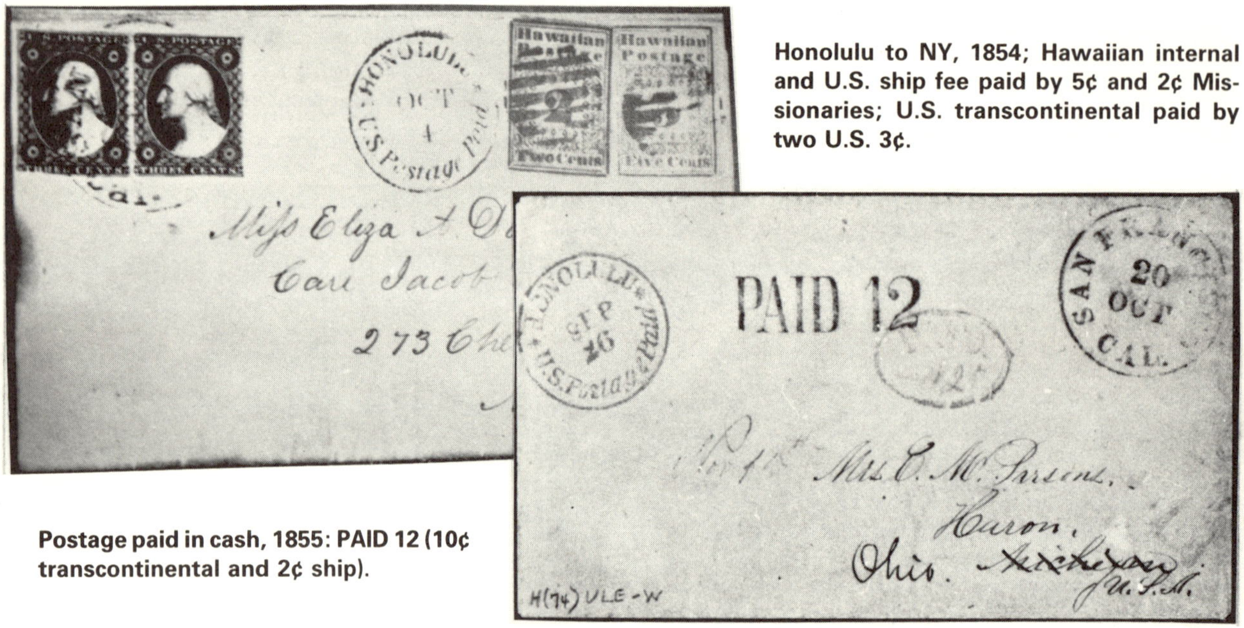

Honolulu to NY, 1854; Hawaiian internal and U.S. ship fee paid by 5¢ and 2¢ Missionaries; U.S. transcontinental paid by two U.S. 3¢.

Postage paid in cash, 1855: PAID 12 (10¢ transcontinental and 2¢ ship).

Postage" was issued in late 1852. The previous year Hawaii began handstamping mail with a circular device reading "HONOLULU U.S. Postage Paid" to indicate cash prepayment of all charges to destination. A similar handstamp "HONOLULU HAWAIIAN ISLANDS" indicated that the letter was sent with U.S. postage to be collected.

A new stamp issue with portraits of Hawaiian monarchs began to appear in 1853. The U.S. rate changes of 1855 meant that the U.S. portion of the rate to the east coast became 12¢: 2¢ ship letter fee plus 10¢ postage for over 3,000 miles or a total combined rate of 17¢. For some years the Hawaiian post office often had a supply of U.S. stamps available for sale, so mixed franking covers are known.

A revision of the Meyer and Harris work on Hawaii is now under preparation under the sponsorship of the Philatelic Foundation and should be published in the near future. It should provide answers to many of the uncertainties regarding Hawaiian postal history, and the mails between Hawaii and the United

Two 1857 covers: Hawaiian postage paid by 5¢ provisional, U.S. postage (10¢ to east, 2¢ ship) paid by 12¢ 1851; Hawaiian postage paid in cash, U.S. by 12¢ stamp.

U.S. postage not paid on 1864 cover; Hawaiian paid in cash. Rated SHIP 6 at SF (two times domestic 3¢ rate).

States in particular.

Prior to 15 October 1867 mail to and from Hawaii was carried by any ship which happened to be sailing when mail bags were ready, although some vessels operated quite regularly in the 1850s. In the decade between 1849 and 1860 more than 200 different ships carried mail; none had a mail contract and each was remunerated by a ship fee per letter. However, by Act of Congress on 2 March 1867, a contract was authorized to be issued to the California, Oregon & Mexico Steamship Company to make 12 round trips per year between San Francisco and Honolulu at $75,000 per year. The *Idaho* began this service on 5 September 1867, followed by the *Montana* on 15 October 1867, to leave San Francisco on the 1st and Honolulu on the 15th of each month, averaging 12 days per trip, thus connecting with mail steamships to and from New York, and to and from China and Japan. Letters from Hawaii had to bear 5¢ Hawaiian postage and 10¢ U.S. postage. Sometimes an oval handstamp "HAWAIIAN/STEAM SERVICE" was applied, very probably by the San Francisco post office, to letters arriving by these steamers during the contract which expired on 30 June 1870.

Despite this regular schedule, Hawaiian postmasters continued to send mail by private ships, if a sufficient quantity had accumulated between the sailing dates of the contract steamers. Such letters were handled in the usual way as ship letters arriving at San Francisco or other U.S. ports.

From Hilo, Hawaii, Dec. 17, 1868. Postage paid by Hawaii 1866 5¢ and U.S. 1861 10¢. Carried by *Idaho* to SF. Red oval Hawaiian Steam Service.

Pacific Mail

On 1 August 1865 the Pacific Mail Steamship Company received an annual contract for $500,000 to provide monthly mail service from California to China and Japan. The company agreed to furnish four large steamers for their transpacific run, and to begin operation on 1 January 1867. Steamers were to leave San Francisco on the 1st of each month and Hong Kong on the 15th. They were to coordinate their San Francisco sailings with the arrival of mail steamers from Panama, and their Hong Kong departures with foreign mail steamers from England and France. Yokohama, Japan, was a port of call, and a branch line serviced other Japanese ports as well as Shanghai, China. The United States Consuls in Shanghai and Yokohama were designated as postal agents to receive and dispatch mail.

To comply with its contract the Pacific Mail Steamship Company diverted the *Colorado* from its San Francisco-Panama route. Nine months later two new steamers, the *Great Republic* and the *China,* were placed in service, followed shortly by the *Hermann.* The steamer *New York* began operations between Hong Kong, Yokohama, and Shanghai. Covers brought to the United States on the first three trips were marked "STEAM/CHINA" or "CHINA/STEAM" at San Fran-

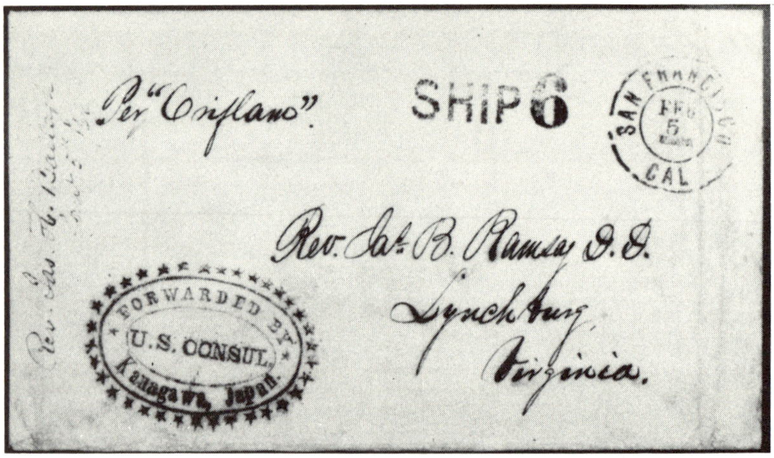

From Japan in 1866. Elaborate forwarding handstamp of U.S. Consul at Kanagawa. Rated SHIP 6 at SF, double domestic rate to Virginia.

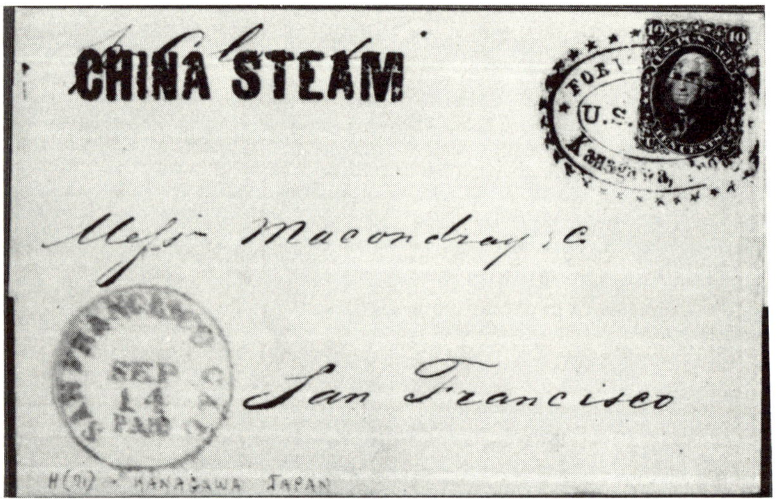

Red CHINA STEAM and SF PAID postmark on letter from Japan "p. Colorado." U.S. 10¢ 1861 tied by consular oval.

cisco. Later covers till the end of 1869 were struck with an oval handstamp "CHINA AND JAPAN/STEAM SERVICE."

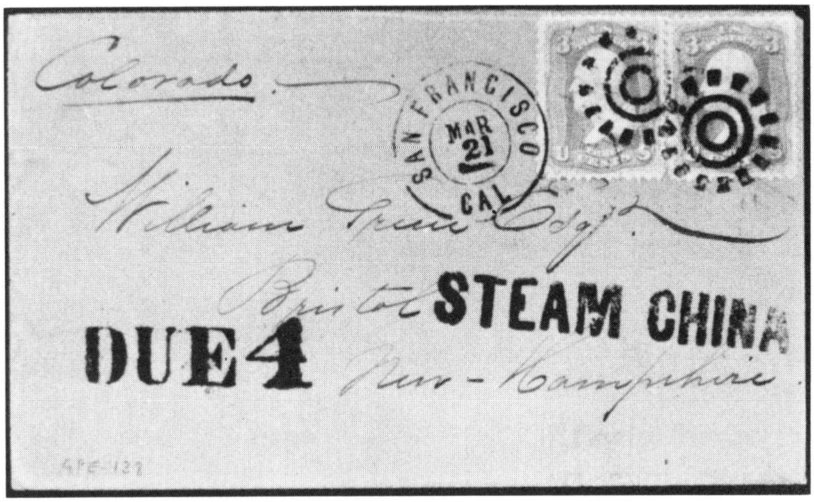

STEAM CHINA and DUE 4 on cover carried by *Colorado.* Two 3¢ 1861, insufficient for 10¢ rate, thus 4¢ due.

Oval China & Japan Steam Service on unpaid 1867 letter from Hong Kong.

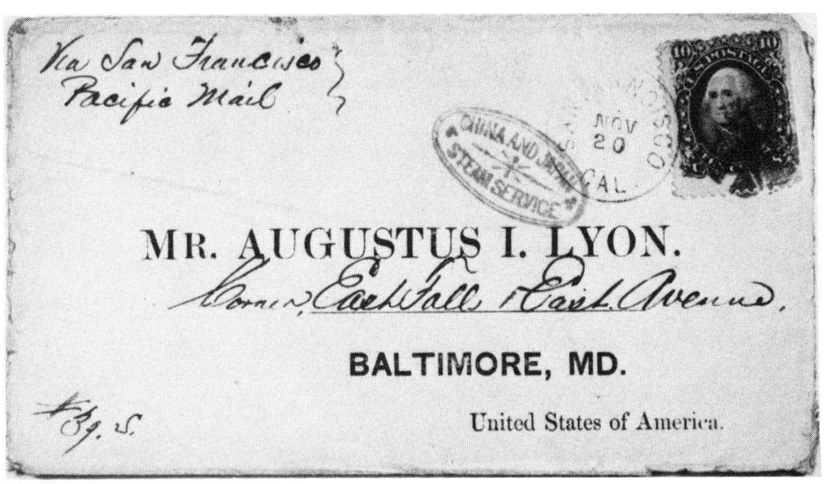

Transpacific 10¢ rate paid by 10¢ 1861. Oval China & Japan Steam Service.

Other Foreign Mail

Although specific mention has been made of certain transpacific mail, Californians corresponded with and received mail from all over the world. Analysis of postal markings on such foreign mail requires specific knowledge of many postal treaties, conventions, agreements, coinage equivalents, and foreign credit, debit and local charge markings. Such detailed information is beyond the scope of this book. The student in this specialized field is referred to George E. Hargest's *History of Letter Post Communication Between the United States and Europe, 1845-1875* and Charles J. Starnes's *United States Letter Rates to Foreign Destinations, 1847 to GPU-UPU.*

For our purpose it is sufficient to know that incoming and outgoing foreign mail was sent through San Francisco which was the principal office for the west coast. From 1 May 1851 San Francisco had made up mail bags for ports on the west coast of South America, forwarding them to the United States Consul at Panama to exchange with the British Packet Agency there for transmission to such ports. Effective 1 October 1853 San Francisco itself became an exchange office for these mails.

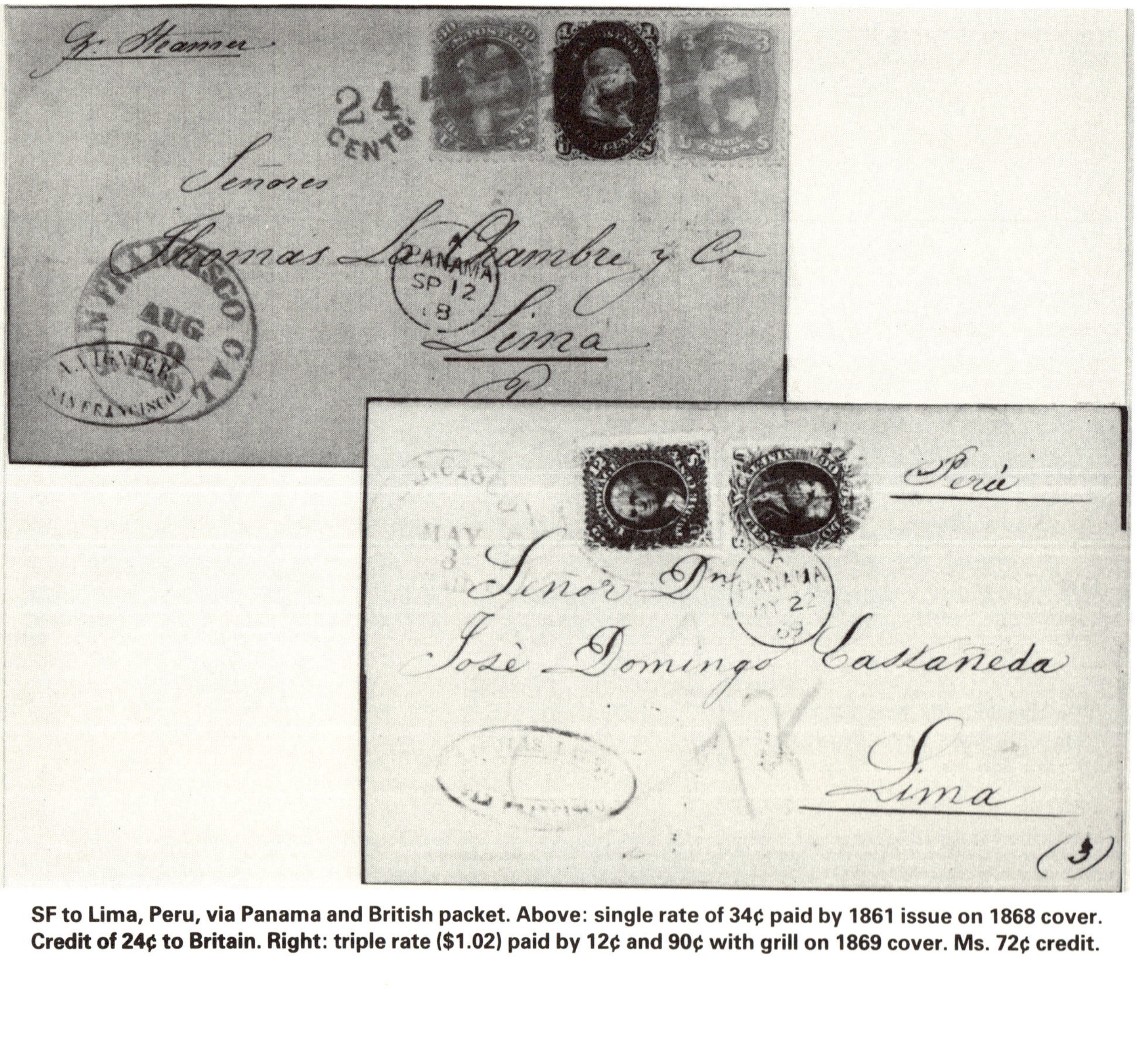

SF to Lima, Peru, via Panama and British packet. Above: single rate of 34¢ paid by 1861 issue on 1868 cover. Credit of 24¢ to Britain. Right: triple rate ($1.02) paid by 12¢ and 90¢ with grill on 1869 cover. Ms. 72¢ credit.

SF to France, 1849. PAID 40¢ for transmission in British open mail by Brit. packet. Due 15 decimes in France.

SF to France, 1856, paid 26¢ in stamps, for carriage in Brit. open mail by Am. packet. Due 8 decimes in France.

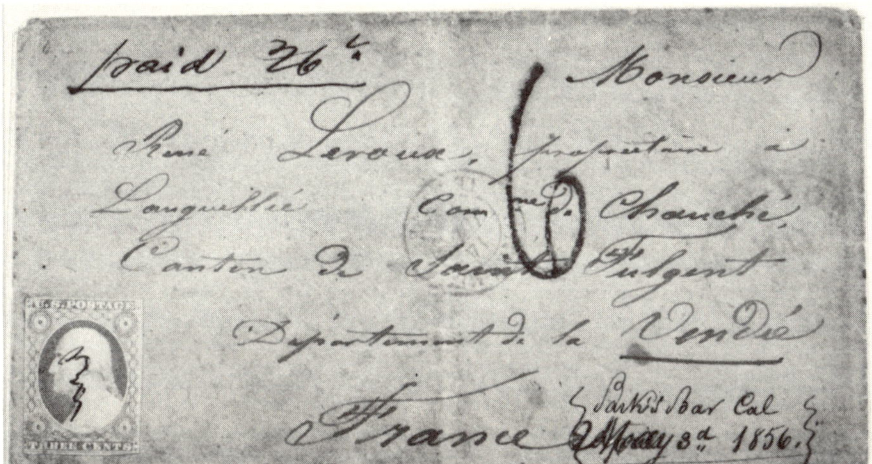

Park's Bar, May 3d 1856 (ms. postmark) to France: "paid 26¢" (presumably 3¢ stamp and 23¢ cash) for Am. Packet, Brit. open mail. Carried instead by private vessel to Le Havre, thus rated 6 decimes due at ship letter rate.

SF to Bordeaux, 1 Dec. 1850. PAID 56 for Am. packet, Brit. open mail (in 40¢ period). Straightline CALIFORNIA may be Brit. marking.

Long Bar, May 18, 1859, to France. A double letter under 1857 French postal convention; 30¢ paid in 1857 issue stamps. SF postmark with "30 Paid."

La Porte, Mar. 4, 1863. A double letter; since the rate was 15¢ per ¼ oz., the 30¢ stamp was enough. The 5¢ stamp was wasted.

By additional articles to the British treaty San Francisco was made an exchange office (paired with Kingston, Jamaica) for mails by the British West Indian packets. These mails were sent from Southampton via Kingston (later St. Thomas) and Panama. During the period 1 April 1857 to 1 April 1861 San Francisco was an exchange office under the postal convention with France and from 19 October 1863 was designated as one under

SF to Switzerland in 1853. PAID 26 for transit in Brit. open mail by Am. packet.

Oregon House, Oct. 29/69, ms. postmark on double letter to Switzerland. Apparently intended for French mail service, overpaid 6¢ (double French mail rate was 42¢), but appears to have been transmitted by Swiss mail. Swiss mail rate was 15¢ per ½ oz.

Los Angeles to Bergamo, Lombardy, May 8, 1863. Prepaid 42¢ in 1861 issue stamps. This amount paid for French service or Prussian closed mail service. At first processed for PCM, then NY postmark struck over and rerated for French mail.

SF to Britain, Oct. 15, 1850. PAID 59, total rate, 19¢ credit to Britain. Liverpool and London PAID postmarks.

Pairs of the 12¢ and 3¢ 1851, from SF to Ireland in 1854; 1¢ overpay of 29¢ rate.

Stampless cover from Jackson, 1857, PAID 29¢ in cash.

SF to Sheffield, 1855, miner's envelope with 6¢ in stamps, and balance "PAID 23" in cash.

the British treaty for *receipt* of mails to California, Oregon, Washington, Hawaii, and British Columbia and Vancouver Island.

Although San Francisco made up mail bags for Europe, it did not use credit or debit markings, even in its capacity as an exchange office under the French and British treaties as outlined above. Mails addressed to or through

Yreka, Sep. 28, 1856, to London. Pair and single of 1855 10¢ overpaying by 1¢.

SF, Sep. 15, 1861, with the 1857 24¢ and 5¢ making the exact 29¢ rate.

Double rate from SF to Scotland, June 1861. Five copies of the 10¢ 1857, plus 3¢ and 5¢, to make the 58¢ double rate.

SF to England, June 6, 1866. Twice the 24¢ rate (from July 1, 1863) paid by 1861 24¢ pair.

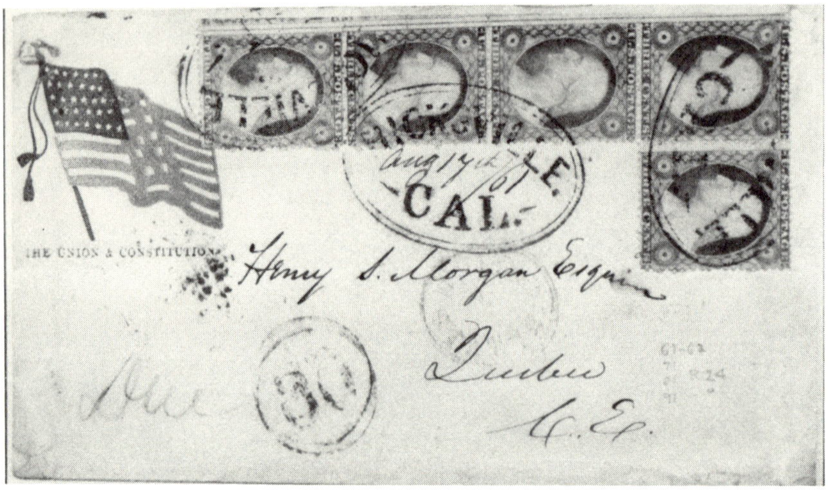

Oval Hicksville, Aug. 17, 1861, to Quebec, with five 3¢ 1857 for the 15¢ rate, but the letter was overweight and marked "Due 30" since part payments were not recognized.

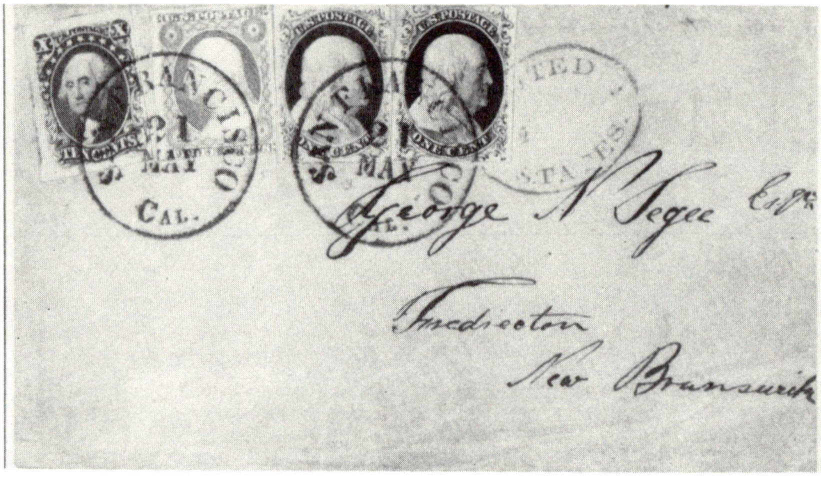

SF, May 21, 1856, to New Brunswick, the 15¢ rate paid by a combination of 1851 and 1855 issues. Oval UNITED STATES marking of Boston exchange office.

SF, Jan. 30, 1860, to Nova Scotia. Rate paid by five 1¢ 1857, and 10¢ 1857.

SF, 25 July 1850, to Bremen, via NY Sep. 6. Rated 64¢ due: 40¢ transcontinental rate and 24¢ U.S. packet rate under first Bremen treaty.

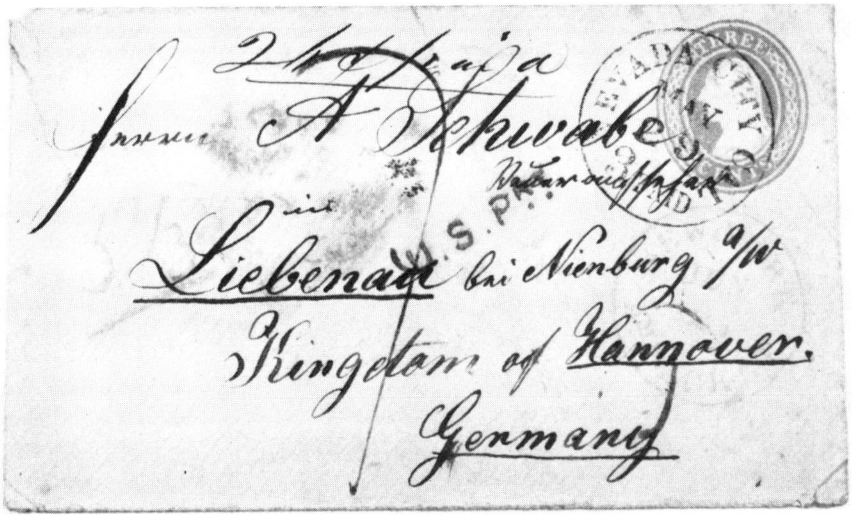

Nevada City, May 29, 1855, to Germany. Ms. "26¢ paid" — rate for Am. packet carriage in Brit. open mail.

Honolulu to Germany via SF, Oct. 28, 1864 (?). Hawaiian postage paid by numeral issue; transatlantic at PCM rate by three 1861 10¢ (2¢ overpayment as prepaid rate was reduced in 1862).

SF to Germany in Prussian closed mail: to Bremen, Aug. 20, 1859, 1857 issue for 30¢ rate. PAID Dec. 30, 1859, on stampless overland envelope. SF, Feb. 17, 1860, strip of three 10¢ 1857, on choo-choo envelope.

Europe were sent to New York for such processing.

From 15 February 1849, the effective date of the U.S.-G.B. Postal Convention, the inland portion of the rate required a 35¢ sur-

(text continued on page 144)

Georgetown to Denmark, May 31, 1861. The 35¢ PCM rate made up by 24¢, 10¢, and 1¢ 1857, all cancelled by fancy black star.

SF to Norway, April 18, 1862. Double letter by PCM. Twice 42¢ rate paid by 1¢ pair and 90¢, 1861 issue.

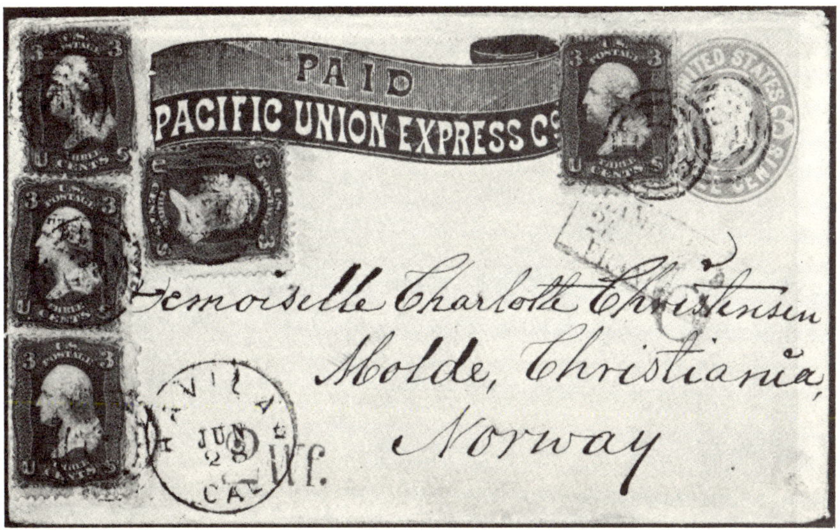

Havilah, June 28, 1869, to Norway. Five 3¢ 1861 on Pacific Union Express franked 3¢ envelope. A 2¢ overpay of the 16¢ North German Union direct rate.

British Columbia to Lombardy via SF, 1862. BC 2½d and 35¢ U.S. postage. By PCM; rate to Lombardy was 42¢, thus marked "FRANCO" only to limits of GAPU. Deficiency probably collected from addressee.

Honolulu Nov. 20, 1852, to Russia, via SF. PAID 28 (it appears) at origin, for transit in Brit. open mail: 2¢ ship fee into SF and 26¢ open mail Am. packet rate. NY Am. packet postmark of Jan. 1.

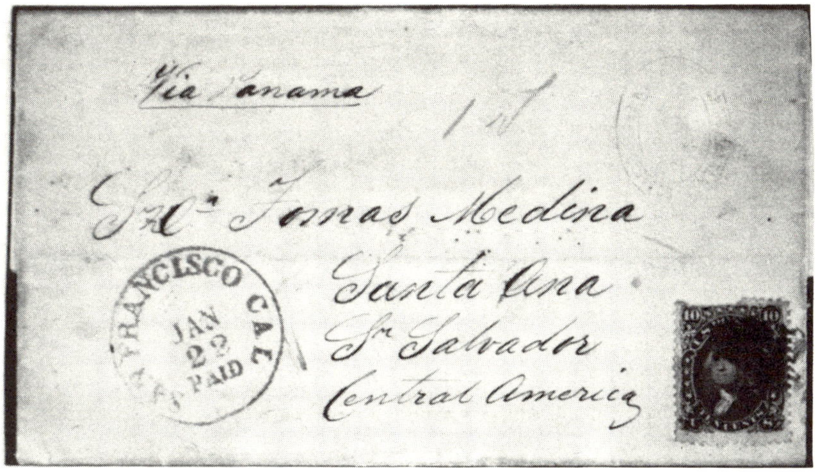

SF to San Salvador 22 Jan. 1869. Postage of 10¢ to Panama exchange office paid by 10¢ 1868 with F grill. SF PAID postmark.

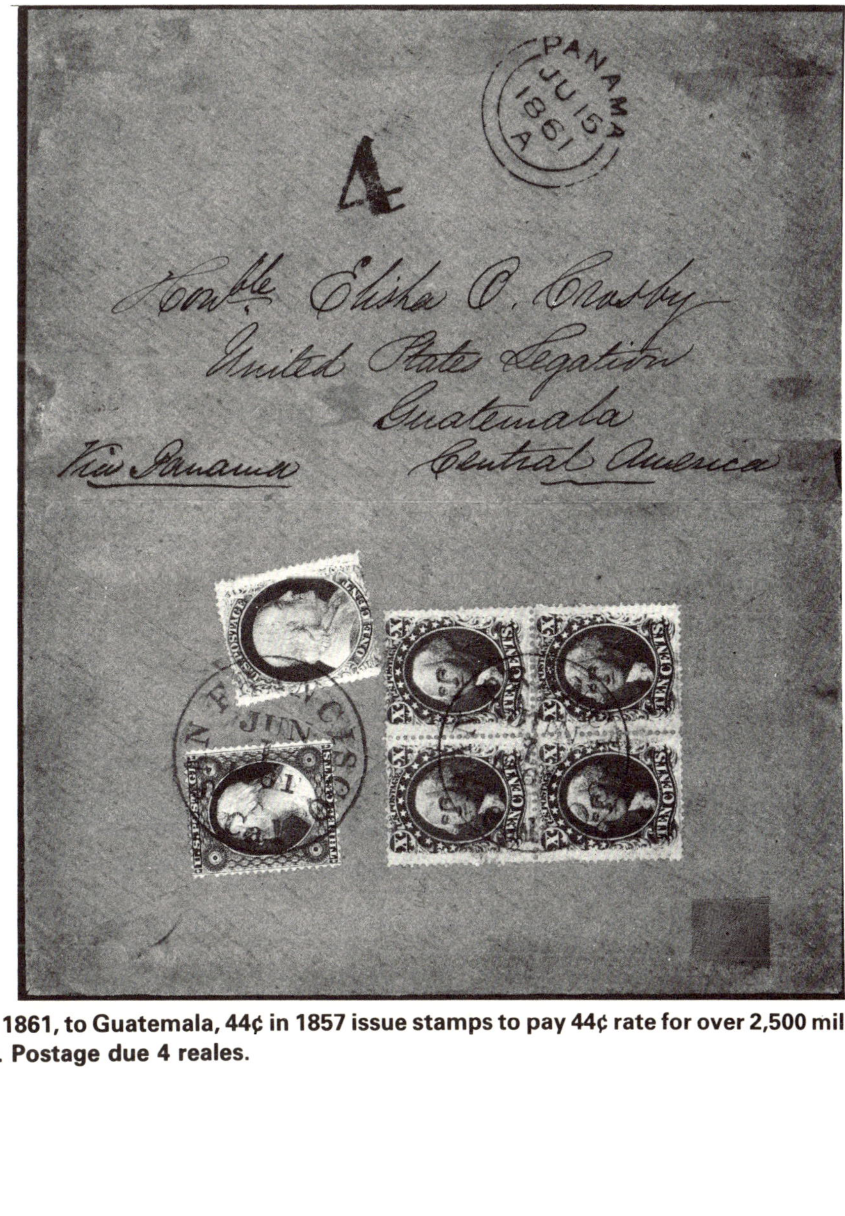

SF, Jun. 1, 1861, to Guatemala, 44¢ in 1857 issue stamps to pay 44¢ rate for over 2,500 miles from office of mailing. Postage due 4 reales.

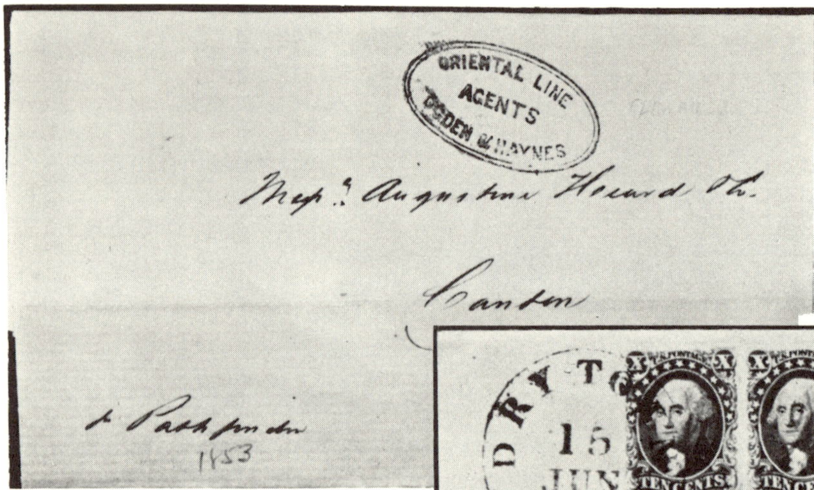

Forwarding oval of Ogden & Haynes on 1853 letter to China outside mails.

Five 10¢ 1857 to pay ½ oz. rate via Marseilles: 45¢ plus 5¢ west coast surcharge.

Pairs of 2¢ Blackjack and 3¢ 1861 from N.H. (Sep. 9, 1869) for 10¢ Am. packet rate to China via SF. Magenta SAN FRANCISCO PAID ALL.

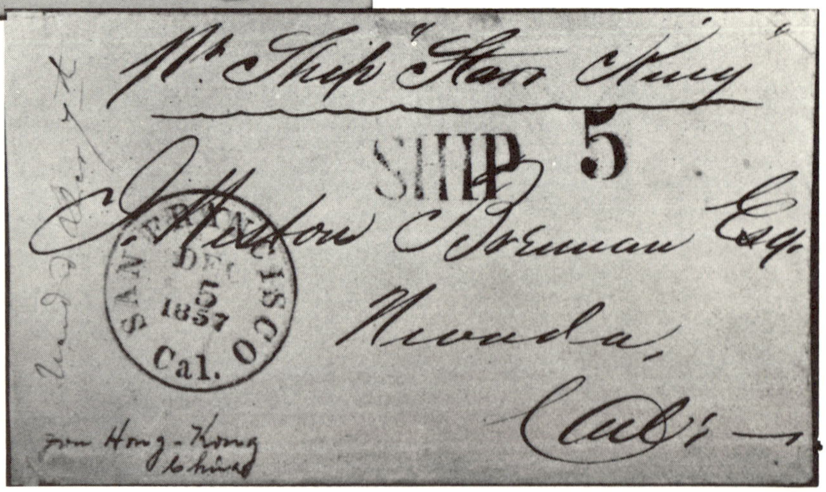

Incoming letter from Hong Kong "pr Ship 'Starr King' " to Nevada City. Arrived SF Dec. 5, 1857; rated SHIP 5 due.

Oakland to Bolivia 13 Dec. 1868. Double rate (2 x 34¢) paid by 1861 issue stamps, via SF PAID, and credit 48 to Britain.

Georgetown to Mexico in 1862. "PAID 20" by 10¢ star die and 10¢ 1857, both cancelled five-pointed star. Rate for over 2,500 miles, Mexican due 2 reales.

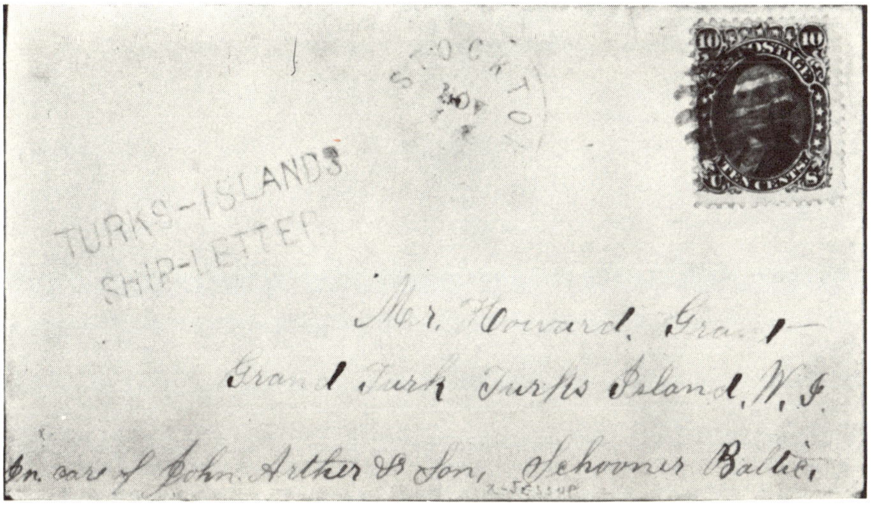

Stockton to Turks I., with 10¢ 1861 for outgoing blanket rate. By private ship and received as TURKS-ISLAND/SHIP-LETTER.

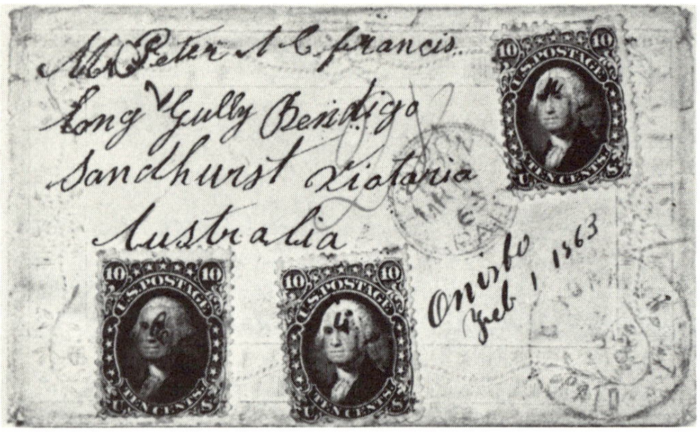

Onisbo, Feb. 1, 1863, ms. postmark to Sandhurst, Victoria, Australia. Only 30¢ postage attached, but correct rate was 38¢ (33¢ in Brit. mails via Southampton + 5¢ west coast surcharge). Balance paid in cash or stamps missing, as credit of 28¢ and NY and London postmarks show payment.

SF to North Queensland, Australia, 5 Sep. 1868. Rate of 22¢ via Panama paid by 10¢ 1861 and 12¢, Z grill. Credit 12 cents to Britain. Transit from Panama by New Zealand packet. Large 10 probably is Australian delivery charge.

SF to London, 18 Sep. 1863, with strip of three 10¢ 1861, plus 90¢, intended to pay five times 24¢ rate. Ms. "over 2 oz." However, until April 1866 no odd rates above single were permitted, so the 2-3 oz. rate was six times. Part payments were not recognized, therefore rated 6/- due in London.

Pescadero to Switzerland 12 Sep. 1867 by PCM. Seven copies of 5¢ 1861 to pay postage. Overpay of 2¢ as PCM rate had been reduced to 33¢ in May 1863. Credit of 12¢ to Prussia.

charge for California (to make up the 40¢ rate to and from the east coast) until 1 July 1851 when the surcharge became 5¢, remaining until 1 July 1863. Specialized postmarks and rating handstamps incorporating the open mail or through rates paid were often used at San Francisco on foreign mail. These include "PAID 59" on letters to Great Britain in the 40¢ rate period, "29 PAID" thereafter, and "PAID 26" on letters by British open mail carried by American packet after 1 July 1851. Postmarks showing "15 PAID" (single) and "30 PAID" (double) were used on French treaty mail to France from 1 April 1857 and were sometimes struck on incoming mail to indicate prepayment, while corresponding markings without "PAID" were used on unpaid letters.

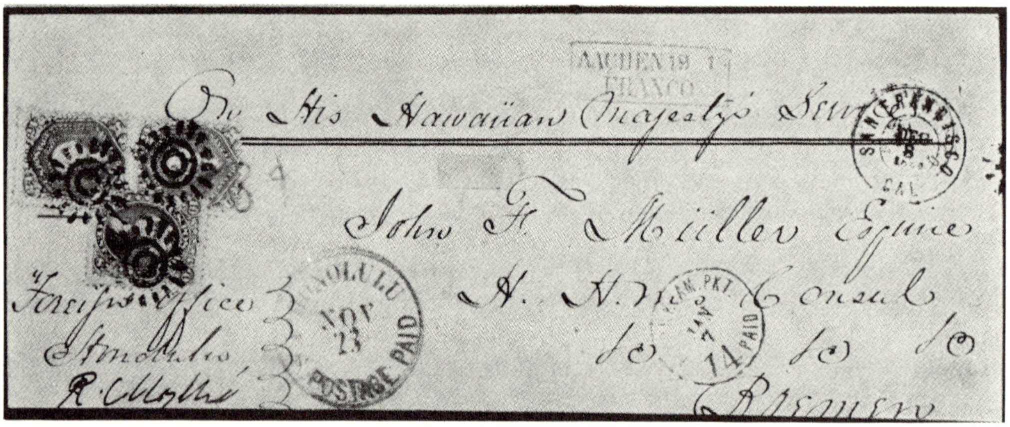

Hawaii to Bremen in Prussian Closed Mail. Official letter, thus no Hawaiian postage. U.S. postage of 58¢ paid by 1861 10¢ and pair of 24¢: two times 28¢ PCM rate plus 2¢ ship fee.

Chapter 5

The Expresses

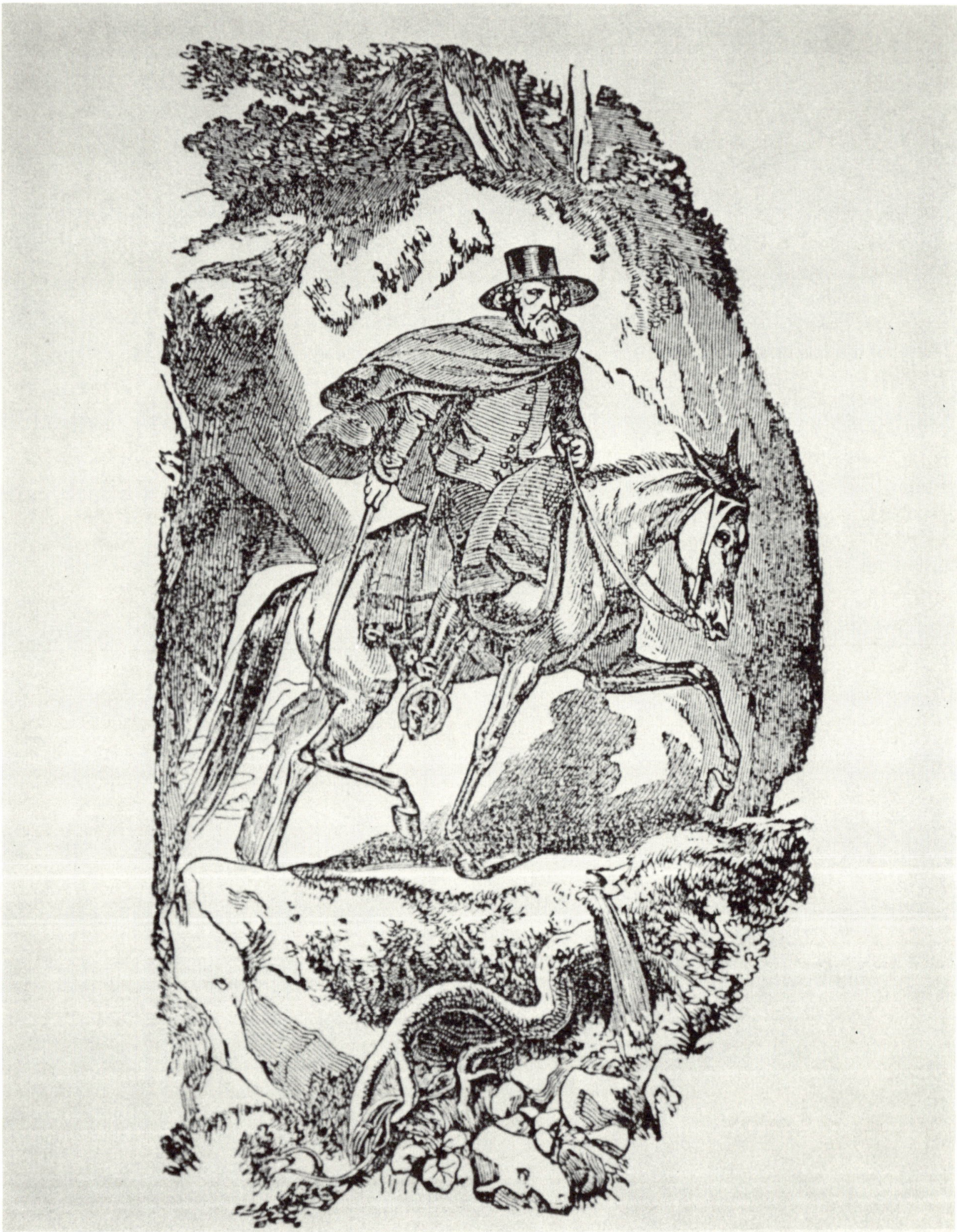

The Mountain Expressman — a woodcut by Nahl, after a drawing by Alonzo Delano.

The Expresses

When most historians write about California, the account of the early-day express companies is an often neglected chapter. The postal historian, however, prizes it. During the early years of the Gold Rush, over 90 percent of all mail was delivered by the express companies because of the inadequacies and delinquencies of the United States postal system. For example, in July 1851 there were only 53 post offices in all of California, and only 25 of these were in the mining regions even though there were several hundred mining camps and towns. While steamers, both subsidized and "outside the mails," were carrying postal material to and from California, there still was the unresolved problem of mail distribution within the state. Letters, newspapers, and other necessities of life reached their recipients through the diligence of the packmule-wagon-and-stage express. The expressmen wrote a vital chapter in postal history.

It is a temptation to distinguish between *freight lines, stage lines,* and *expresses.* Obviously the freighting companies specialized in the transport of food, clothing, mining equipment, and other physical necessities. The stage lines were organized to carry passengers, and they added the transport of gold and mail both to increase their profits and because there was no other agency to do it. The early express companies saw themselves first as mail and package carriers, and then they had the transportation of raw gold thrust upon them by the miners with whom they had such intimate contact. But the distinction becomes erroneous when it is evident that all three services ultimately, to some extent, duplicated all of these functions. Since the definition of an express is "a means of rapid conveyance of valuable articles," the term "expresses" will be used in a generic sense to describe all of the land transportation agencies which met the needs of the miners and other inhabitants of California until the transcontinental railway made them obsolete in 1869. Expresses still served remote areas after that date, but their heyday had passed.

It was these expresses, stages, and freight lines that nurtured the physical, spiritual, aesthetic, and emotional needs of the Far West. They carried people, food, clothing, household utensils, merchants' wares, farm supplies, mining equipment, bank and government documents, gold and currency — and, above all else, letters and newspapers. Until the telegraph line was completed in October 1861 Californians had no current knowledge of world events, of Congressional and Presidential laws and acts which affected their lives. Mail not only meant contact with loved ones and information about national affairs, but it also conveyed currency, bills of exchange, bank drafts, and business documents. Mail and newspapers literally were worth much more than their weight in gold. "The Express companies that brought their mail, delivered their letters to each camp, bar, creek or gulch . . . were almost as essential to their existence as the food they ate and the air they breathed," wrote Wiltsee in his study *The Pioneer Miner and the Pack Mule Express.*

It must also be remembered that the argonauts already were familiar with the concept of "express companies." Back in the States, expresses had been operating since 1834. W. F. Harnden's Express, which began in 1839 between Boston and New York, had grown until it had offices all along the Atlantic seaboard as well as agents in England, France, and Germany to facilitate its transatlantic business. Adams Express Company covered the entire eastern United States, and also had foreign offices. Handstamp markings on letters and parcels thus were familiar to the California argonauts.

In the beginning most California-bound mail simply was addressed to San Francisco. Later it was addressed to the "base towns" of Marysville, Sacramento, or Stockton according to the instructions of the recipient. As more towns, in turn, became supply centers with post offices, mail was addressed to them. There was, of course, no delivery service by these post offices. It was the responsibility of the peripatetic miners to come to these post offices and claim their mail. Oftentimes this was inconvenient or impossible for the miners — and thus the need for the California expresses was emphasized.

Expresses in California

According to Bancroft's *History of California,* the first express in California was operated in 1847 by C. L. Cady between San Francisco and Sutter's Fort, this being verified by an announcement in *The Californian* for 24 July 1847, stating that C. L. Cady offered a weekly service. Nothing more is known about it except it was short-lived and existed prior to the discovery of gold. Samuel Brannan also "pioneered" in the express area by organizing an overland mule train delivery of 2,000 copies of his newspaper *The California Star* to Independence, Missouri. According to an announcement in the *Star* on 15 January 1848 this express would carry private letters, 50¢ each. This was a one-time affair, however.

On 18 August 1849 the Sacramento *Placer Times* voiced the universal complaint and asked the right question: "The regular mail is a regular humbug . . . who will establish an express? And who will not give a dollar for every letter promptly delivered?" The answer came from plenty of enterprising men willing to supply packmules and wagons, and

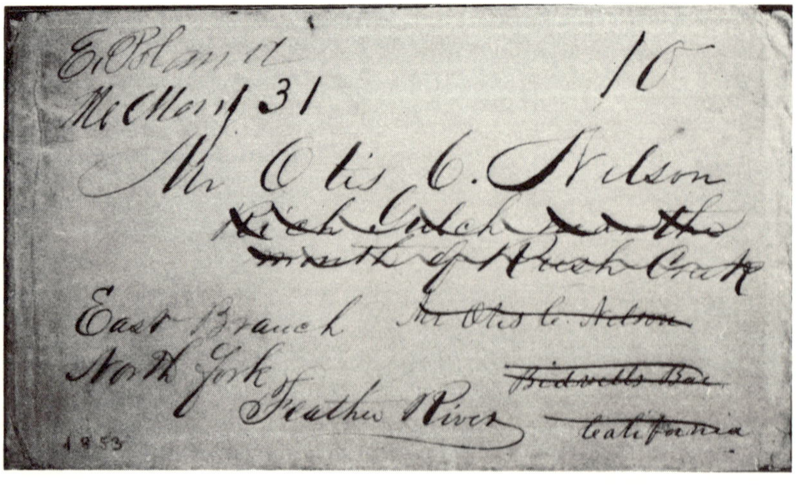

An unpaid 1853 letter from E. Poland, Me., to Otis Nelson, variously addressed to Rich Gulch, Bidwells Bar, and to East Branch North Fork Feather River, where presumably the letter was finally delivered.

plenty of people eager for mail at any cost.

Some time later, in a "looking back" article, *The California Chronicle* of 4 May 1854 commented "It soon became a matter of vital necessity that Expresses should be started." Evidently such awareness was common quite early, for on 17 January 1850 the *Alta California* noted: "There are so many express companies daily starting that we can scarcely keep the run of them."

On 27 July 1850 an editorial in the *Alta California* paid tribute to the work of the expresses: "We scarcely know what we should do if it were not for the various express lines established, enabling us to hold communication with the mines, with the present defective mail communications . . . . Our enterprising express agents, however, supply the deficiency . . . ."

Alonzo Delano, who in 1850 had gone to California to be a miner, quickly abandoned that grueling task and returned to earning more money by making pen sketches and publishing his illustrated accounts of life in the mining camps. To him we owe a vivid description, from his *Pen Knife Sketches,* of the importance of the expressmen:

> At every camp where there is a store a letter memorandum is kept by some merchant where miners who desire to receive letters can enter their names. These are taken by the expressman who makes it his business to get from the post office all letters and papers directed to his subscribers, and deliver them to the individuals at their diggings . . . . There is scarcely a gulch which he does not visit. 'The Express has arrived!' Every pick and shovel is dropped, every pan is laid aside . . . and they crowd around . . .with eager inquiries 'have you got a letter for me?' . . . Without him, the miner would be shut out from the world, and next to the trader who furnishes the means of sustaining life, the Expressman is of most importance.

The expressman arrives with "Letters from Home."

Often quoted are the words of Ernest A. Wiltsee: "How few realize that seven years after the discovery of gold, the leading journal of the Pacific Coast, the San Francisco *Alta California,* declared that the United States postal service was practically useless throughout the interior, and that the State was saved from business chaos by the then all prevalent express companies!"

As late as 2 July 1861, even though there were then about 500 post offices in the state, we read the following accurate account: "Mails in this state are so uncertain that all important letters are carried by private express in government envelopes . . . the Wells Fargo mail often being larger than the government mail . . . ." The writer added that it was not uncommon for a letter to take 18 days between San Francisco and Monterey if sent by the post office, whereas an express or stage line would deliver it in one day.

Which was first? Todd, Ballou, and Kimball would all claim the priority. Nevertheless, Kimball & Co.'s Express for Sacramento would be the first to advertise. On 4 October 1849 the following notice appeared in the *Alta*

California: "The undersigned have made arrangements with the schooner JAMES L. DAY and the steamer SACRAMENTO, by which Charles P. Kimball will run regularly between this city San Francisco and Sacramento, and will attend personally to the transmission of parcels, bundles, orders, etc."

Alexander Todd usually is given credit for organizing the first real express service in California. Arriving in San Francisco in the summer of 1849 after a 170-day trip around Cape Horn, Todd proceeded to "the diggings." A few weeks of wading in icy streams and finding no gold convinced him that he should seek an easier way of making a living. He observed that the hundreds of miners around him were desperate for mail. This was his opportunity for a new vocation. Since all mail lay in San Francisco 150 miles away, he would become a mail carrier for the miners. He went from camp to camp along the southern end of the Mother Lode, listing the names of miners who would pay him to carry their outgoing letters for $2.50 each; for inquiring at the post office, $1.00; for bringing back a letter, one ounce of gold. Hundreds of miners were eager to sign the lists, and it required a pack-horse to carry the outgoing mail. Todd rode into Stockton, which was then a sprawling tent city, hired a small boat to carry him downstream — and found a group of merchants asking him to deliver $150,000 in gold dust to San Francisco. Todd was willing, at a toll of 5 percent of the gold, which he packed into an empty butter keg. Arriving at the San Francisco post office Todd found great heaps of unsorted mail.

Confronting the frantic postmaster, Todd explained his subscription list, was sworn in as a postal clerk, and was allowed to sift through the unsorted letters for the names of his clients. To maintain a good future working relationship, Todd paid the postmaster 25¢ for each letter he claimed. Returning to the mining camps he was "the man of the hour," especially since he also brought back bundles of New York newspapers which he sold for $8.00 each. Soon he had over 2,000 names on his list and was in business. He now was an expressman and before long business was so good that he could purchase equipment, hire needed help, and acquire partners. He had handstamps made so that he could mark the letters he carried, thus advertising his express companies.

Todd's idea may not have been original, but he was a pioneer. His success created an avalanche of competition. During 1849 at least 20 other express companies appeared on the scene. Only the secretive past knows exactly how many other expresses may have operated and disappeared unrecorded. Not only was there an unending need for food and supplies, but the mania for news from home was an obsession. Letters from miners poig-

Todd & Bryan's shield on 1850 cover to Missouri. Bryan was Todd's first partner.

Hauling supplies by mule and wagon.

nantly express this common sentiment: "The letter carryer goes to San Francisco one per month for letters and he charges $2. That is cheap enough for the country. I shall not care if it is fifty so long as I get a letter. Tell them to write again. I would give an ounce anytime for a letter from any of them."

While Todd was carrying mail and newspapers, Daniel Dancer was carrying freight by mule trains. From Marysville on the Feather River to Downieville 55 miles away, Dancer's trains of 150 mules, each loaded with 300 pounds, were carrying everything from barrels of flour and salted pork to stoves and iron safes, through the canyons and along the mountain creeks and over the precipitous ridges. On his return trips, miners on the move replaced the barrels on muleback, and packsaddles were filled with mail to be forwarded to a post office and gold dust to be safely deposited. Since Dancer charged 5¢ per pound for everything, he could average more than $2,000 per trip. As trails became primitive roads, heavy wagons were used and freighting expanded rapidly. The *Alta California,* 2 December 1851, reported that from the single supply town of Marysville 794 wagons, each capable of carrying 1½ tons, departed in one week.

While Todd and Dancer and others were carrying mail and freight, James Birch had decided that people would rather ride in a wagon than walk or suffer the saddle sores of horse-or-mule-backing. So he began a primitive stage line that was an immediate success. All these, and others like them, were expressmen, an array of entrepreneurs shuttling goods, people, money, and mail.

It has been noted that one of the essential functions of the expresses was the transportation of gold dust. Therefore it was inevitable that some of them would end up in the banking business. The first rudiments of a banking system in California had begun in the hide-and-tallow period when merchant agents extended credit. Similarly, the first bankers of the Gold Rush again were merchants, primarily because they had safes and more fireproof buildings. Also, in many cases, they had coins to exchange for raw gold, and connections with mercantile firms in the East which afforded a means of exchanging gold for merchandise.

The express companies got into the banking business because of the unusual conditions of life in the mining camps and diggings. A miner lost valuable time and risked loss of his claim if he personally had to carry his earnings down to San Francisco, or even to the town supply centers in the Valley. He also faced the possibility of being robbed and murdered. Since the expresses followed the miners into the canyons and river bars, and established offices in the larger communities, it was inevitable that some express agencies would accept the miners' raw gold and issue a certificate of deposit for it, charging a small percentage of the amount for the services. These certificates could be cashed for their face value and often circulated as currency. Again, if a successful miner decided to "go back to the States" it was less hazardous for him to deposit his treasure with an express company and receive "a bill of exchange" drawn on a correspondent banker in any large city in the East. The bills of exchange were payable in gold at their destination, which increased their value if exchanged for paper currency.

Adams & Company Express is an example of this activity. This express had been founded in the East in 1839 and by 1845 had branches in almost every important city in the eastern United States, and also agent offices in many foreign countries. In 1849 it decided to enter into California operations. Its agent, Daniel H. Haskell, arrived in San Francisco on 31 October 1849. Nine days later he had a building and office and was advertising in the *Alta California* that Adams & Company would buy and transport gold to any large eastern city; also that letter bags would be sent on all steamers, and that both mail and treasure would be under the protection and care of a special messenger.

To insure that its internal sources were maintained, Adams quickly opened offices in the California towns that became terminal points for the smaller expresses which served the camps and diggings. Thus Adams & Company became the banking institution for other expresses and their clients, as well as the transmitter of their mail and treasure to San Francisco and beyond. By 1853 it was the leading business house in California. Its decline began in 1854 as the 1852 newcomer, Wells, Fargo & Company, became a fierce competitor. In the financial panic of 1855 the eastern owners of Adams & Company stripped their California connection of its assets, and Adams & Company in the West closed its doors.

ADAMS & CO.'S

California and Atlantic States

EXPRESS.

Our Atlantic States Express leaves San Francisco on the 1st and 15th of each Month, by the

PACIFIC MAIL S. S. COMPANY'S STEAMERS,

And the Treasure Crosses the Isthmus under charge of a Strong Guard.

The Treasure forwarded by us to the Philadelphia Mint, is always deposited there previous to that sent by any other conveyance. Our rates are *lower* than those offered by any other House with *the same security*. We also forward Treasure, on the 1st and 15th of each month, to England, by the P. M. S. S. Co.'s Steamers to Panama, and from Chagres by the West India Mail Steamers.

Our small Package Express goes forward in charge of our regular Special Messengers, who, by constantly traversing the Isthmus, have the information and facilities not possessed by any other person. Our Express being the most extensive in the United States, we can guarantee greater facilities than can be offered by any other concern. We do a Banking Business of deposit only, special or otherwise. We draw

BILLS OF EXCHANGE

on any of our Houses in the following places:

BOSTON, NEW YORK, PHILADELPHIA, BALTIMORE, WASHINGTON, ST. LOUIS, CINCINNATI, NEW ORLEANS, LONDON, &c. &c.

In the Northern Mines we run Expresses in our own name, always accompanied by faithful Messengers, to and from the following places:

San Francisco, Benicia, Sacramento City, Marysville, Shasta, Nevada, Grass Valley, Coloma, Placerville (or Hangtown,) Greenwood, Georgetown, Mormon Island, Salmon Falls, Auburn, and every other part of El Dorado and Placer Counties. Also, to Jackson, Dry Town, Volcano and Sutterville, in Calaveras County. Through Langton & Co., to and from Downieville, and all other places on the Yuba and Feather Rivers.—Through Cram, Rodgers & Co., to Yreka, Weaverville, &c. To and from Sacramento and Stockton, via. Benicia. In the Southern Mines, we run an Express in our own name, always accompanied by faithful Messengers, to and from San Francisco, Stockton, Sonora, Moquelumne Hill, Columbia, Mariposa, &c. &c. By Brown's Express, from Stockton to all the Camps in the Southern Mines.

Our Bills of Exchange can be procured at and Treasure forwarded to us for shipment, from any of the above places. In all the above places, we have Brick Vaults or Iron Safes for the security of Treasure entrusted to us, and on board of Steamboats, on any of the above routes, we have Iron Safes and Messengers for the security of all valuable Packages transported by us.

There are two unusual sagas of express history worth noting to illustrate the versatility of the early expressmen to satisfy the miners' hunger for mail. In dealing with the vagaries of weather and terrain, expressmen resorted to some strange means of conveyance. When heavy snow closed the trails, roads, and passes to mule or horse or foot travel, some expressmen put on snowshoes and carried on. Even this solution was too

Fenton Whiting and his dog team.

slow for Fenton Whiting of the Feather River Express, which operated across the Magalia Ridge between Gibsonville and Quincy. In the winter of 1858 he hitched dogs to a snowsled, loaded 600 pounds of mail and packages onto the sled, and launched a "dog team express." He did this each winter until 1865.

Another ingenious express was initiated by John Thompson. In 1856 snows closed the road between Placerville and Genoa, Nevada, 110 miles away. Thompson, a Norwegian immigrant, had used skis in his native land. He carved and fashioned a set of long skis, and with an 80-pound sack of mail strapped to his back, Thompson crossed the Sierras and in five days was in Genoa. He continued to make these winter runs carrying mail, sometimes medicine, and in 1858 the font of type for Nevada's first newspaper. The miners delightedly dubbed him "Showshoe Thompson," and each winter for some years he was the only communication link across the Sierras.

The majority of the expresses were established to serve the mining camps and communities. In the beginning of the Gold Rush most of these areas were in the Mother Lode district, usually referred to as the Southern and the Northern mines. The Northern mines were on the slopes drained by the Feather, Yuba, American, and Consumnes Rivers. Sacramento and Marysville were their main supply bases. The Southern mines were along the Mokelumne, Stanislaus, Tuolumne, Merced, and Mariposa Rivers. Stockton was the main supply center for this area. Several years later the Trinity-Klamath mountain ranges, drained by the Klamath, Trinity, Pit, Eel, Scott, and Salmon Rivers, would become a distinct mining area with Shasta and Yreka as its main supply towns.

Mel C. Nathan, *Franks of Western Expresses,* states that about 1,900 known expresses existed in the western half of America, but that the franks or handstamps of fewer than 400 have been catalogued. The others, thus far, are known only from newspaper advertisements and articles or from reference to them in letters and journals. Future discoveries, of course, will add to the catalogued list.

Stage Lines

Stage lines were a logical development from the earlier packmule and wagon expresses. As roads were improved, bridges

Snowshoe Thompson carrying mail across the mountains on skis.

built, camps developed into established towns, and population continued to increase, there was a persistent demand for better means of communication and transportation. Stage lines were the answer, and many of the expressmen gravitated naturally into stage-line operators. As indicated earlier, the stage lines performed the functions of the early expresses with the additional ability of transporting passengers.

Urban expansion is a familiar story today, but the Gold Rush gave it its first impetus in the West. San Francisco in four years grew from 1,000 inhabitants to over 50,000. Sacramento grew from Sutter's Fort into a town of 12,000 in less than two years. Marysville had a population of 300 in 1850 and 8,000 in 1855. Stockton, which did not exist in 1848, was a town of over 5,000 by 1850. Everywhere throughout the state, to a lesser degree, towns had become supply and agricultural centers. It was the stage lines which linked them together, until the coming of the railroads.

From the testimony of journals and remi-

Cover with corner card of San Francisco-Alviso packets.

niscences of pioneers, John Whistman is generally credited with operating the first stage line in California, just as Alexander Todd is believed to be the first of the expressmen. John Whistman in late August or early September 1849 started a stage line from San Francisco to San Jose. San Jose had become the temporary capital, and also was on one of the routes to the southern mining areas. His stage line was supposed to follow the old mission trail of *El Camino Real,* but that trail never had been a wagon road. The Mexican Californios traveled on horseback, and their only wheeled vehicle was the ox-drawn cart (*caretta*) with its solid wooden wheels turning on a wooden axle, used only for short hauls. Also, *El Camino Real* was not a single road but a network of paths crossing lowlands in the dry summer and moving up into the foothills in the rainy winter. Nevertheless, by the autumn of 1849 enough travel had marked out a road of wheel ruts that could be used the year around, weather permitting. So John Whistman, using spring wagons and an old French omnibus, began an irregular nine-hour run. If heavy rains turned the northern end of the road into a bog, then he took his passengers to Alviso on the southern tip of San Francisco Bay and they completed their trip by boat.

To determine who was first to establish a California stage line, an advertisement in the *Alta California* in June 1849 must be noted:

> MAURISON & CO.'S EXPRESS AND MAIL LINE. For the transportation of Passengers and Baggage from Stockton to the Stanislaus Mines. Through in Twelve Hours! — The undersigned respectfuly inform the public that they have established a line of Stages between Stockton and the Stanislaus Mines, for the accommodation of passengers and baggage.
>
> A stage will leave Stockton every other day for the mines, at 4 o'clock, A.M., and arrive at the other end of the route in 12 hours. Returning, a Stage will leave the mines at the same hour on the intermediate days, and arrive at Stockton at 4 o'clock, P.M.
>
> MAURISON & CO. Stockton, June 25, 1849

Since this was Maurison's first venture into staging, he too probably used spring wagons and operated as a one-man line. The distance covered was only about 45 miles. The "and Co." could have been "window dressing," for this became a common practice among early expresses and stage lines in the belief that it enhanced their prestige and public image.

In the early spring of 1850 Maurison, in partnership with Ackley, decided to begin an opposition line between San Francisco and San Jose. The *California Star* carried an advertisement and announcement that it would be called The United States Mail Line, though it had no such government contract. On 12 April 1850 the *Alta California* carried as a news article:

> STAGES TO SAN JOSE. We are glad to learn that a line of Stages has been established between this place and San Jose, by Messrs. Ackley & Morrison. The Stages leave the St. Francis hotel, at 7 A.M., three times a week, and go through in about nine hours. The proprietors have placed us under obligations by taking charge of despatches from our reporter and promptly delivering them.

Several months later a capable rival appeared. Warren F. Hall and Jared B. Crandall, who had had staging experience in Mexico, bought the John Whistman operation, changed the name of the line to Berford & Co.'s Line, and began daily runs between San Francisco and San Jose with a twice-weekly run on to Monterey via Gilroy. Their advertisements are headed "Berford & Co., Hall and Crandall, Agents." However, though not so named, it actually was the Hall-Crandall Stage Line, appropriating the prestige of a better-known nationwide express which it serviced on this route. Hall and Crandall knew how to operate a successful stage express line. They purchased a ranch in the Santa Clara Valley for breeding and training good horses, paid their drivers $300 per month and allowed them the perquisites of charging 25¢ for each letter deposited in a post office, or $1.00 if hand-delivered to an addressee, and 1 percent of any monies given

A typical stagecoach underway.

into their keeping for transmission. In May 1851 Hall and Crandall were awarded a four year mail contract at $6,000 per annum to carry mail bags three times per week between San Francisco and San Jose.

The first of the famed Concord Coaches had arrived at San Francisco by clipper ship around Cape Horn on 24 June 1850. Crandall had looked it over and decided that it could be improved for usage on the poor roads of California. He went to Concord, New Hampshire, to Abbott Company which built the coaches, and persuaded them to build a revised version of their coach and purchased a number of them. This Crandall improvement, later to be known as the "Wells Fargo Coach," became *the* stage coach of California.

While Hall and Crandall were growing wealthy on their peninsula traffic, James Birch was doing equally well in the Sacramento Valley and the Sierra foothills. Arriving in California with his friend Frank Stevens by immigrant train in August 1849, both he and Stevens quickly realized that mining was not their *forte.* Each, though only 21 years old, had been a stage driver in Rhode Island. While Stevens decided to go into the hotel business, Birch went back to his known profession. On 1 September 1849 James Birch, driving a springless ranch wagon with boards nailed across the top and pulled by four mustangs, was at the Sacramento landing. A small river boat was unloading argonauts from San Francisco. Birch called out: "All aboard for Mormon Island and the forks of the American River!" He loaded his wagon at a fare of $32 per person and was in business. As fast as he made money he improved his equipment, bought the best horses available, hired men to build bridges and level the road. He hired only experienced drivers. By 10 February 1850, six months after his first trip, Birch was advertising daily service from Sacramento to Mormon Island and on to Coloma.

Success soon breeds competition, and there was need for it as mining camps and towns sprang up through the whole Mother Lode area. The amount of freight needed to

Some hazards of stagecoach travel.

supply the towns and miners was enormous. The freighters had no choice but to improve their roads in order to carry heavier loads. During the 1850s a network of rough but passable roads existed all through the mining area, and the stage lines and expressmen were using them.

As the stage lines increased and expanded their coverage, these roads were improved by blasting and scraping. Since state and federal funds were not available, towns would often appropriate monies for road improvement. The stage lines sometimes used their own profits for road and bridge building. In 1850 the state, in lieu of appropriations, offered franchises to private individuals to build roads and bridges, provide ferries for crossing the larger rivers, and to charge tolls for passage over all of them. At the end of 1858 there were 64 such toll roads within the state.

By the fall of 1850, 12 stage lines were operating out of Sacramento alone, most of them as one-man one-stagecoach businesses. There was one conspicuous exception, however; William A. Beeks had come to California for the sole purpose of staging. He brought a caravan of horses, mud wagons, and coaches overland, sold some of them to Hall and Crandall, and used the rest to begin a daily service between Sacramento and Nevada City. On 19 October 1850 the following advertisement appeared in the *Sacramento Transcript:*

> SACRAMENTO AND NEVADA CITY EXPRESS LINE. THROUGH IN 12 HOURS. The undersigned, having a very choice Lot of Spring Wagons and a splendid stock of Horses, have es tablished a Daily Line of Stages between Sacramento, Rough and Ready, and Nevada City, and confidently rely upon the patronage of the traveling community. Being well aware that a single trip will convince one and all that, in point of safety, comfort and speed, this line is unequaled in the New State of California. Stage leaves Iowa House at 6¾ o'clock A.M. and Crescent City Hotel at 7 o'clock each day.
>
> WM. A. BEEKS, Proprietor, R.O. SELFRIDGE, General Agent

In the spring of 1851 Frank Stevens decided to give up the hotel business and go back into staging. Like his friend James Birch, Stevens bought the best horses and equip-

Pioneer Stage Co. coach.

ment available. He then began as the Pioneer Stage Line, running from Sacramento to Hangtown (Placerville). By 1853 Stevens and Birch were giving stage service to the mining towns and camps along the North Yuba, the South Yuba, and the American Rivers.

Meanwhile, California's two senators were complaining in Congress about the lack of adequate communication between their state and the rest of the Union. They were demanding an immediate railroad connection, equal mail service, federally constructed postroads between California's main towns, and the awarding of substantial mail contracts. Though unable or unwilling to meet all these demands, Congress did authorize the giving of generous mail subsidy contracts to California stage line companies.

Birch, Crandall, Hall, and a few other operators realized that those lucrative mail contracts would be awarded only to companies operating between important towns on fixed schedules with first-class equipment. On 18 February 1851, the *Sacramento Transcript* verified that Birch had decided which towns were likely to be mail centers:

> NEW LINE OF STAGES. Mr. James Birch, extensively known in this vicinity, has established a line of stages between Nicolaus and Marysville, in connection with the steamer Gov. Dana. He has fifty of the finest horses and the best coaches in California, and the way he puts through his passengers is a caution to slow teams. He has an abundance of passengers as the Gov. Dana has carried one hundred passengers daily for the last two weeks. Mr. Birch, we believe, is the first man that established a stage line in this country. He is endowed with great perseverance and integrity, and if any person deserves success it is James Birch.

Two months later the *Sacramento Union* announced that Birch had established the Telegraph Line of U.S. Mail Coaches, "running between this city and Nevada City, the largest and most important mining town in California." The Telegraph Line would pass daily through Rough and Ready and Grass Valley on its way to Nevada City.

Birch was not alone in his analysis of the situation. By midsummer of 1851 Sacramento had become the stagecoaching capital of the state. The *Alta California,* 12 November 1851, reported that out of Sacramento six lines were running to Marysville, one to Nevada City, two to Coloma, and one each to Placerville, Auburn, Stockton, Dry Town, and Jackson. Twenty-seven mail contracts had been awarded. Hall and Crandall had bought out and re-equipped a line between Sacramento and Marysville. Baxter and Monroe had opened a stage line to Shasta City, 180 miles north of Sacramento, improving the roadway that followed the trail blazed by Jedediah Smith in 1828. Other stage lines followed suit, for the combination of paying passengers and a mail subsidy was highly profitable.

In April 1853 Hall and Crandall sold their San Francisco to Monterey line to Dillon, Hedge & Company, and then immediately established a passenger service between Marysville and Yreka. At Redding their road left the Sacramento River, went northwest 20 miles to French Gulch and ended there. For the next 60 miles across the Trinity and Scott mountains, passengers and mail traveled by muleback to Callahan's Ranch near Mt. Shasta. From there a semiroad again led to Yreka. Also in April 1853 Charles Greene began the first stage line that cut directly across from Sacramento to Sonora, via Ione Valley, Jackson, and Columbia.

By late 1853 the competition of the many stage lines had resulted in ruinous rate-cutting wars. As a solution, James Birch, Frank Stevens, Warren Hall, and Jared Crandall persuaded many of the other operators to consider a consolidation. Consequently, in December 1853 the California Stage Company was organized, with a capitalization of $1 million, with its headquarters in Sacramento. It officially started on 1 January 1854, and soon operated over a total of 1,400 miles of roads.

Not included in the new consolidation were Wells Fargo & Company and the stage lines of southern California. Phineas Banning in 1850 was operating between Los Angeles and Wilmington on San Pedro Bay. In 1852 he had joined with D. W. Alexander to establish a stage line from Los Angeles to San Diego, and again in 1853 to combine a freighting and stage line from Los Angeles to Ventura, Santa Barbara, and San Luis Obispo. In 1854 when the gold rush to the Kern River began, it was Banning and Alexander who opened a wagon road across the Tehachapis through the Tejon Pass and down the Grapevine into the San Joaquin Valley. When the Kern River rush ended, the firm had 500 mules, 40 huge freight wagons, 15 stagecoaches, and herds of horses. So they branched out again, running stages to San Bernardino, and hauling army supplies across the Mojave Desert to Fort Yuma on the Colorado River, and north to Salt Lake City.

On 18 February 1855 the steamer *Oregon* brought the news that Page & Bacon, the largest banking institution in St. Louis, had closed its doors. A run began upon its bank in San Francisco and on 22 February it was bankrupt. A financial panic ensued and frightened depositors stormed the other banks in the city. Adams & Company in San Francisco could not meet its obligations and also closed its doors. One by one its other banks in California towns failed. Of all the major banks, only Wells Fargo bank survived. Scores of firms of all kinds were ruined by the loss of their monies in the failed banks.

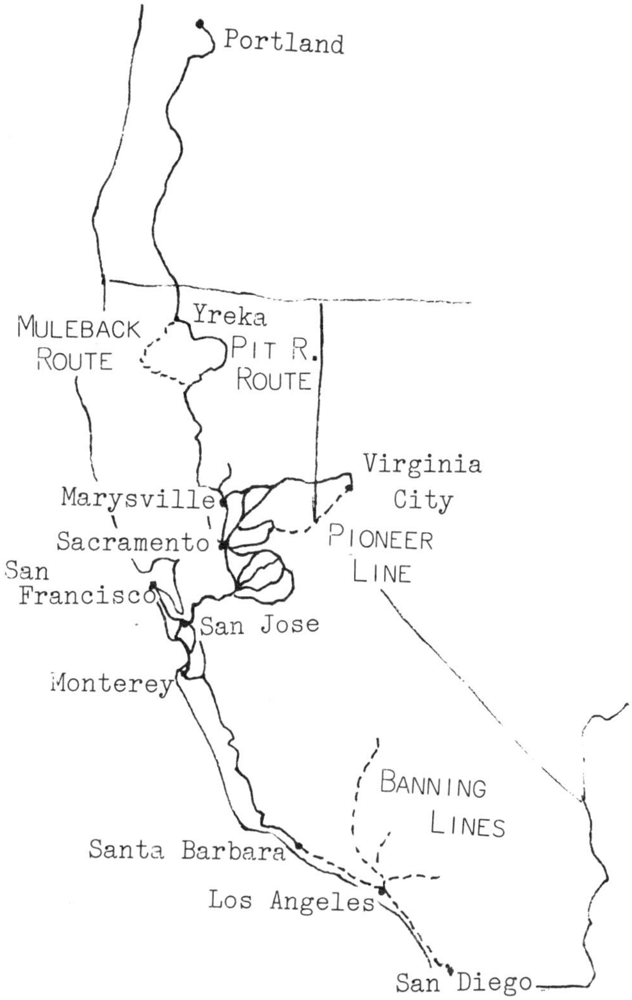

California Stage Co. routes.

The California Stage Company, however, was not hurt by the panic. In that same year it purchased two lines that had not been part of the consolidation: the Dillon and Hedge line and the Talmadge line from Oakland to San Jose. It expanded its line south to Los Angeles, then to San Diego. It went north along the Hall and Crandall road to Redding, and had its northern terminus at Shasta City. In 1857 the company began blasting, scraping, and bridging a new road across the Cascade Mountains to reach Yreka. It was completed in 1861, and by absorbing Oregon companies the California Stage Company operated from San Diego to Portland, Oregon. It was then the largest stage line in the West, operating daily over 1,970 miles of roads.

It was not a monopoly, however, for small feeder lines continued to serve another 1,000 miles of roads in more remote areas. Then the company lost its mail contract in 1865 and many of the partners withdrew and re-established their own firms. In 1866 the Oregon Stage Company took advantage of the disintegration and began service from Portland to Sacramento.

Now Wells Fargo began to show its strength. Its great banking rival, Adams & Company, was gone. The mammoth California Stage Company had fallen apart. Since 1852 Wells Fargo had been building quietly for this moment, and it moved quickly to become the greatest stage line in the United States.

Wells, Fargo & Company

The Wells Fargo record in western mail service demands the use of superlatives. Its success was legendary, its honesty and reliability were proverbial. The miners and merchants swore by it. Entering the California scene in 1852, by the mid-1860s it was the most important express agency in the West, the richest bank, the farthest-ranging stage line, and one of the largest freighting agencies. Almost every town in California had a Wells Fargo building, a combination bank and stage agency. Its green mail boxes stood alongside the red U.S. postal boxes, and became the largest depositories of mail because of an awareness that Wells Fargo could and would provide better mail service than the United States postal system. It had agents in Mexico, Panama, Hawaii, and other foreign nations. Within 15 years of its founding in San Francisco in 1852, it had absorbed every one of its major rivals, and was the most powerful institution in the West.

It began as a connection of an already established successful express and stage company in the East. Henry Wells and William Fargo, operating out of Albany, Buffalo, Cincinnati, and Chicago, earlier had challenged the government for a short time by delivering mail for 6¢ per letter when the postal system was charging 25¢. In 1850 they merged with another large eastern express firm controlled by John Butterfield to form the American Express Company. Almost overnight the new company became an economic giant in the eastern half of the continent.

Then its eyes turned westward. From 1849 to 1851 millions of dollars in gold dust and bullion had been shipped from California to the U.S. mints in Philadelphia and New Orleans to be converted into coinage. The western expresses and banks had reaped a rich harvest from this lucrative expediting service, charging 3-5 percent of the value of each shipment as their service fee. Wells and Fargo wanted the American Express to enter the California competition. The other directors were reticent to this expansion, and, unable to convince them, Wells and Fargo moved on their own.

On 18 March 1852 Wells and Fargo, with support from seven other eastern financiers, organized a new express company specifically to serve the Western frontier, capitalized it with $300,000, and christened it "Wells, Fargo & Company." Four months later two carefully selected agents, one to oversee the banking business and the other to organize the express services, opened the first Wells Fargo office on Montgomery Street in San Francisco. A policy of establishing branch offices in California towns began immediately. Slowly Wells Fargo & Company began buying out other express companies, but only if they served major routes over good roads. Wells Fargo was content to allow the small expresses to continue the packmule-type service into the more inaccessible mountain areas. It carefully avoided the reputation of being a monopolistic consumer of weaker ri-

Interesting Wells Fargo markings: oval Sacramento with Gothic lettering; fancy STEAMBOAT; oval Messenger Southern Coast; SF shield.

vals. It expanded quietly, patiently, and continuously. By the end of 1854 Wells Fargo had 24 branch offices, had purchased Gregory's Express in November 1852, Todd's Express in September 1853, Hunter's Express in July 1854, and now owned a network of express lines covering much of California. It had only two major rivals then: the powerful Adams & Company and the California Stage Company. Adams & Company was a tough competitor, and for two years it had out-maneuvered and out-shipped Wells Fargo. But when the financial panic of 1855 destroyed Adams & Company, Wells Fargo survived and paid off its depositors, thus emerging with heroic stature as a company of honesty, integrity, and reliability.

In 1856, Louis McLane, a former naval officer who had participated in the conquest of California and then had resigned his commission to engage in the shipping business, was appointed general manager of all California operations for Wells Fargo. It was a brilliant choice of a visionary but experienced man. When the U.S. government struck at the private mail service of the express companies by legally requiring all letters to bear U.S. postage stamps, McLane simply bought stamped government envelopes by the thousands, printed the Wells Fargo frank on them, and sold them at a profit in all of its branch offices. By the end of the decade, envelope sales were averaging $15,000 per month. Other express companies followed the Wells Fargo practice. Pacific Express Company began using government envelopes in November 1855, Everts & Company in April 1856, and soon numerous expresses and stage lines were printing their company franks on government envelopes. None, however, reached the volume of Wells Fargo which operated almost as if it were a branch post office. Wells Fargo records indicate that in 1863 that company purchased two million 3¢ envelopes, 15,000 6¢, and 30,000 10¢ and 12¢ envelopes. In addition, it bought 70,000 3¢ stamps and 12,500 12¢ stamps.

By 1859 there were 126 branch offices, each marking mail with the familiar Wells Fargo handstamp containing the name of the company and the town office. As the West grew, Wells Fargo grew with it. When other gold rushes occurred, Wells Fargo followed the miners. Its agents smelted down the gold dust and ore and shipped the treasure East or carried the ingots to the San Francisco Mint, which had opened in 1854, to be converted into coinage and returned to the miners. The company not only carried the mails, but also packages of everything from cigars to grapevine cuttings. It used every available route and mode of transportation: river boats, coastal steamers, ships around the Horn, the Panama passage, express wagons to haul freight, mule trains, and men on snowshoes — but its major identification was with the stagecoach.

An interesting anomaly emerges here. Logically, it would be assumed that a huge enterprise such as Wells Fargo would by 1860 have owned an empire of its own stage lines. On the contrary, except for buying a few feeder lines to give it better access to the Mother Lode gold camps, Wells Fargo had been content to send most of its express on the coaches of other established lines. When James Birch put together the mammoth California Stage Company, Wells Fargo avoided competition. In 1860, however, events occurred which would lead Wells Fargo into its own massive stage operation.

First, its California agent, Louis McLane, and his brother bought out the Pioneer Stage Company which ran from Placerville to Carson Valley, Nevada, servicing the famed

Early "Through California" frank to Phila.; fancy STEAMBOAT handstamp; oval COLLECT 50cts. and WF label.

WF Nevada oval and FREE in red on cover from SF to Nevada, and on to Sacramento.

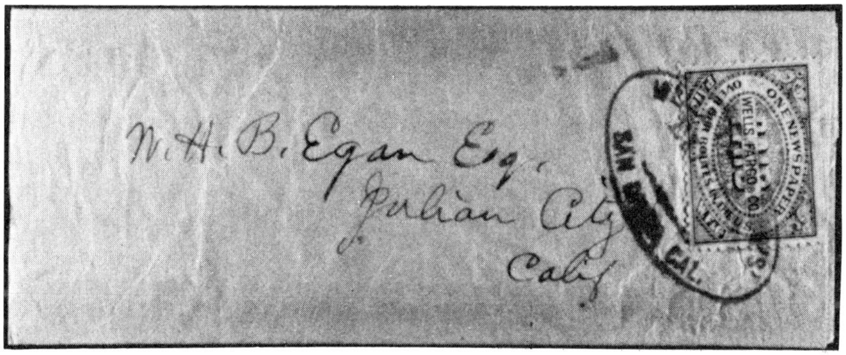

WF newspaper stamp on wrapper.

Ladies' envelope from Neosha, Wis., to Crescent City, unpaid 10, with WF Crescent City and $1.00 rate.

Comstock Lode mines and camps. McLane immediately improved the road, making it usable the year around. The 140 miles were run by daily stages, taking two days for the trip. He still retained his position as head of Wells Fargo operations, and Wells Fargo utilized his new stage line for shipments. However, on 12 December 1864, Wells Fargo quietly bought the line from McLane for $175,000, though retaining its name as the Pioneer Stage Company.

The next event of 1860 really put Wells Fargo into staging. The Overland Mail Company, running over the famed Southern or Oxbow route from California to the Mississippi River, was in deep financial trouble. Wells Fargo was a principal creditor, and in March 1860 it took active control of the Overland Mail Company, forced its owner, Butterfield, to resign, and put in a new president. A year later, with the outbreak of the Civil War, Washington directed that the stage line move north to the Central route. With this move, Wells Fargo found itself the prime operator of the western end of the only transcontinental stage line, the Central Overland California and Pikes Peak Express Company (C.O.C. & P.P.E.). Wells Fargo controlled the 966 mile route from San Francisco to Salt Lake City, and the Russell-Majors-Waddell company (C.O.C. & P.P.E.) ran the eastern part of the line from Salt Lake City to the Missouri River towns. During the desperate though romantic saga of the transcontinental Pony Express, Wells Fargo operated the western half of that operation. It was a pleasant cooperative enterprise until Ben Holladay forced Russell, Majors, and Waddell out of control, taking over that portion of the C.O.C. & P.P.E. to protect his loans. Holladay was a "rough and tumble" businessman and it was with great relief that in 1866 Wells Fargo accepted Holladay's offer to sell them the eastern half of the line (The Holladay Overland Mail & Stage Express Company) for $1,800,000. This purchase gave Wells Fargo a monopoly on long distance staging and mail service west of the Missouri River.

Perceptive as they were, they did not realize that Ben Holladay had outsmarted them. Holladay had watched the rapid progress of the building of the transcontinental railroad, and knew that its completion would end transcontinental stagecoaching. Wells Fargo was strangely blind to the potential effects the railroad would have, and this set the scene for the next climactic event in Wells Fargo history.

Lloyd Tevis, a lawyer from Kentucky, had come west with the 1849 gold rush. He accumulated wealth in a variety of successful business ventures. He became friends with "the Big Four" (Crocker, Stanford, Hopkins, and Huntington) who were building the Central Pacific railroad to meet with the Union Pacific in 1869. He invested heavily in the stock of the Central Pacific during its most difficult financial days. Tevis realized that when the transcontinental railroad was completed, the right to carry mail and treasure on it would be worth millions. In May 1867, two years before the railroad's completion, Tevis organized the Pacific Express Company. (Many writers, including Nathan: *Franks of Western Expresses,* have named this as the Pacific *Union* Express. According to Wells Fargo records, the Pacific Union Express was a separate and different company, and Tevis's Pacific Express Company was a "shell" company, organized simply as the holder of the railroad contract, and never actually operated. It should be noted, however, that other historians claim that Tevis did operate in Nevada and engaged in a "rate war" with Wells Fargo.) Tevis persuaded the Big Four to grant him an exclusive contract to use the services of the railroad after its completion. He did not announce that he possessed the contract. Meantime, the value of Wells Fargo stock dropped from $100 to $13 as the railroad neared completion. At this depressed price, Tevis quietly began buying it. On 4 October

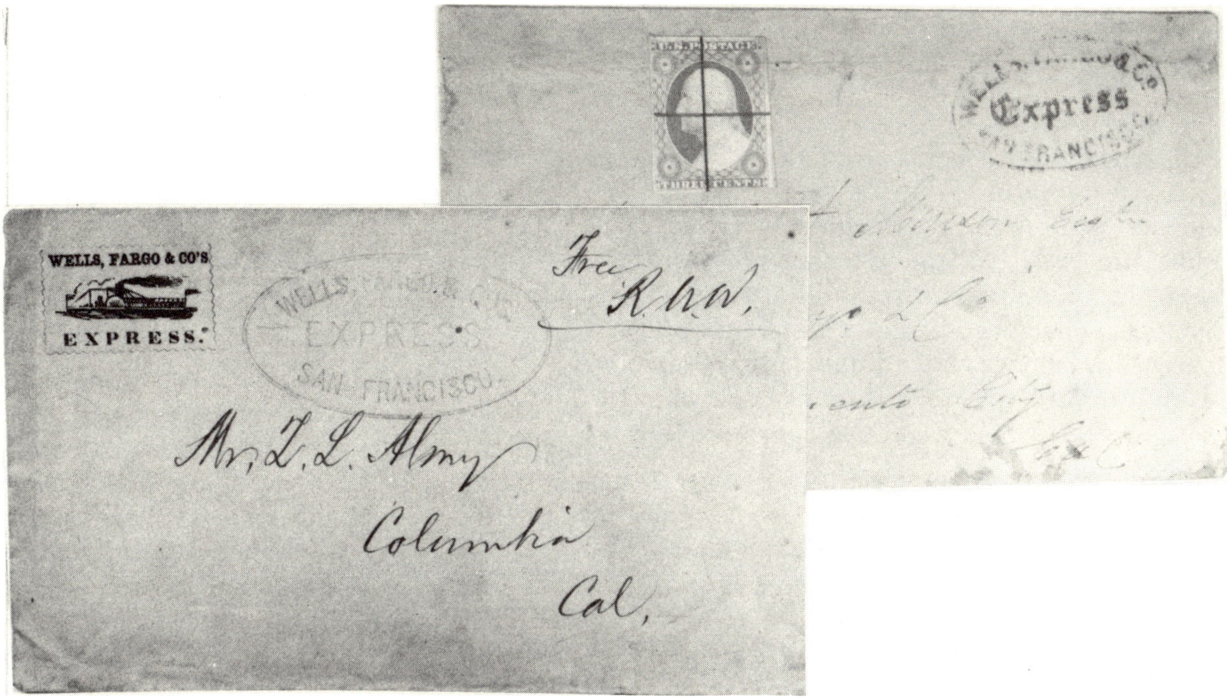

Oval SF and ms. precancelled 3¢ 1851 to Sacramento; another type of SF oval on WF corner card marked "Free."

1869 Tevis met with the directors of Wells Fargo in Omaha, Nebraska. No details of this meeting have ever been revealed; however, we can observe the results. At the meeting evidently Tevis presented Wells Fargo with a copy of his exclusive contract and informed them of his ownership of a huge block of their stock. Tevis's demands were accepted and for $5 million he agreed to terminate the Pacific Express Company and give the railroad contract to Wells Fargo. Wells Fargo had survived again, and later under Tevis's leadership as president (beginning in 1872) it continued to grow. By the close of the century Wells Fargo had almost 3,000 branch offices and operated over 38,000 miles of express routes, and had become a premier banking institution. It had helped to create the new West, and it served well what it had aided to create. (A direct continuation of the banking arm of this firm exists to this day as the Wells Fargo Bank.)

SF oval to Sacramento with W. F. & (Co.) in blue on 3¢ 1851.

Wells Fargo's Virginia City Pony Express

An interesting chapter in the story of Wells Fargo was its operation for almost three years of a miniature Pony Express. It did not compare with its famed predecessor where riders traversed hazardous terrain day and night for 2,000 miles. Also, unlike the original, the Wells Fargo Virginia City Pony Express of 1862-65 was not a financial disaster. On the contrary, it was a highly profitable venture, over a good road, and covered only 150 miles.

The story begins with the discovery of the great Comstock Lode in the Washoe Mountains of Nevada in 1859. In 1860 more than 10,000 Californians stampeded to Nevada. As miners disemboweled Mt. Davidson in their search for silver and gold, the new towns of Virginia City, Silver City, and Gold Hill were born. Carson City became a supply center for the new towns which housed the miners, merchants, and businessmen who were part of this new gold and silver rush. Their economic ties, however, principally were westward to San Francisco and Sacramento, and the Comstock area was virtually a California colony. For supplies, equipment, capital, and manpower, the isolated Nevada mines depended on California. Freight, passengers, mail, and treasure made their way across the Sierras. The Pioneer Stage Company already was serving the Placerville to Carson Valley route and, with the explosive growth of the new communities, it was operating 75 stagecoaches. The California Stage Company and the Nevada Stage Company were also running from Downieville to the new Nevada towns. Wells Fargo for some years had been operating the western end of the Central Overland from Placerville to Salt Lake, a line which passed through the Carson Valley. The stages, however, were too slow to satisfy the desires of businessmen and their mail, when fortunes could be made or lost by the delay of a letter.

So it was that on 11 August 1862 Wells Fargo initiated a pony express which would leave San Francisco and Virginia City each evening, cross the Sierras at night, and 14 hours later be at its destination.

Wells Fargo was not a novice in the pony express business. During most of the runnings of the famous Pony Express of 1860-61, Wells Fargo had been responsible for the western half of that epic drama. Now, as before, they limited their new pony express to the carrying of letters and documents. They required their customers to use 3¢ government envelopes bearing a Wells Fargo frank, with a 10¢ brown Wells Fargo adhesive stamp affixed if the weight did not exceed ½ ounce. Each additional ½ ounce required an additional 10¢ pony adhesive and 3¢ more postage. In the early spring of 1863 the rate was raised to 25¢ per ½ ounce and the color of this adhesive pony stamp was blue. In the summer of 1864 the color of the 25¢ adhesive was changed to red. Each envelope also was marked by a Wells Fargo handstamp bearing the name of the town of origin. For a short time Wells Fargo also impressed their pony stamps on envelopes, but very few of these are known to have survived for collectors to prize. As a special service, Wells Fargo would carry any letter bearing the pony adhesive to any town in California or Nevada where they had an office, without any additional charge.

By 1865 the stage lines had improved their service until they almost could equal the running time of the ponies. So, in March 1865 the Virginia City Pony Express discontinued its operation.

The history of Wells Fargo has been covered briefly in this chapter. No attempt is made to list all of the Wells Fargo franks. The classic work of John F. Leutzinger: *The Handstamps of Wells, Fargo & Co., 1852-1895* is already in the libraries of most serious collectors. Nevertheless, some of the rarest of the Wells Fargo covers are illustrated in this chapter.

WF envelope with Virginia City impressed stamp.

Virginia City 10¢ brown on WF envelope to SF, 1862.

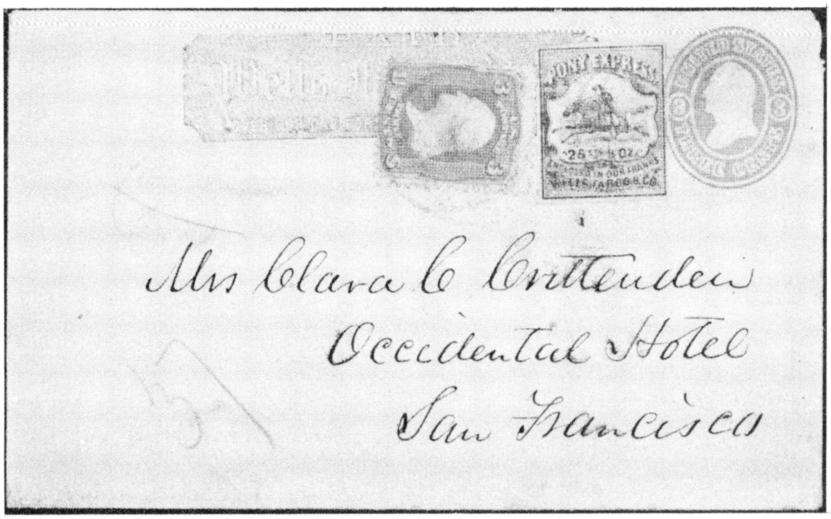

Virginia City 25¢ red and 3¢ 1861 on double letter to SF, 1864.

Wells Fargo shield in red tying pair 3¢ 1851 to 1855 cover from NY to SF. Fancy red and white corner card of "Shippers Line of California Packets."

Wells Fargo & Co. S. Fco. double circle in blue tying 10¢ 1857 to pink allover WF illustrated envelope to NY. Oval COLLECT and straightline McINTIRE'S POST 3 WALL ST, local delivery in NY.

Chapter 6

Franks of California Expresses and Stage Lines

New York to Stockton, FEB 11/40. The transcontinental rate paid by a strip of four of the 1847 10¢. Straightline handstamp of REYNOLDS & CO.'S/SONORA EXPRESS. This company operated from Sacramento and Stockton to the Southern Mines. Oval forwarding handstamp of Reynolds Express — the letter was carried by Reynolds to the addressee in a camp somewhere in the Sonora area.

Franks of California Expresses and Stage Lines

In this chapter the term "frank" will be used to describe any marking, manuscript or printed, which was placed on a folded letter, envelope, paper, or circular by a non-government postal agency to designate the carrier or transmitter of such mail. By this definition a frank may be a manuscript inscription, a handstamp, or a printed corner card. The existence of more than one express company handstamp on a single cover (often called "compounds") simply indicates that expresses serving a particular limited area would pass along a letter to another express until it reached a post office, or, in some cases, until it was delivered to the addressee by an express company. In the latter case, such letters never entered the postal system, thus being described as carried "outside the mails."

Of course, a printed corner card does not always represent express use since the envelopes were sometimes used outside the chronological period or geographical area of the express named.

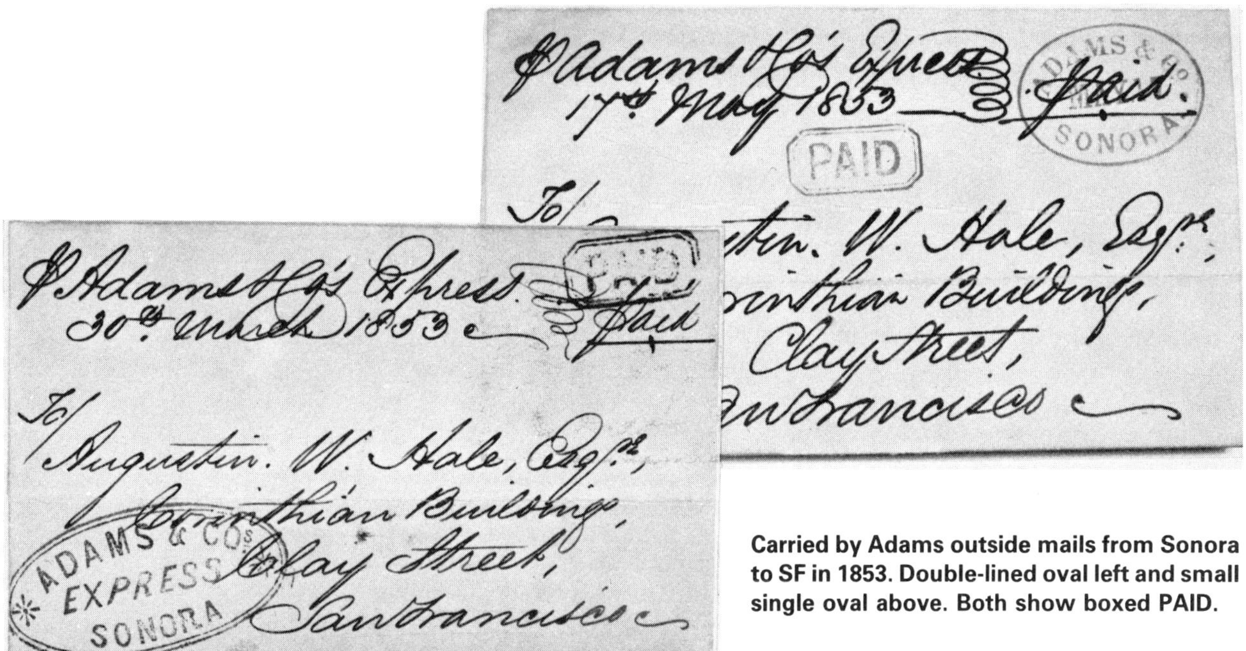

Carried by Adams outside mails from Sonora to SF in 1853. Double-lined oval left and small single oval above. Both show boxed PAID.

Adams Express covers (clockwise from lower left): Boston PAID 6 to SF, there "Fd 5" to Mormon I., hence by Adams to Murderers Bar; boxed frank on cover to Alabama; SF rope oval and "collect 50¢" to Mormon Gulch; Sacramento double oval to SF; shield corner card of SF office "free" to Mok Hill; Rattlesnake Bar oval and boxed PAID to Mormon I.; Stockton double circle and 25¢ adhesive to North Branch; SF to Don Pedro's Bar, forwarded with latter oval to Columbia; SF double circle tying 3¢ 1851 to San Jose; oval Chinese Camp tying 3¢ 1851 to Sacramento; Sonora oval to Benicia.

Adams & Co.

Dates: 8 November 1849-23 February 1855.

Routes: All of California and Oregon, with offices in foreign countries.

Adams & Co. was the western branch of a large eastern express. Until its failure in the financial panic of 1855, it was the largest and most successful express company in the West. At least nine different handstamps, corner cards, and adhesive labels were used by Adams & Co. (The Adams adhesive labels also were printed in sheets on cardboard-type glazed paper, for use as currency to pay postage or make small remittances. They were popularly called "poker chip" stamps.)

American Express

Dates: Not yet determined, but known to have been in operation on 13 March 1857.

Routes: California and Oregon.

Owned by McCombe Co. No connection with the famed American Express of the east-Co., and Berford's Express. Nine different franks have been identified for this company. *(See also p. 230.)*

Alta Express Co.

Dates: July 1857-November 1858.

Routes: Northern California, Oregon, Washington, and into British Columbia.

Connected with the following expresses: Nichols & Co., Freeman & Co., Whitney &

American Express frank.

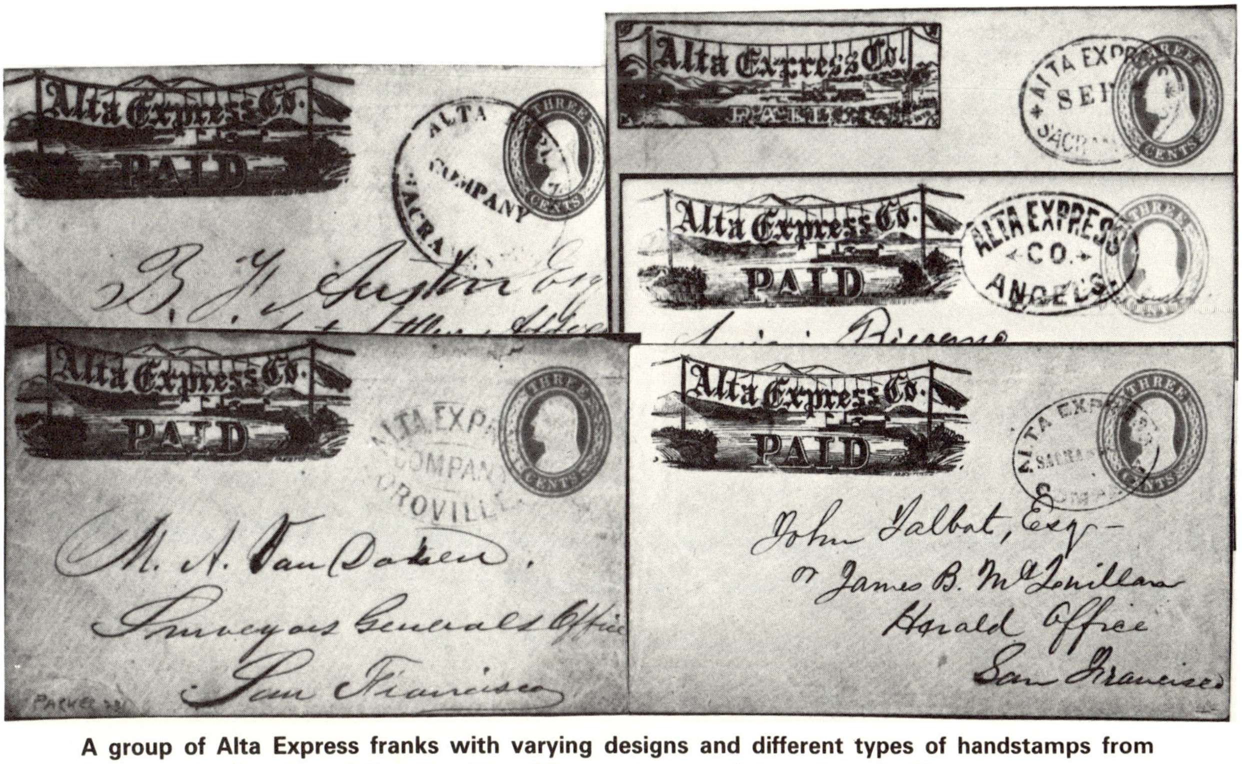

A group of Alta Express franks with varying designs and different types of handstamps from Sacramento, Angels and Oroville. All to SF, except upper right to Downieville.

ern United States. Not known to have used any frank until after it purchased Nichols & Co. Express, at which time it began using the same unique printed frank of the Nichols company.

Angiers' Express
[One-man express]
Dates: 1861-64.
Route: Humbug through mining towns to North San Juan.
Connected with Wells, Fargo.

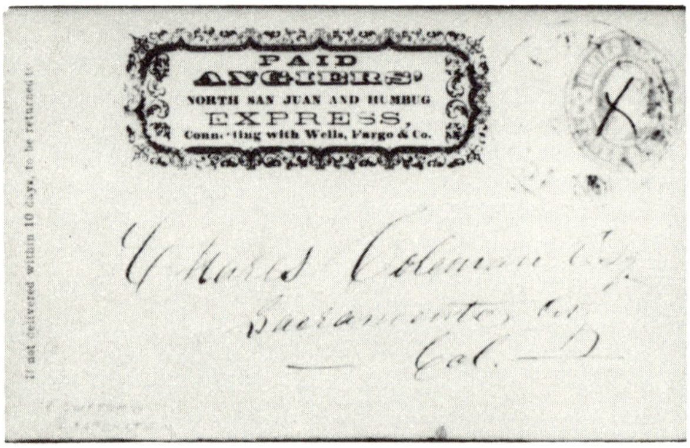

Anthony & Co.'s Northern Express
Dates: Early 1852-mid-1853.
Route: San Francisco to Northern Mines via Benicia, Sacramento, Marysville, and Nevada City.
Connected with Berford's Express.

Auburn Express
[One-man express]
Date: 1865.
Route: Auburn to neighboring mining camps.
The only known cover is in the Wells Fargo History Room. Handstamp is almost indecipherable.

Bamber & Co.'s Express
Dates: 1858-69.
Route: San Francisco to Contra Costa and Alameda Counties.
Connected with Wells Fargo for service to other areas. The Bamber covers may be the most common of all express covers. At least 12 varieties are known.

Various Bamber franks and handstamps.

Bamber covers from Oakland and San Francisco, with oval CONTRA COSTA, diamond COLLECT 12½ Cts., and "Answer by Bamber" slogan.

Bamberger Six Horse Wagon

Many early collectors regarded these printed corner cards to be the frank of a mystery express. Recent research reveals that they are illustrated advertising covers of Booth & Co., wholesale grocers, with offices in Sacramento and San Francisco. Extant covers are dated in 1853, 1855 and 1860. The same printed illustration is found as a logo on invoices of this company. Booth & Co. was not an express, and the inclusion of these covers is to correct an oft-repeated error.

Phineas Banning Express

Dates and Routes:

1851: Los Angeles to San Pedro.

1852: A stage line from Los Angeles to San Diego.

1853: Freight and stage line from Los Angeles to Ventura, Santa Barbara, and San Luis Obispo.

1854: Stage and freight line from Los Angeles through Tejon Pass to Kern River

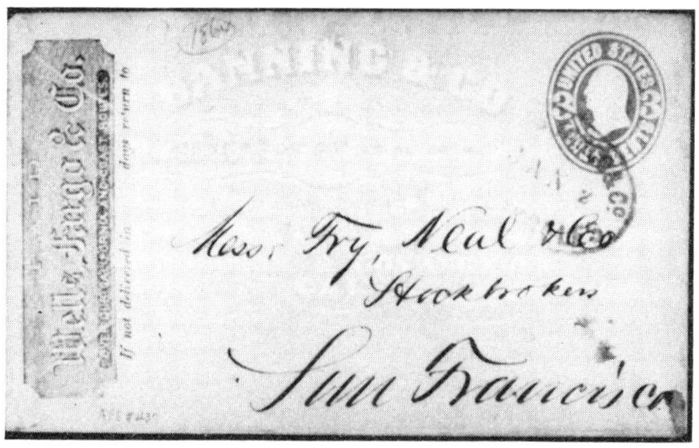

Banning & Co. all-over printed envelope.

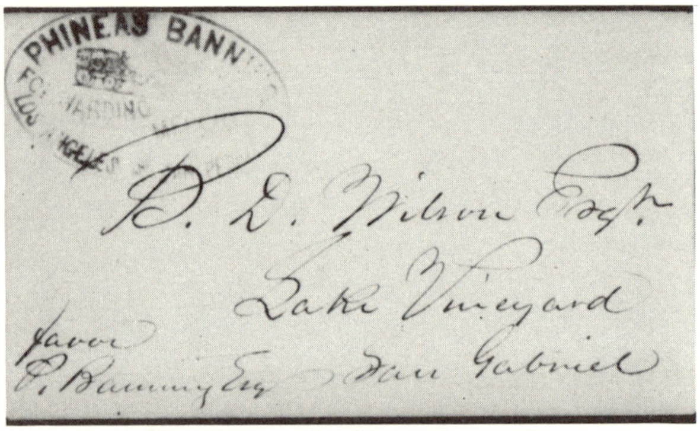

Banning oval with stagecoach.

mining camps.

1855: Stage line to San Bernardino and freight line through Cajon Pass to Fort Yuma.

Banning is not known to have used any express company handstamp except for a blue oval reading "Phineas Banning/Forwarding Merchant/Los Angeles & San Pedro" with a pictured stagecoach and horses in center of oval.

Beck's Express

[One-man express]

Dates: 1853-1 July 1854.

Route: Downieville to Gibsonville and all intermediate points. Only one cover known.

Becker & Co.'s Express

[One-man express]

Dates: 1852-53.

Route: Marysville to French Corral via the various bars of the Lower Yuba River.

Connected with Gregory's Express and later with Wells Fargo.

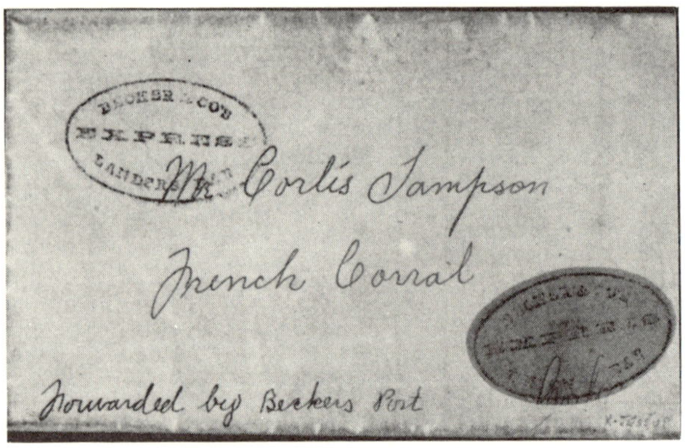

Becker oval, Landers Bar.

Beekman's Express

Dates: Summer of 1855-1863.

Route: Jacksonville (Oregon) to Crescent City and Yreka, and later to Portland (Oregon).

Connected with Wells Fargo.

Beekman oval on cover to SF.

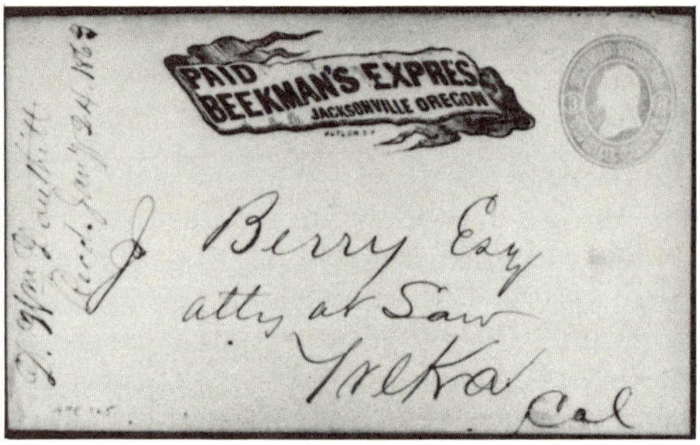

Beekman printed frank used to Yreka, 1863.

J. M. Benton's Stable

Dates: Late 1860s into early 1870s.

Berford handstamps: double oval SF; oval Atlantic & Pacific on 10¢ envelope to Boston; shield and single oval California, all to SF. Last two with COLLECT marking.

Route: Carson City (Nevada) through Virginia City and Gold Hill to Lake Tahoe.

Scant information is available about this company. Although primarily a stage line, the Wells Fargo & Co.'s label indicates it also carried express materials.

Berford's Express

Dates: Late 1849-summer 1854.

Routes: Through their contacts with other stage lines they covered most of California, and carried mail and treasure to the Atlantic states under the protection of their own agents. They functioned much as Wells Fargo would later on.

Connected with Todd & Co. and Anthony & Co. Berford used at least seven different franks.

F. W. Blake & Co.'s Express

Dates: Spring 1852-late 1855.

Route: Shasta to Weaverville, through intervening camps in Trinity and Shasta counties.

Connected with Wells Fargo in Shasta. Counterfeits of Blake covers are known. *(See also p. 231).*

Blake Weaverville oval.

Bowers Express

Dates: 1849-52.

Route: Sacramento to Nevada City via Grass Valley, and Rough and Ready.

Connected with Freeman & Co. in Sacramento. *(See p. 71.)*

Brooks Express

Date: 1855.

Route: Scott River camps.

According to Konwiser and Mason an oval blue frank was known on an 1855 cover; not available for illustration.

Brown's Express

Dates: Early 1850-1855.

Route: Stockton to the Southern Mines.

Delivered by Brown from Stockton to subscriber at a mining camp for $1.

Connected with Todd & Co. and Newell & Co. in Stockton. After Brown was killed in April 1853, Adams & Co. took over the route and continued as Brown's Express until shortly before Adams failed in February 1855. Three varieties of Brown's franks are known.

Carried by Brown for $1.50 charge.

Brown & Maguire's Copper City Express

Dates: 1858-possibly to 1861.

Route: Shasta to Copper City.

Other than the existence of the printed frank illustrated in Nathan, little is known of this express.

J. Butts Diamond Springs Express

[One-man express]

Date: 1853.

Route: Diamond Springs to Placerville.

Little is known about this company except for the existence of this cover. Enclosed in the cover is a letter, dated in 1853, in which the writer says that he will give it to "Butts the expressman."

Circular handstamp of J. Butts.

Byam's Express

Dates: 1851-55.

Route: Sacramento to the Consumnes River through Cook's Bar, Michigan Bar, then to Fiddletown and Volcano.

Connected with Wells Fargo at Sacramento. Byam began by writing the name of his express in manuscript. Later his handstamp was an oval "Forwarded by/Byam's Express/Fiddletown."

Byam's manuscript frank.

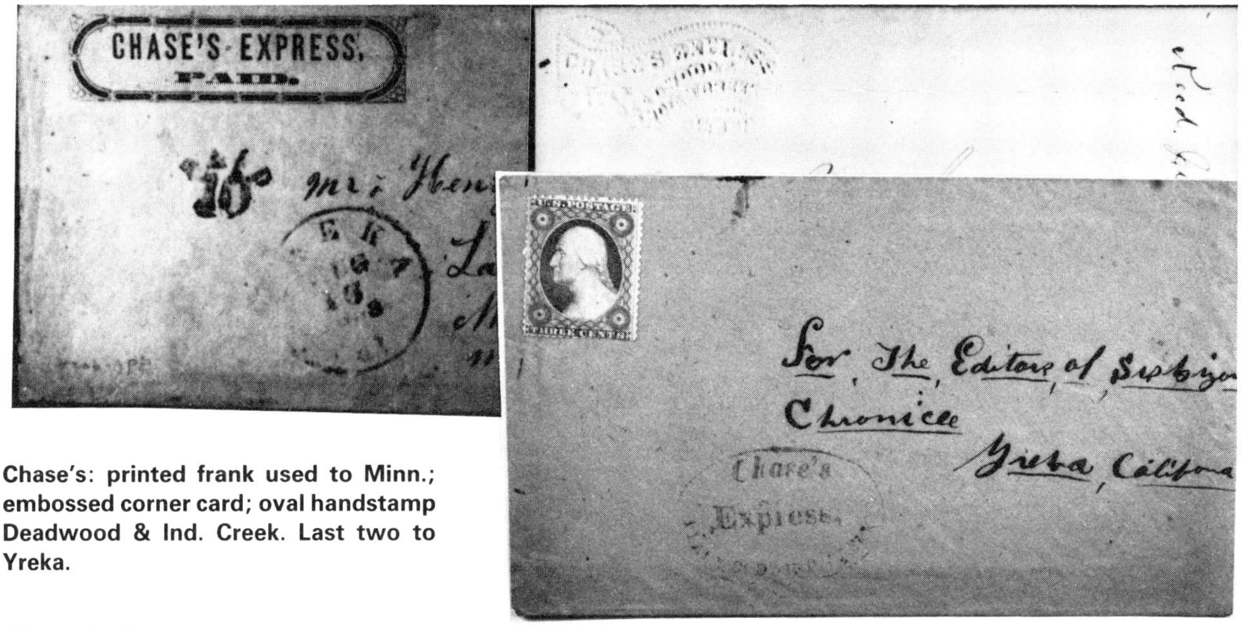

Chase's: printed frank used to Minn.; embossed corner card; oval handstamp Deadwood & Ind. Creek. Last two to Yreka.

Chase's Express

Dates: 1856-63.

Route: Yreka to Deadwood, Humbug, and Indian Creek.

Connected with Wells Fargo at Yreka. There may be a very rare third frank of a horse and rider with "Chase's" in front and "Express" in back.

Cherokee Express

Date: 1863.

Route: Oroville to Cherokee, a distance of about 14 miles.

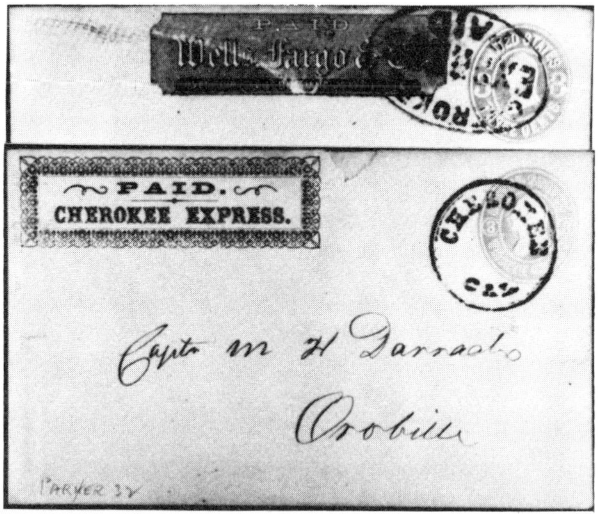

Cherokee: oval handstamp; printed frank. Both to Oroville.

Clippinger's Express

Dates: Between 1852 and 1855.

Route: In the area of Yankee Jims.

Little is known about this express. Only one cover is known to exist.

Clippinger oval on cover from Winchester, Ill., to Yankee Jims.

Coast Line Stage Co.

Dates: 1 January 1868-late 1870s.

Route: San Francisco to Monterey, Salinas, San Luis Obispo, Santa Barbara, Los Angeles, San Diego.

This firm used a variety of corner cards showing stagecoaches.

Colby's Nevada & Dutch Flat Express

Dates: 1860-61.

Route: Nevada City to Red Dog, You Bet, and Dutch Flat.

One cover reported.

Condon's Express

Date: 1850.

Route: Probably from Sacramento to the Northern Mines.

Almost nothing is known about this express since no mention of it has been found in newspapers or other records. Only one cover is known, a folded letter dated 15 June 1850.

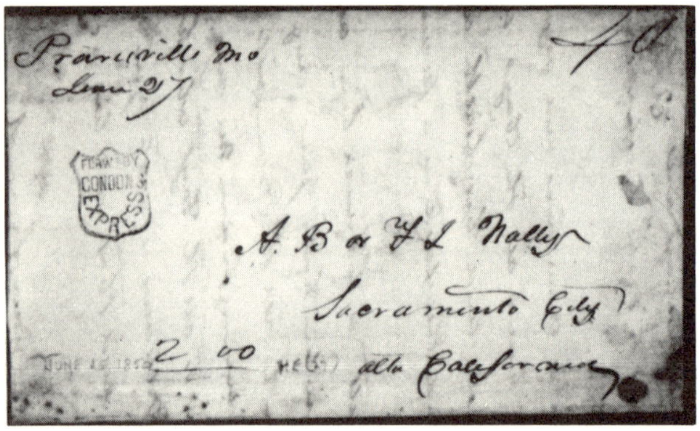

Copley & Co.'s Miners Express

Dates: 1855-56.

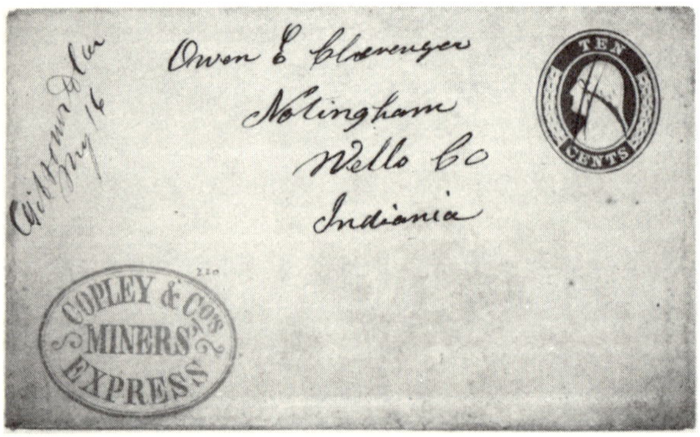

Route: Marysville to Gibsonville Ridge and Feather River camps.

Connected with Wells Fargo at Marysville.

Cornwall's Inland Express

Date: 1850s.

Route: Crescent City to Klamath River camps.

A cover is reported with the express name in a black oval.

Cramer's Express

Dates: 1867-70.

Route: Susanville to Fort Bidwell in northeast corner of state near Oregon border.

Only known frank was a two-inch-long dark ribbon scroll. The black printing of the company name on the dark background is almost impossible to read; no cover was available for portrayal.

Cram, Rodgers & Co.'s Express

Dates: 1851-27 March 1855.

Route: All of the territory of the Siskiyou-Klamath-Cascade mining areas, and continuing into Oregon to Jacksonville.

Connected with Adams & Co. and Newell & Co.

Cram Rogers SF and Shasta ovals to Siskiyou Co., Scott's Valley, Shasta, last with ADV and ms. 2.50.

Crawford's Middle Fork Express

Date: 1855.

Route: In El Dorado County from Placerville to Greenwood (Lewisville), crossing the South Fork and serving the mining camps near the Middle Fork of the American River. *(See also p. 231.)*

Crooks' Express

Dates: 1855-February 1860; and again in 1862.

Route: Yreka to the Scott and Klamath River mines; and in 1862 Yreka to Red Bluff to Sacramento.

Crooks' Scotts River oval.

Davis & Brothers Express

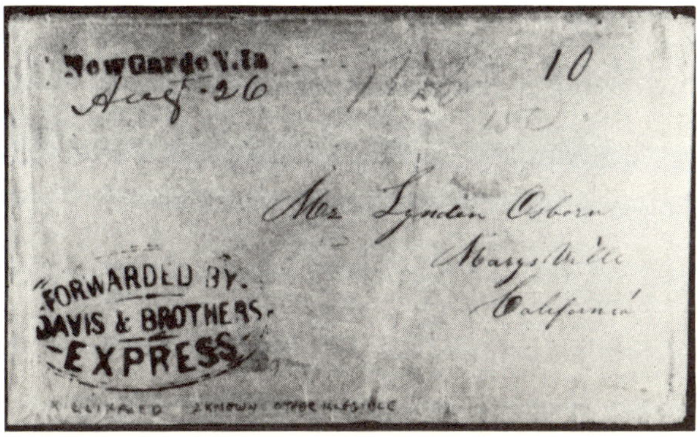

Dates and Routes: 1858: From Quincy to Honey Lake.

1862: From Oroville to Susanville.

The cover shown, the only known one with a legible Davis & Brothers handstamp, has a New Garden, Iowa, straightline postmark. It is stampless with a manuscript "10" and a manuscript "1.50" which may be the express delivery charge. There is no indication of how or why Davis & Brothers picked up the letter in Marysville for delivery to someone on their subscription list.

Deming & Wall's Union & Eureka Express

Date: Believed to be 1856 or 1857.

Route: In Humboldt County between Union (Uniontown) and Eureka.

This company is known only through the existence of a stampless cover addressed to Bucksport, a town near Eureka; carried "outside the mails."

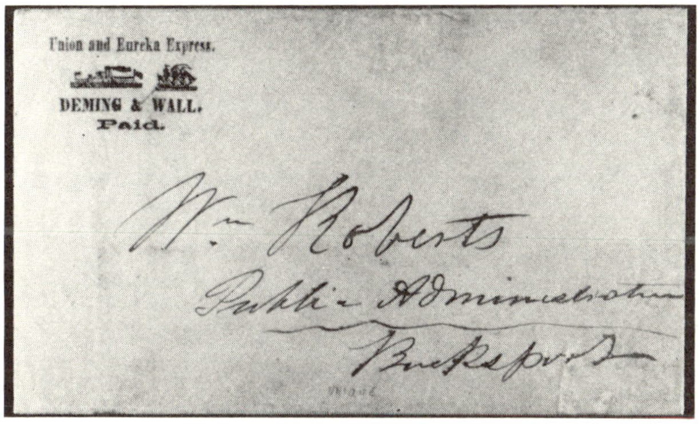

Dewey's South Yuba Express

Date: 1863.

Route: In the region of Omega, Washington, Alpha, and North Bloomfield on the South Fork of the Yuba River in Nevada County. Little is known about this small express. Its frank was a printed adhesive label. No cover available.

Dillon & Co.

Dates and Routes:

1853-54: Stockton to Sonora, serving the Southern Mines en route.

1856-57: Stockton to San Jose, Santa Clara, and Monterey.

Connected with Wells Fargo at Stockton.

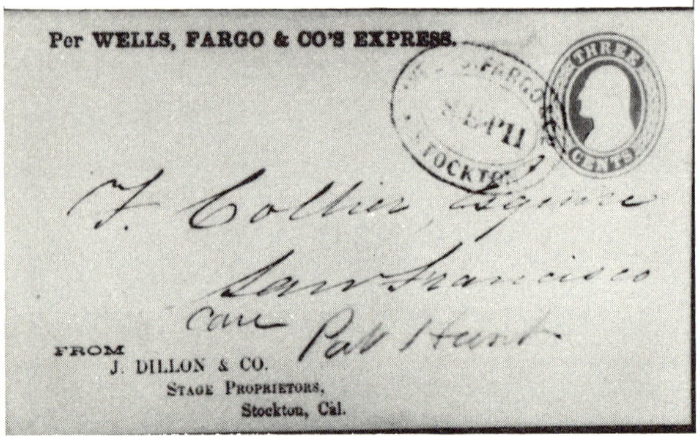

J. Dillon & Co. on envelope to SF with Wells Fargo oval.

Dodge & Co.'s California Express

Dates: 1849-52.

Routes: This company operated overseas and worldwide as well as in California and to Portland, Oregon.

Connected with Mumby & Co. for service to the mining areas.

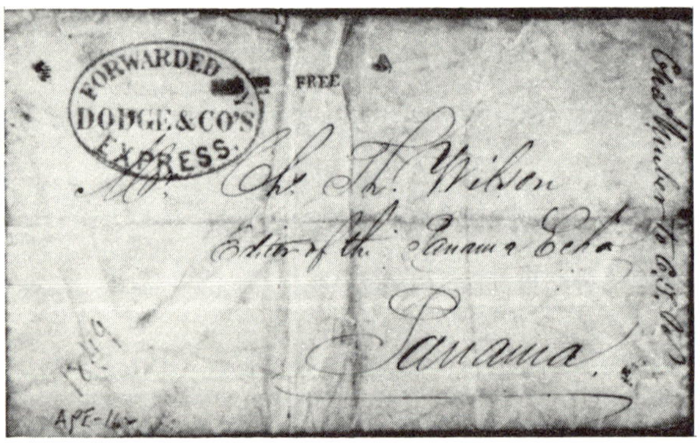

Dodge oval on 1849 letter to Panama.

Doherty's "Halfmoon" Bay Express (Doherty & Martin's Half Moon Bay & Pescadero Express)

Date: 1865.

Route: San Francisco to Half Moon Bay and Pescadero on the San Mateo County coast.

Doherty & Martin Pescadero oval to N.Y. state, 1864: oval Doherty "Half Moon" Bay.

Downieville & Howland Flat Express

Date: 1865 (lasted only a few months).

Route: The 12 miles between Downieville and Howland Flat in Sierra County.

Two types of printed frank are known with the same wording but different style frames.

Dugan & Wall's Express

Dates: 1863-64.

Route: Crescent City to Jacksonville (Oregon) and to Del Norte City (California).

They are said to have used two handstamps. The author has never seen one in the many collections he has examined.

(J. R.) English Express

Date: 1865-66.

Route: Nevada City, Alpha, Omega and Washington.

Connected with Wells Fargo at Nevada

City.

Only one cover of this company is known at present.

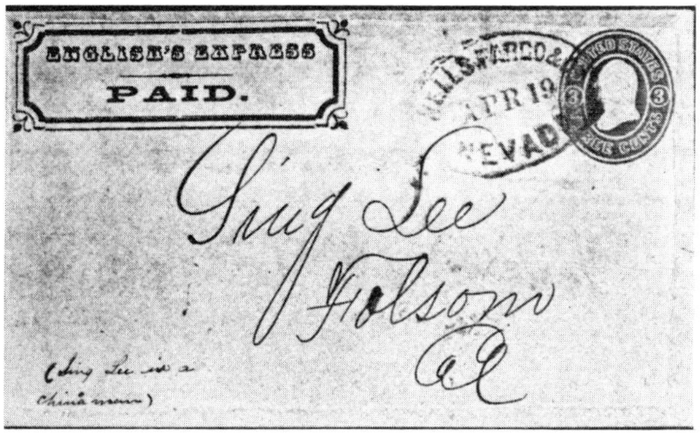

English & Wells Moore's Flat & Eureka Express

Dates: 1863-69.

Route: Nevada City to Moore's Flat, Eureka South, and Emigrant Gap.

English & Wells had not only a black printed frank but also used a blue oval handstamp, both of which appear on the cover illustrated. Connected with Wells Fargo at Nevada City and Emigrant Gap.

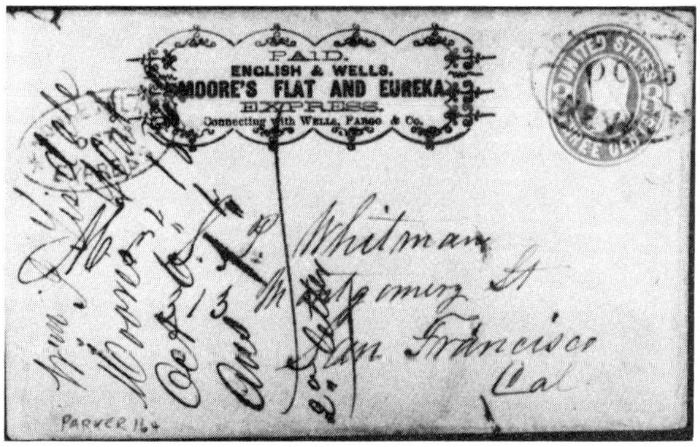

Estill & Co.'s Express

Date: 1850.

Routes: Transcontinental.

Colonel J. M. Estill tried a short-lived experiment of carrying mail along the emigrant trail for a fee of 50¢ per letter. He picked up mail from westbound emigrant trains and carried it to Weston, Missouri, where he placed it in the post office. Estill used a manuscript "Estill & Co./Express." Only three covers are known, all with manuscript franks; two are in the Wells Fargo History Room.

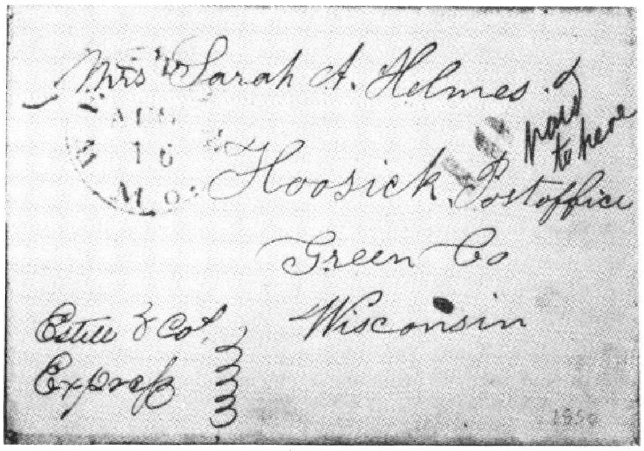

Letter carried by Estill Express to Weston, Mo., and there mailed, 1850.

Eureka, No. Bloomfield Express Co.

Date: 1857.

Route: North Bloomfield to Eureka South in Nevada County. The cover illustrated here, stampless, no postal markings, is addressed to Mohawk Valley (Plumas County).

Evans & Co.'s Jacksonville & Crescent City Express

Date: Early 1860s.

Route: Crescent City (California) to Jacksonville (Oregon).

Various Everts & Co.: Feather River on cover from Wis.; boxed Forwarded from Nelson Creek, into mails at Marysville; oval Marysville on short paid Quincy to Ky.; Forwarded from Ind.

Everts & Co.'s Express

[One-man express]

Dates: 1850-51.

Route: The mining camps between Marysville on the Yuba River and LaPorte on the South Fork of the Feather River, and over the Gibsonville Ridge.

Connected with Adams & Co. at Marysville. Frank D. Everts, while operating by himself, is known to have used one printed frank and three different handstamps. His was to become the largest express company in his area, operating for a total of 12 years during which he formed six companies with varying partnerships, all of which served the same area. (For purposes of clarity, the six companies will appear in chronological order.)

Everts, Snell & Co.'s Express

Dates: 1851-25 July 1854.

In 1851 Everts's business became too large for a one-man express, so he formed a partnership with a Mr. Snell. This company used three different handstamps. A shield handstamp is probably the scarcest of all Everts handstamps. Connected first with Adams & Co., then with Wells Fargo.

Everts Snell and Evert Snell ovals.

F. D. Everts & Co.

Dates: 25 July 1854-February 1855; again February 1855-10 May 1855.

In July 1854 when Everts dissolved his partnership with Snell, he took in his brother Henry Everts, and moved his headquarters from Marysville to LaPorte. This company lasted less than a year when they sold out to Adams & Co. When Adams failed a few days later, the Everts brothers took back their business. No franks or handstamps of this company are known and it is assumed that they used Frank Everts's original 1850 Everts & Co.'s franks and handstamps.

Everts, Davis & Co.

Dates: 10 May 1855-1856.

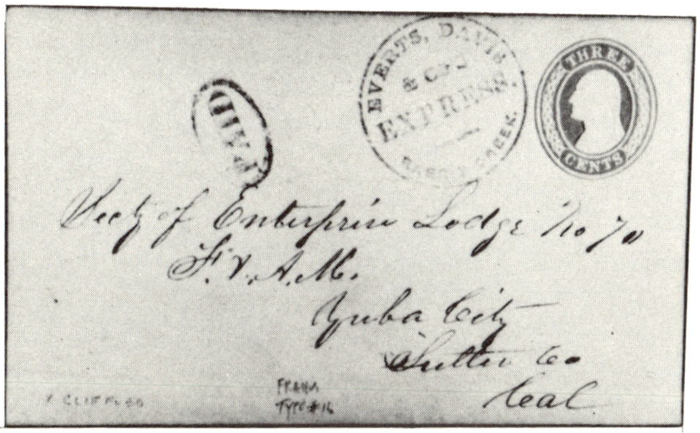

Five partners formed this company and moved their headquarters back to Marysville to connect with the Pacific Express Co. This company used a printed frank and two handstamps, one circular and one oval, each including the name of a town or camp.

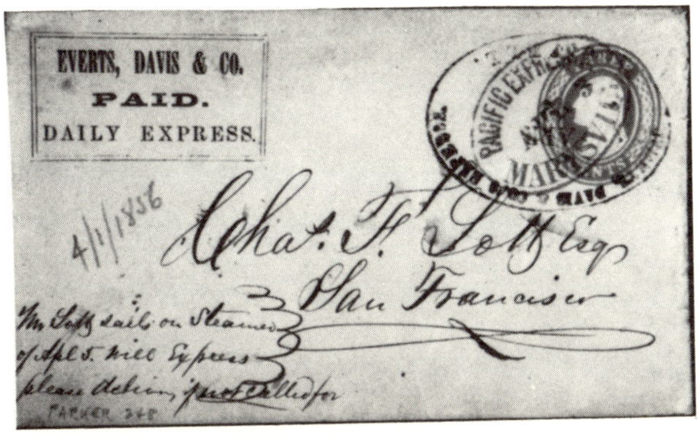

Everts, Davis frank.

Everts, Hannon, Wilson & Co.

Date: 1856.

Sometime in 1856 Davis dropped out of Everts, Davis & Co. and the remaining partners formed Everts, Hannon, Wilson & Co. The frank of this company is one of the scarcest of the Everts franks. Connected with Pacific Express and W. E. Singer & Co.

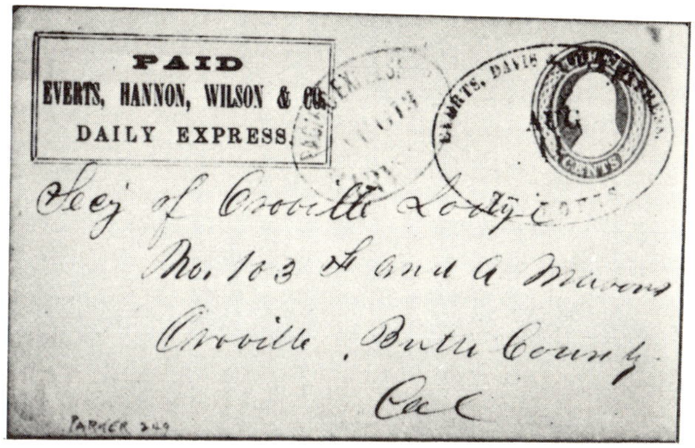

Everts, Wilson & Co.'s Daily Express

Dates: 1856-62.

Within a year Hannon dropped out of the partnership. When the Pacific Express went out of business in 1857, Everts, Wilson & Co. then connected with Wells Fargo and continued this arrangement until Everts left the

Three varieties of Everts, Wilson & Co. franks, two with circular handstamp.

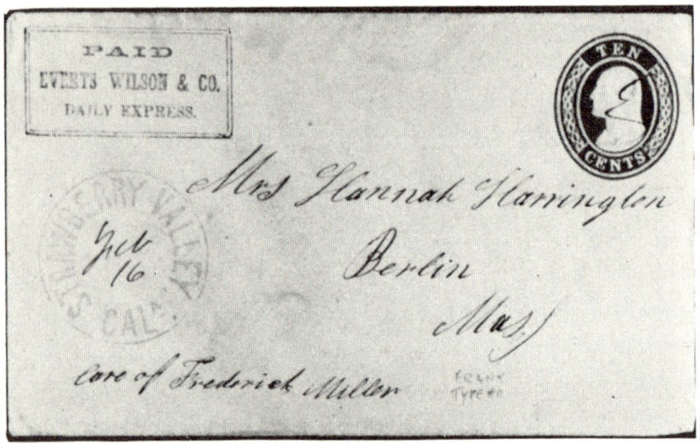

express business in 1862. This company used seven varieties of corner cards and two handstamps.

Fargo & Co.'s Feather River Express

Dates: 1852-53.

Route: Marysville to Bidwell's Bar and other North Fork Feather River camps.

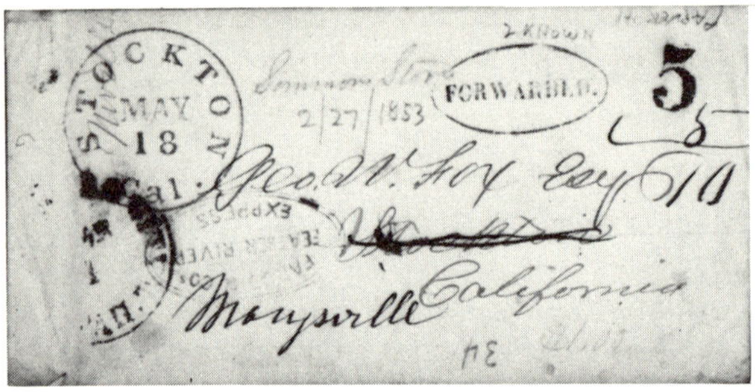

Farley's Express

[One-man express]

Dates: 1855-57.

Route: Camptonville to mining camps around Howland Flat via Ira Creek and Youngs Hill.

Ford & Co.'s Express

Dates: 1851-52.

Route: Sacramento to Greenwood in El Dorado County.

Ford circle and oval handstamps.

Wm. Ford's Express

Dates: Early 1850s-late 1862.

Route: From LaPorte to mining camps on the Gibsonville Ridge along Hopkin's Creek

Farley handstamps: Ira Creek, Camptonville.

Wm. Ford frank on paste-up.

and Poorman's Creek.

Connected with Holland & Morley Express at LaPorte.

Chester P. Fox's Half-Moon Bay & San Francisco Express

Date: October 1862.

Route: San Francisco to Half-Moon Bay and intermediate towns.

This express began operation in October 1862 according to an advertisement in the *San Mateo County Gazette* of 25 October 1862, using a printed corner card frank.

R. Francis' Express
Francis & Co.'s Express

Dates: November 1859-1864.

Route: LaPorte to Quincy via Gibsonville, Onion Valley, and Nelson Point.

Connected with Whiting & Co. at Quincy. In July 1862 Robert Francis took in J. S. Rump as a partner and changed the name to Francis & Co.'s Express, which lasted until 1864. In his first company he used two printed

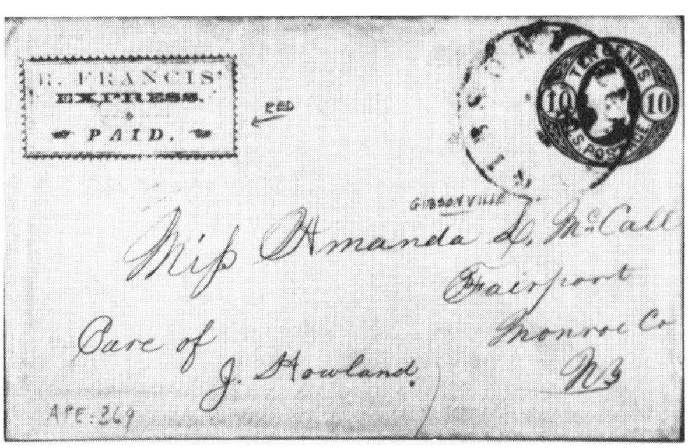

R. Francis frank.

Two varieties of Francis & Co. franks, lower with Gibsonville circle.

franks, in one of which the initial "R" is deleted. Printed frank of his second company is known in both black and red.

Freeman & Co.'s Express

Dates: November 1850-14 November 1851; and again 1855-60.

John M. Freeman was one of the most active early California expressmen. By November 1850 he had purchased Hawley's Express running between San Francisco and Sacramento.

In June 1851 he bought out Mumby & Co. which gave him access from Sacramento to the Northern Mines. Five months later he sold out to Adams & Co. Freeman served as the Adams agent in Sacramento, then became the Adams agent in Panama until the failure of Adams & Co. in 1855.

Freeman then returned to San Francisco to organize a new Freeman & Co. This company served foreign countries and also extended through all the mining regions of California, even opening offices in British Columbia from April to November 1859. Sometime after April of 1860 he sold out to Wells Fargo and retired from the express business. During different periods he had connected with Alta, Berford, Bowers, and Wells Fargo expresses.

Freeman markings (from lower left): SF oval; oval and COLLECT $1 to San Jose; oval to SF; franks with Coulterville, Oroville, and Sac. Mess. ovals; SF oval to Mass., frank to Phila.; circle to SF; slogan "Reply by Freeman & Co.'s Express."

During his express career Freeman used a great variety of printed franks, handstamps, adhesive labels, and corner cards. They exist in red, green, black, and brown.

Freeman corner cards and label.

J. C. Fuller's Express

Dates: 1858-60.

Route: North San Juan, Railroad Hill, Camptonville.

This express is reported to have had a three-line oval blue handstamp used on an 1858 cover.

Garland's Express

Dates: 1854-72 (maybe not continuous).

Routes: Richard E. Garland started in the express business in 1854 along a route between Oroville and Quincy. In 1864 he began operating from Quincy to Taylorville and Crescent Mills along the east branch of the North Fork of the Feather River. In the same year he bought out J. P. Wharton's Express which gave him access to the mining camps from LaPorte to Gibsonville to Jamison City.

Garland also operated under the names of Quincy & Indian Valley Stage Line, and Quincy & Crescent Mills Stage Line. By 1872 Garland had sold his entire express business to Wells Fargo. As far as is known, Garland had only one frank.

W. T. Gibbs' Express

Dates: 1854-59.

Route: Coloma, Georgetown, and Spanish Flat.

Connected with Wells Fargo at Georgetown. Gibbs used both an oval handstamp and a printed straightline.

W. T. Gibbs printed straightline and oval handstamp.

Gilbert & Hedges Express

Date: 1855.

Route: San Francisco, Monterey, and other coast towns south to Santa Barbara.

Connected with Wines & Co. Express.

Grammer & Co.'s Express

[One-man express]

Date: Late 1849.

Route: San Francisco to mining camps as far north as Georgetown in El Dorado County.

Joseph Grammer was a pioneer expressman. He used a subscription list from these Northern Mines to obtain mail from incoming steamers. It is known that he was still operating in December 1851.

Gray's Cal Express

[One-man express]

Dates: 1850-51.

Route: San Francisco to Sacramento.

The cover illustrated is a stampless letter, postmarked "Oswego, N.Y.," manuscript "40" rate, red stamp "Paid," and Gray's delivery fee of "1.10." Addressee was probably on Gray's subscription list; letter picked up in San Francisco, delivered in Sacramento.

Gray's Express

Dates: 1862-63.

Route: Coultersville to various towns in the Southern Mines in Mono and Alpine Counties.

Little is known about this company except that it connected with Wells Fargo and used Wells Fargo envelopes on which it printed its own frank in different colors.

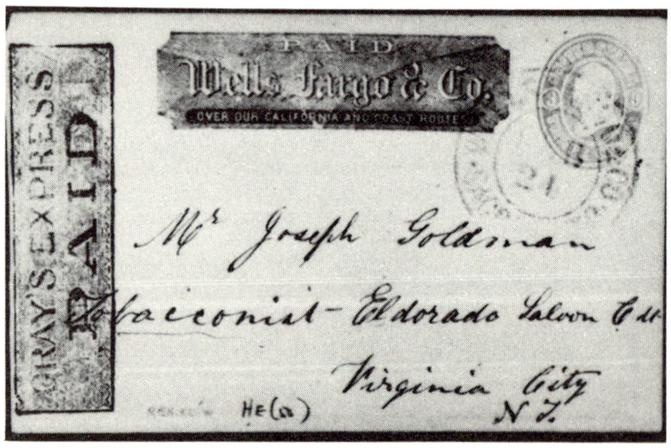

Greathouse & Slicer's Express

Dates: Fall 1855-1864 or 1865.

Route: Shasta to Callahan, Yreka, into Oregon and the Boise Basin of Idaho.

From 12 August 1854 to the fall of 1855, brothers George and Henry Greathouse operated a mule train express between Shasta and Callahan's Ranch (Callahan). No frank is known to have been used.

In the fall of 1855 H. Slicer became a

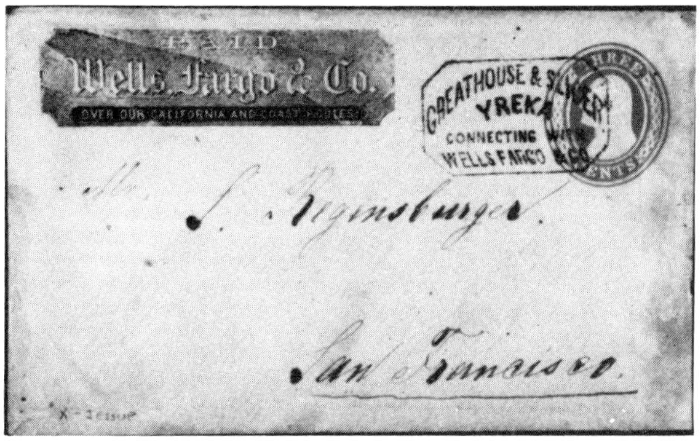

partner and the express company of Greathouse & Slicer extended its operations through Yreka into Oregon and over to the Boise Basin of Idaho. They connected with Wells Fargo. They are known to have used one handstamp. Counterfeits of this frank exist.

Great Pioneer Express

Date: 1858.

Route: From Placerville to the mining camps on the North Fork of the Middle Fork of the American River to Michigan Bluff.

Connected with Tracy & Spears Express at Georgetown. Many of the most important mining camps of the Mother Lode area were on this route. It is unfortunate that more information is not available about this short-lived express. It is reported to have used a black oval handstamp.

(Thomas) **Greene's Copper City Express**

[One-man express]

Dates: 1864-65.

Route: Copper City (at the junction of Pit and McCloud Rivers) to Shasta to Weaverville.

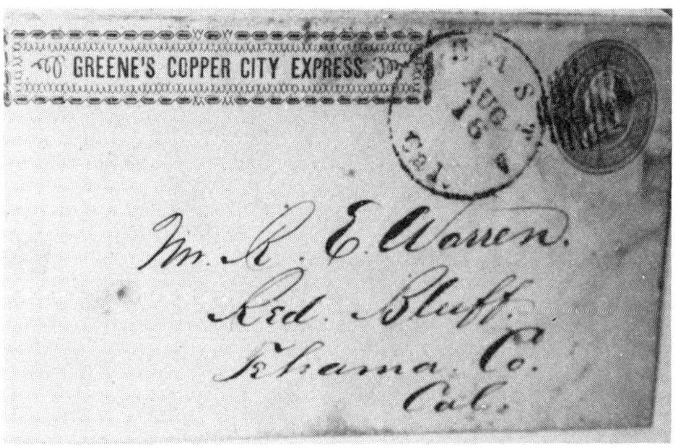

(J. A.) **Green's Express**

[One-man express]

Dates: 1850-54.

Route: Nevada City to Red Dog, You Bet, Lowell Hill, Alpha, Omega, Washington, Moore's Flat, and Eureka South, all in the area of the South and Middle Forks of the Yuba River.

Greenhood & Newbauer Northern Express

Dates: 1857-August 1867.

Route: Weaverville to Shasta and Crescent City.

Connected with Wells Fargo at Shasta and Weaverville. There exists a rare error in their printed frank in which Newbauer is spelled Newba*n*er.

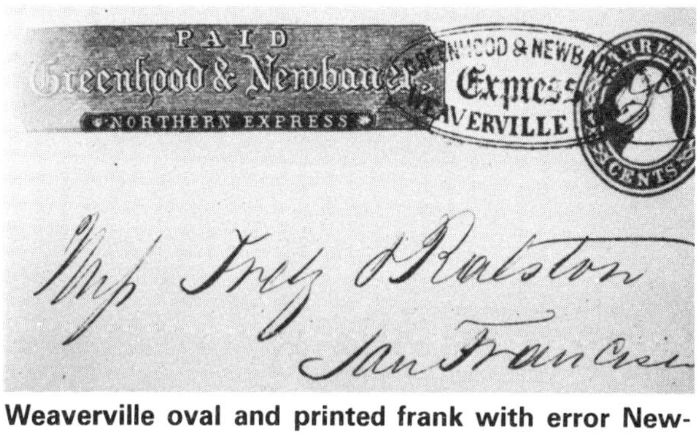

Weaverville oval and printed frank with error Newbaner.

Greenhood & Newbauer printed franks with Weaverville oval, double circle, and Northern Express corner card.

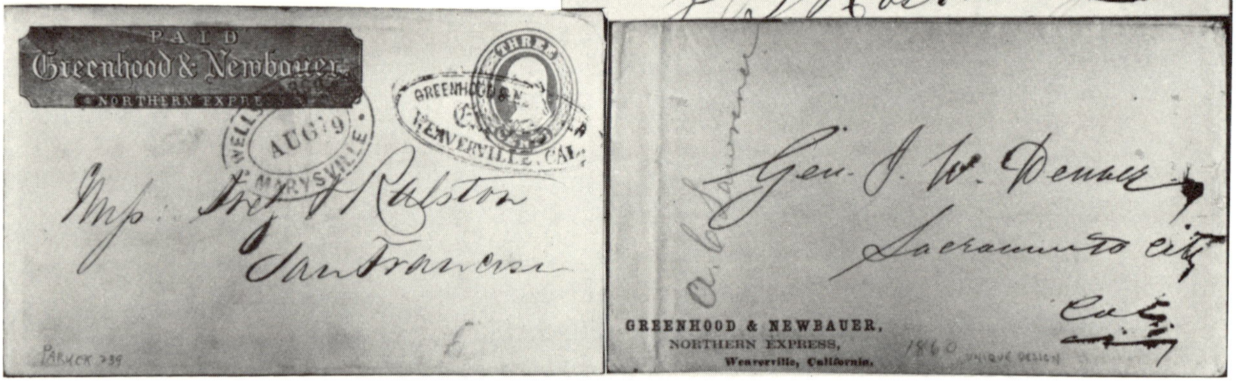

Gregory's Express

Dates: Early 1850-May 1853.

Joseph W. Gregory created an express system which ran not only from San Francisco to Sacramento and the towns of the Northern Mines, but which also operated nationwide as well as into foreign countries. He operated under many Gregory titles and used at least 22 different handstamps in black, blue, and red.

Various Gregory markings: Atlantic & Pacific to Hawaii, 1851; fancy oval New York & California, 1850; two different SF straightlines to N.Y.

Gregory handstamps from SF, Sacramento, Marysville.

Gregory & English's Moore's Flat & Eureka Express

Date: 15 April 1866.

Route: Nevada City to North Bloomfield, Lake City, Woolseys Flat, Moore's Flat, and Eureka South.

Connected with Wells Fargo at Nevada City. Although in business only a few months, this company used five different printed franks, primarily variations in the frank frame line.

Gridley's Express

[One-man express]

Dates: 1859-62.

Revel Colt Gridley was among the first to offer the mining communities in the High Sierras between Oroville and Honey Lake Valley a letter mail service some years prior to the establishment of any post offices in the area. He operated north from Oroville along the west branch of the Feather River to Inskip and then east to the North Fork of the Feather River to serve the communities of Humbug Valley and Honey Lake Valley. Connected with Wells Fargo at Oroville. Little information is available about this one-man express who traversed a lengthy and difficult trail. His only known frank is shown in Nathan.

Hall & Allen's Express

Date: 1864.

Route: To mining camps in the Dutch Flat area.

Connected with Wells Fargo and Bamber & Co.

Hanford & Co.'s Express

Dates: 1857-58.

Route: Almost the entire length of northern California, from Yreka via Shasta, Marysville, and Sacramento to San Francisco.

Connected with Alta Express and Berford & Co.

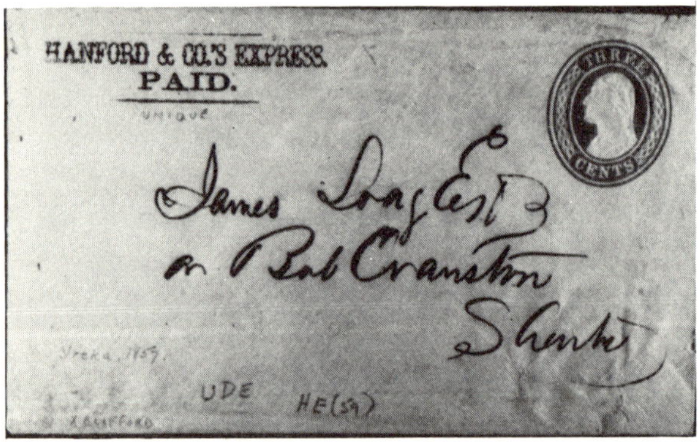

D. W. Harrier's Express

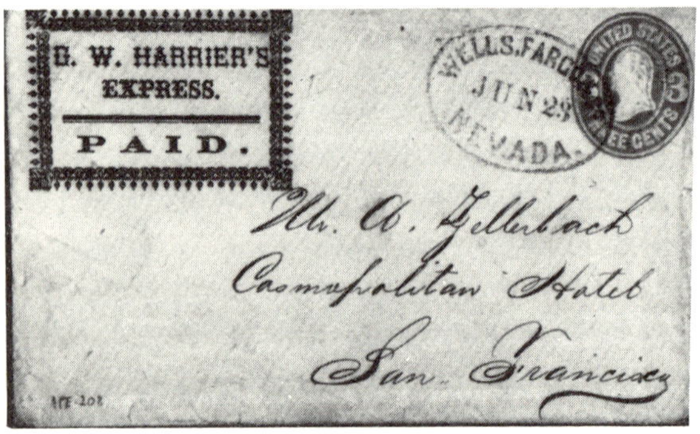

Dates: 1862-65.

Route: Nevada City to Moore's Flat, North Bloomfield, and intermediate camps.

Connected with Wells Fargo at Nevada City. Three printed franks are known to have been used by this express.

Haven & Co.'s Express

Date: 16 July 1850.

In September 1849 J. P. Haven organized an express company with C. Livingston called Haven & Livingston's Foreign & Inland Express, which ran from Stockton to San Francisco, and had an arrangement with the banking and express firm of Livingston, Wells & Co. in New York to receive treasure. No covers or franks of this company are known at present.

On 16 July 1850 the partnership of Haven & Livingston was dissolved and Haven immediately established Haven & Co.'s Express. He now operated from Sacramento to San Francisco, and again forwarded letters and treasure to New York.

Haven & Co. is known to have had two types of frank: a handstamped oval and a manuscript.

Haven & Co. "free to NY."

J. Hawes & Co.'s Express

Dates: May 1853-1857.

J. Hawes, who in May 1853 had purchased the remnants of Gregory's Express, established a company called J. Hawes & Co.'s New York & San Francisco Express whose operation was similar to that of Haven

& Co.'s Express. Hawes evidently ran from Sacramento to San Francisco, and from San Francisco provided service to New York via steamers carrying treasure and mail.

His one known frank, usually found on the back flap of a cover, is illustrated here.

Hawley & Co.'s Express

Dates: January to November 1850.

Route: San Francisco, Benicia, and Sacramento.

Hawley was an early expressman, beginning Hawley's Express in October 1849 with four partners, all of whom had formerly been associated with Weld & Co.'s Express. No franks of Weld & Co.'s Express nor of this original Hawley's Express are known to exist.

In January 1850 Hawley bought out his partners and changed the name of the express to Hawley & Co.'s Express. The two known handstamps are illustrated here. In November 1850 Hawley sold out to Freeman's Express.

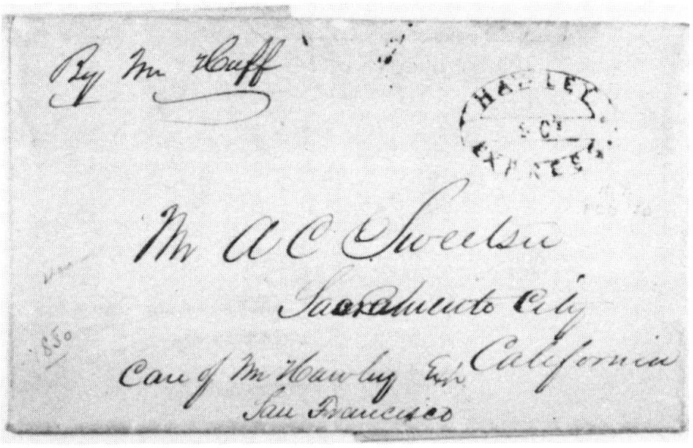

Hawley oval on 1850 letter of N.Y. origin.

Hawley shield on letter from SF.

Haydon's Express

Dates: 1855-57.

S. C. Haydon, a merchant in Petaluma, established his express to forward mail from Sonoma and Marin County towns to San Francisco.

His only known frank, an oval, is illustrated on the cover shown here, which also has a printed corner card.

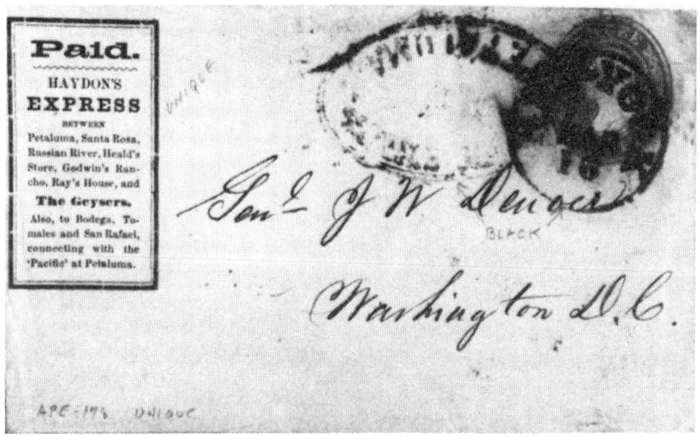

Haywood Express Co.

Date: 1864.

Nothing is known about this small express company except that it operated in 1864, connected with the Pacific Union Express Co., and soon sold out to them.

Its only known frank is a straightline handstamp printed over the Pacific Union ribbon frank.

Henderson & Co.'s Express

Date: 1854.

Route: San Francisco south to coastal towns.

An 1854 cover is reported with a block letter blue printed frank.

W. F. Herre's Contra Costa Express

Dates: 1855-56.

Route: Contra Costa County towns to San Francisco.

An 1855 cover is reported with red oval handstamp.

S. Hervey's Express

[One-man express]

Dates: 3 November 1858-4 April 1860.

Route: Nevada City to Lowell Hill, Red Dog, Dutch Flat, and intermediate points.

Samuel Hervey's only known frank is on the cover of *Hutching's California Magazine* of May 1859, discovered in 1978. *(p. 232.)*

Heston's Express & U.S. Stage Line

Dates: 1857-June 1859.

Route: Hornitos to Visalia via Millerton and King's River.

Connected with Wells Fargo at Hornitos.

W. L. Hice's Express

Dates: June 1859-1860.

Route: Hornitos to Visalia via Millerton and King's River.

After purchasing Heston's Express & U.S. Stage Line in June 1859, Hice continued serving the same route. Hice's only known frank is a manuscript on cover.

J. W. Hoag & Co.'s Contra Costa Express

Dates: 24 October 1855-1858.

Route: Oakland south to San Leandro, Mission San Jose, Warm Springs, and intermediate points.

Connected with Wells Fargo and the Pacific Express. Hoag, John Bamber, and Charles D. Gable formed this partnership. In 1858 Bamber bought out his partners and took over the express route, forming Bamber & Co.'s Express. Hoag & Co. used a double oval handstamp in both red and blue colors.

Hoag ovals on COLLECT and PAID covers.

Hoffman & Co.'s Express

Dates: 1850-51.

Route: Picked up mail in San Francisco, but operated from Sacramento to Weberville and Placerville. Later Hoffman ran an express line north into Oregon.

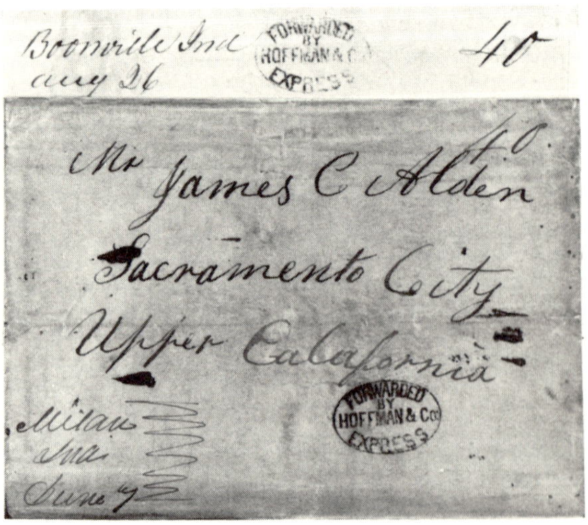

Two early covers from Ind. to Sacramento with Hoffman ovals.

Holland, Morley & Co.'s Express

Date: 1862.

Route: Marysville to LaPorte to the Gibsonville Ridge and the Feather River mining camps.

Holland and Morley purchased the express business of Everts, Wilson & Co. and their partnership lasted until Morley sold his interest to Sam Wheeler, which led to the formation of a new company. The Holland and Morley company had two printed franks and a black oval handstamp.

Holland & Wheeler's Express

Dates: Late 1862-December 1863.

Route: Marysville to LaPorte to the Gibsonville Ridge and the Feather River mining camps.

Connected with Wells Fargo at Marysville and Whiting & Co. at Quincy. Holland and Wheeler used three printed franks and two handstamps, one a double oval, the other a large double circle.

Holland & Wheeler franks to N.Y. and SF.

Hopkinson's Express

Dates: 1862-64; again 3 February 1865-1867.

Route: Nevada City to Dutch Flat via Red Dog and Little York camps.

Connected with Wells Fargo at Nevada

Hopkinson corner card and printed franks.

City and Dutch Flat. Hopkinson used a variety of printed franks with different frame lines and town names, also a rare handstamp.

Horsley & Brastow's Express

Dates: 1855-56 (10 months).

Route: Shasta, Yreka, and Scott River country, and also southern Oregon.

Hunter & Co.'s Express

Dates: Early 1851-July 1854.

After the Gold Rush began, Alexander Hunter started out at first as a one-man express, picking up and delivering miners' mail. Then early in 1851 he formed Hunter & Co.'s Express. Using Sacramento as a base, he operated from the Mokelumne River to the Bear River, reaching most of the mining camps in

Hunter: straightline El Dorado to N.Y. state; Placerville oval; Mud Springs shield; fancy double oval with horse, and Mormon I. double oval, both from the east to Murderers Bar.

between, connecting with Gregory and Adams expresses. Hunter & Co.'s franks were numerous and varied.

Indian Creek Express

Dates: 1865-66.

Route: Between Yreka and Indian Creek, about 20 miles.

G. F. Johnson's Inland Express

Date: 1858.

Route: From Crescent City to Jacksonville, Oregon.

Connected with Dugan & Wall in Crescent City and Beekman in Jacksonville.

Kennedy & Co.'s Express
Kennedy, Long & Co.'s Express & Transfer Co.

Kennedy & Co. and Kennedy & Long franks, one with PAID.

Dates: M. G. Kennedy and F. W. Utter: 1862-64. M. G. Kennedy and Mr. Long: 1864-66.

Route: San Francisco to Half Moon Bay and Pescadero.

Lamping & Co.'s Express

Dates: 1861-65.

In 1861 P. A. Lamping began running a small express from Downieville north. Early in 1865 he bought Samuel Langton's routes from Langton's widow and covered the Yuba River area from Marysville to Downieville to Meadow Lake. Within the year he sold out to Wells Fargo. Lamping's frank and handstamp are on the cover shown here.

The Langton Expresses

Samuel W. Langton was the most vigorous of all the independent expressmen. He operated many companies under successively different names. He used at least 20 varieties of printed franks and handstamps.

Langton was an expressman for 14 years, beginning as a one-man express: **Langton's Express.** His route ranged up the Yuba River from Marysville to Downieville including towns and camps en route. Six months later, with William T. Ballou as a partner, Langton organized **Langton's Yuba River Express.** Its route covered the same area. After seven months Ballou sold out to Langton, and in March 1851 Langton formed a new partnership with Samuel L. Clapp and E. White: **Langton & Co.'s Express.** When the partnership dissolved in November 1853, Langton and his brother, W. T., formed **Langton & Bro.'s Express,** which they operated until 21 February 1855. In February or March 1855 Samuel Langton took as partners N. W. Wilkinson and Charles P. Stone, forming **Langton's Pioneer Express** and extending his route to San Francisco and Sacramento. In January 1860, after the discovery of the Com-

Various Langton Downieville ovals, including error Downeville on *The Mountain Echo.*

Langton's (from lower left): label with ms. Unionville; Cherokee oval; Columbia Hill oval; Forest City oval; small Pioneer oval; straightline Forwarded Nevada (City); Unionville oval with adhesive label; ms. Dayton oval; Marysville oval.

stock Lode in Nevada, Langton formed **Langton's California & Nevada Express** running from Downieville to towns in Nevada. He continued to use the Langton's Pioneer Express printed frank on this route. One month later, Langton added a branch to his Nevada operations to run to the Humboldt Mines in the northern part of the state: **Langton's Humboldt Express.** For this express Langton created a rare adhesive stamp to indicate an additional fee of 25¢ because of the difficulty of the route. On 24 August 1864, Langton was accidentally killed at Silver City, Nevada, and in 1865 his family sold out to Lamping & Co.

LaPorte Express Co.

Date: 1865.

Route: From LaPorte to the mines on the Gibsonville Ridge.

Connected with Wells Fargo at LaPorte. This little-known express had a printed frank and an oval handstamp, both shown.

Latta's Mountain Express

Dates: 28 February 1866-1867 or 1868.

Route: Nevada City to Alpha, Omega, and Washington, California.

Latta used two similar printed franks.

Latta: two different franks, both Washington, Cal., to SF.

Leland's Express
(Leland's San Jose Express)

Date: 1855.

Route: San Francisco to San Jose.

Leland & McCombe's Express
(John McCombe)

Dates: Late 1853-28 June 1854.

Routes: From San Francisco to Crescent City and towns on the Humboldt Bay; express sent by ships. Their secondary route was over-

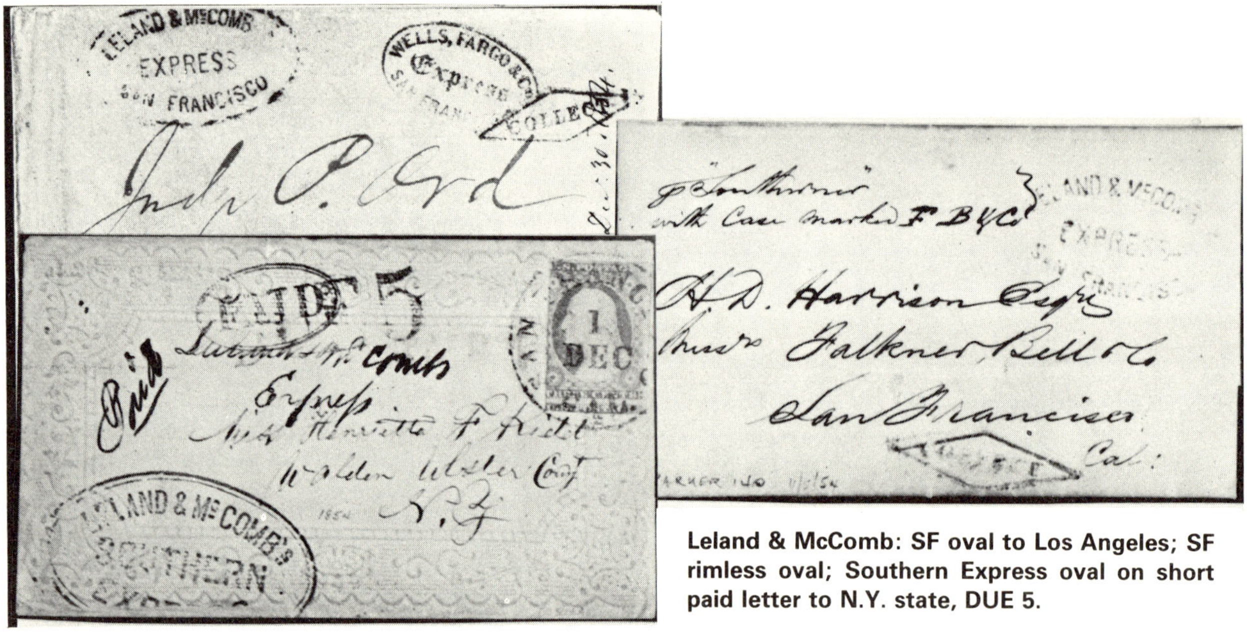

Leland & McComb: SF oval to Los Angeles; SF rimless oval; Southern Express oval on short paid letter to N.Y. state, DUE 5.

land from San Francisco to Santa Clara to San Jose.

Leland & McComb's Express
(Leland & McComb's Southern Express)
(David McComb)

Dates: August 1854-October 1855.

Route: San Francisco, Monterey, Santa Barbara, San Pedro, Los Angeles, San Diego.

Leland continued using his Leland & McCombe's Express handstamp with the "E" removed from McCombe. A later handstamp was created reading "Leland & McComb's/Southern/Express."

E. Lobenstein's Pony Express

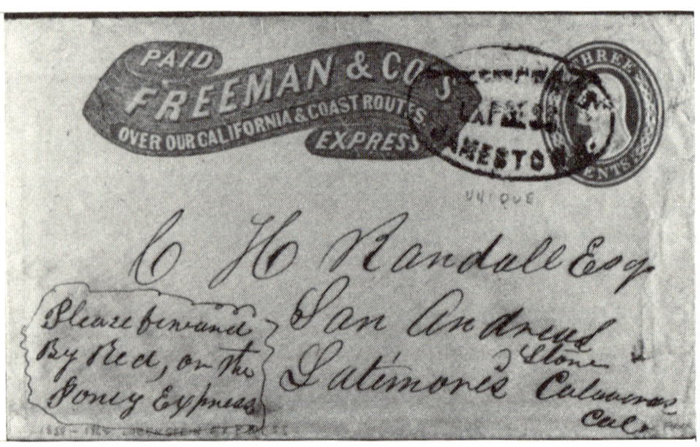

Dates: 1858-62.

Route: From San Andreas to Calaveritas and Old Gulch.

E. "Red" Lobenstein had neither a frank nor a handstamp for his pony express. He depended on a manuscript notation such as appears on the cover here: "Please forward by Red, on the Poney Express."

McCann & Morison's Express

Date: Early 1860s.

Route: Placerville area.

C. McLaughlin's U. S. Mail Line

Dates: 1856-58.

From April to July 1853 Charles McLaughlin and Mr. Mann operated McLaughlin & Mann's Express from Marysville to Downieville and intermediate camps. No franks are known to have been used. In July 1853 Mann left the partnership, and McLaughlin continued to operate alone under the name of U.S. Mail Line. His route covered much of the Yuba River area. As far as is known, no franks were used by this firm either.

Three years later in 1856 McLaughlin moved his U.S. Mail Line into the Santa Clara

Valley and operated from San Francisco to San Mateo, Santa Clara, and San Jose; also from Oakland to Union City, Warm Springs, and San Jose. Two months later he extended his route to include Santa Cruz and Monterey. This Santa Clara Valley company, C. McLaughlin's U.S. Mail Line, used a printed corner card.

The last word on McLaughlin is in 1858 when he sold half-interest in his firm and changed the name to C. McLaughlin & Co. There are no known franks of this latter company.

Mann & Co.'s Express

Date: 1856.

Route: Between Marysville and Oroville.

Little information is available about this express, but it is reported to have used a black oval handstamp.

Mann's Inland Express

Date: 1855-59.

Route: Crescent City via intermediate points to Jacksonville (Oregon).

A very few Mann's Inland Express covers with handstamped town postmarks exist, but most Mann's covers are unused. Those with manuscript town postmarks are highly suspect as being fraudulently created as *used* by a known forger who obtained a number of these unused covers.

Mead & Clarke

Dates: 1865-68.

Route: From Snelling's Ranch (Snelling) to Fort Miller (Millerton) northeast of Fresno, over their Clear Creek route, possibly extending to Visalia and Havilah.

Mead & Davis Letter Express

Date: Late 1860s.

Route: From Visalia to towns on the Clear Creek route.

Connected with Wells Fargo at Visalia. This is probably the same Mead who operated Mead & Clarke, but with a new partner. The frank is an oval handstamp.

Miller & Co.'s Express

Date: 1853.

Route: Chinese Camp, Montezuma, Jackson.

An 1853 cover with black double-line oval handstamp has been reported.

Moore's Flat Express

Dates: 1868-69.

Route: Nevada City, Moore's Flat, and Eureka South.

Connected with Pacific Union Express.

Morley, Caulkins & Co.'s Express

Dates: 1853-61.

Route: LaPorte to the Gibsonville Ridge mining camps.

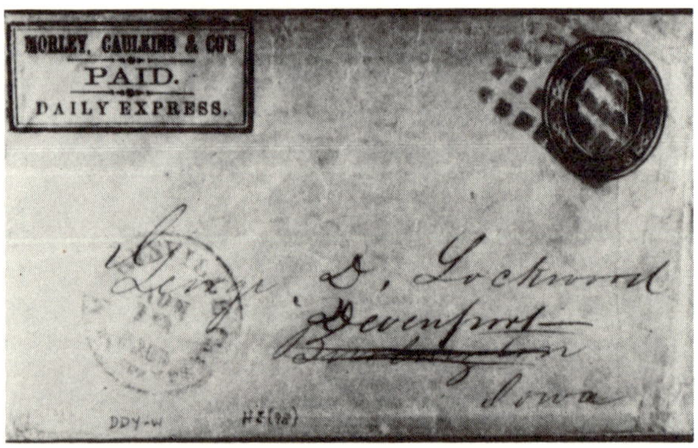

Mumby & Co.'s Express

Dates: Spring 1851-June 1851.

Route: San Francisco to Sacramento and Marysville by boat, and then to the Northern Mines.

The handstamp is almost indecipherable on the only known cover.

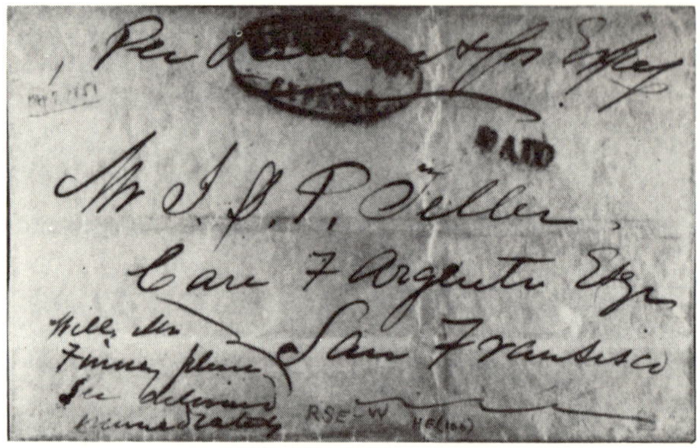

Mumby & Co. oval.

Nevada City & Meadow Lake Express

Dates: 1861-64.

Route: Between Nevada City and Meadow Lake in the Yuba River area.

Newell & Co.'s Express

Dates: October 1851-July 1853.

Routes: They utilized two routes: one from Stockton to Sonora, and another from San Francisco to Portland (Oregon).

Connected with Brown, Adams, and Cram, Rodgers expresses.

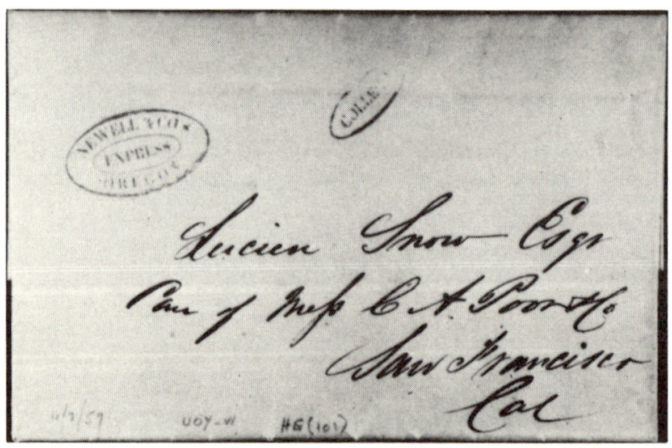

Nichols & Co.'s Express

Date: Spring 1858.

This was a busy express which began in 1858, running from San Francisco to San Jose. They also ran a semi-monthly express to Oregon via Crescent City. They connected with the Alta and Freeman expresses. Later they sold out to the American Express (California), which was already in operation, and which now began using on their printed adhesive frank the interesting picture from the Nichols & Co. printed frank illustrated here.

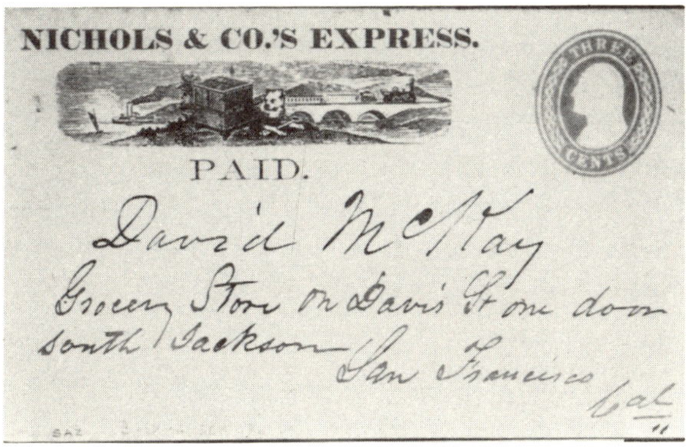

Nils Express

[One-man express]

Date: 1850.

Route: Foster's Bar on North Fork of Yuba River to Marysville to San Francisco.

The letter enclosed in the cover illustrated states that the writer will give the letter to the expressman "who will carry it down" and bring back letters in about four weeks, indicating that Nils made a monthly round trip to San Francisco.

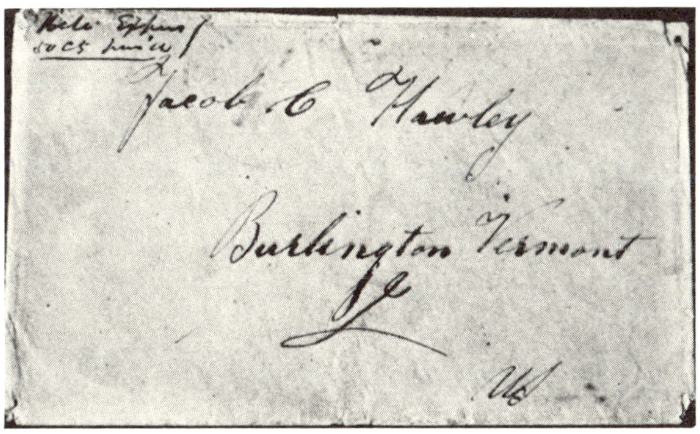

G. H. Norman's Express

Date: 1864 (short-lived).

Route: Feather River district from Gibsonville to mining camps on the Ridge.

Within a few months G. H. Norman (Nohrman) and N. O. Pauly formed a partnership to be known as Pauly & Nohrman's Express.

Norman & Grant's Express

Date: 1868.

Route: Gibsonville to camps on the Ridge.

When the partnership forming Pauly & Nohrman's Express dissolved in 1868, Nohrman (Norman) took a new partner named Grant. It is not known how long this company existed.

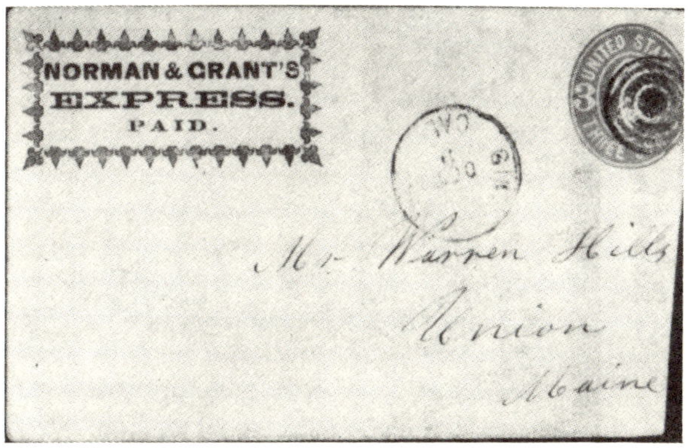

Oregon Stage Co.

Dates: 1 July 1866-August 1870.

Route: Between Portland (Oregon) and Lincoln (California), 15 miles west of Auburn. By this time Lincoln was connected with Sacramento by railroad.

Organ & Tibbetts' Excelsior Express

Dates: 1865-68.

Route: Nevada City through the Excelsior District to Meadow Lake.

Because of the severe terrain, it is possible that this was a pony express. On the cover illustrated here, the spelling of "Tibbetts" would appear to be in error. This is the same William Tibbetts who operated Tibbetts' U.S. Mail Line (for which no franks are known) from 1859 to 1864, and Tibbetts & Co.'s Excelsior Express from 1864 to sometime in 1865. Two varieties of the printed frank of Organ & Tibbetts' Excelsior Express are known.

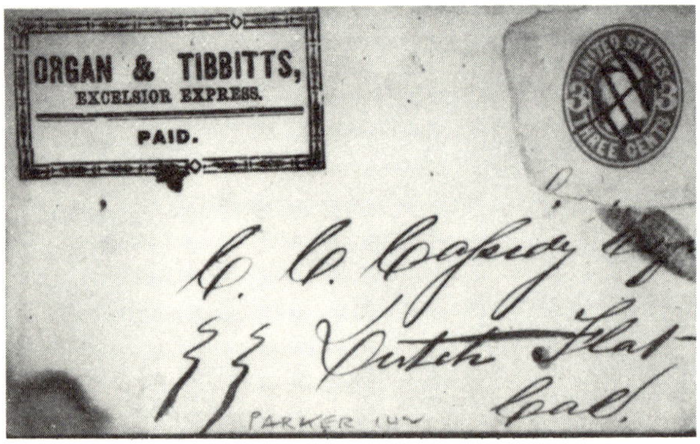

Overland Mails

Careful distinction must be made between the various overland mails. Some of the confusion is caused by the use of similar corner card franks showing stagecoaches and horses, a device used by at least two completely different companies. Further confusion ensues when a cover bears the simple manuscript "per Overland."

The **Overland Mail Company** was the Butterfield Line, also known as the Southern or Oxbow Route, operating from 1858 to 1861.

The **Central Overland** was the Central Overland California & Pikes Peak Express Company operating from February 1860 to March 1862 across the central part of the western United States. This company utilized the route of the earlier Overland mails operated by Woodson, Chorpenning, and Hockaday.

The **Overland Mail Stage** was operated by Sanderson, Parker & Co. from 1867 to 1872, running from Placerville, California, to Salt Lake, Utah. This company used a stagecoach-and-horses corner card frank very similar to that of the Overland Mail Company.

The confusion can be resolved by careful attention to dates and routes indicated by towns, also by the important fact that the Overland Mail Company (Butterfield) in its stagecoach-and-horses frank always used the words "via Los Angeles," whereas the Overland Mail Stage frank always included the words "via Placerville and Salt Lake."

The Overland Mail Company and the Central Overland are considered in the chapter on the Overland Mails.

Overland Mail Stage

Dates: 1867-72.

Route: Placerville to Salt Lake.

This stage line is the subject of controversy among postal historians. Some have insisted that it followed the route of the old Butterfield line south through California to Los Angeles, then north through the Carson Valley of Nevada and east to Salt Lake. This circuitous route is suspect, both because of its geographical difficulties and because of the absence of covers with town postmarks along this route. Most authorities have accepted the printed corner card "Placerville to Salt Lake" at face value, even though this route was dominated by Wells Fargo stages from 1867 to 1869, and from 1869 onward mail and passenger service was provided by the transcontinental railroad.

If the Overland Mail Stage did operate from Placerville to Salt Lake (1867-72) in competition with Wells Fargo and the railroad, then newspaper advertisements should have carried their schedules. Extensive research fails to reveal such evidence. This lends credence to those postal historians who insist that this was a stage line "which never ran," and that after abandoning their plan, Sanderson and Parker sold their attractively printed corner card envelopes to stationery stores for resale to any customers sending letters via the overland route. While future discoveries may resolve the controversy, nevertheless these covers are very rare.

Pacific Express frank on 6¢ Nesbitt with 1¢ and 3¢ 1851 added to make up 10¢ rate from Knight's Ferry to Mass. shortly after transcontinental rate was increased.

Pacific Express Co.

Dates: 2 March 1855-April 1857.

Like Adams & Co., this company attempted to conduct an express and forwarding business throughout California and Oregon, as well as to the Atlantic states by steamer. For a company in business only two years, it created a prolific number of printed franks and handstamps in black, blue, and red, at least nine in number. *(See also p. 231.)*

Pacific Express ovals: Sacramento, Nevada, SF, Marysville, El Dorado, Napa.

Pacific Express SF and Stockton ovals, and SF double circle.

Pacific Express Company (#2)

(Printed frank of a railroad mail car)

Postal historians raise questions about this company. An advertisement in the *Daily Morning Message* of Gold Hill, N. T., 24 June 1864, announced the formation of a company with this name and a route from Gold Hill and Virginia City in Nevada, to San Francisco via Nevada City, Grass Valley, San Juan, Auburn, and Sacramento. Confusingly, a 2 January 1869 notice in the *Quincy Union* announced the resignation of the Pacific Express local agent, but Quincy is over 75 miles north of the announced route. Since the original Pacific Express had ceased existence in April 1857, the resigned agent could not have represented that company. Further, the printed franks depicting a railroad mail car imply existence after completion of the transcontinental railroad in April 1869. In addition, no used covers with this frank are known prior to the 1870s. Therefore, despite the newspaper items, many historians maintain that this company never operated, and the railroad car franked envelopes were later used as illustrated covers for letters carried over the transcontinental railroad. Large numbers of these unused envelopes are offered in auctions, but have little value and most collectors regard them as curiosities.

Pacific Stage & Express Co.

Date: About 1864 (short-lived).

Route: From San Francisco and Sacramento to Virginia City (Nevada) via Auburn, Grass Valley, Nevada City, and Eureka South. *(See p. 229.)*

Pacific Union Express Co.

Great controversy exists among postal historians about this company. Nathan states that it was formed by Lloyd Tevis and associates on 1 July 1868 to compete with Wells Fargo in California and Nevada, and operated until December 1869 when Wells Fargo purchased the company. However, Wells Fargo authorities dispute these facts, claiming that their records indicate that Tevis and associates formed a "paper company" simply as the holder of the express contract on the soon-to-be-completed transcontinental railroad. Their records indicate that Tevis's company was named "The Pacific Express Company," and that it never operated. (Possibly this could be Pacific Express Co. #2).

Regardless of who owned it, there was a Pacific Union Express Co. operating in 1858 throughout California and with an office in New York. Its franked and handstamped covers show its agencies in coastal towns, in the San Joaquin and Sacramento valleys and

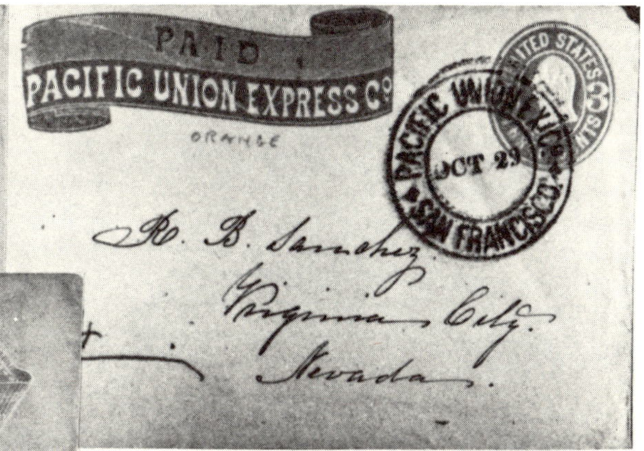

Pacific Union Express: Woodland and SF double circles; fancy ribbon frank.

along the Russian River. A very few covers indicate that it had a route into Virginia City and Treasure City, Nevada.

Palmer & Co.'s Express

Dates: 1850-51.

C. S. Palmer started his company in New York on 7 March 1850. Some weeks later offices opened in San Francisco, Sacramento, and Marysville. It connected with other expresses which served the mining areas. It was a transcontinental express in that it sent shipments east via Panama under the protection of a special messenger. In an advertisement in the *Sacramento Union,* 19 March 1851, the firm is referred to as "Palmer & Co.'s California Express Company." This advertisement is the last known reference to Palmer's express, indicating that it may have gone out of business shortly thereafter. There is reason to believe that it actually operated in California for only about seven months.

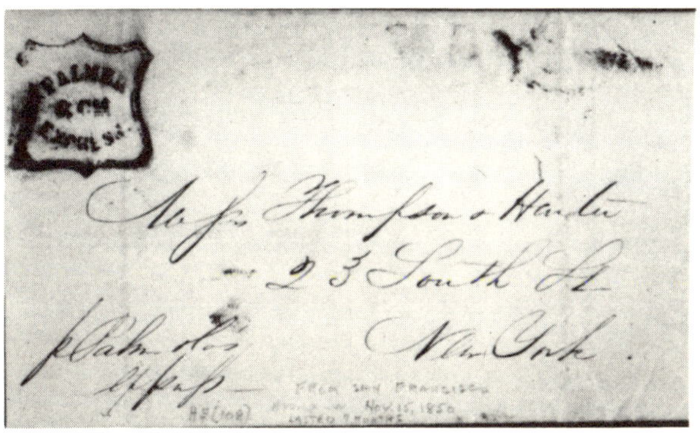

Only one type of Palmer handstamp is known, in red and black. There are, however, manuscripts "per Palmer & Co. Express."

Pauly & Nohrman's Express
N. O. Pauly's Express

Dates: N. O. Pauly and G. H. Nohrman (Norman): 1864-68. N. O. Pauly: 1868 (short-lived).

Route: From Gibsonville to mining camps on the Gibsonville Ridge.

In 1864 both N. O. Pauly and G. H. Norman were operating weekly expresses in the same area. Pauly's company was called N. Pauly & Co.'s Express, for which no franks are known. Late in that year the two expressmen combined their operations into one company: Pauly & Nohrman's Express. In 1868 Pauly bought out Nohrman and operated again under his own name, this time as N. O. Pauly's Express. Soon after this he ceased operation. The six printed franks of these

Pauly & Nohrmann, N. Q. Pauly franks, last with tiny Gibsonville straightline.

companies are similar with varied frame lines.

Pescadero & Half Moon Bay Stage Co.'s Express

Date: 1865.
Route: From San Francisco to Half Moon Bay and Pescadero.

Petaluma & San Francisco Express

Dates: 1865-68.
Route: Between Petaluma and San Francisco.

Peterson's Lower California Express

Dates: Spring 1861-summer 1863.
Route: From San Francisco to San Diego and La Paz, Mexico, via steamers.

Only two Peterson covers are known.

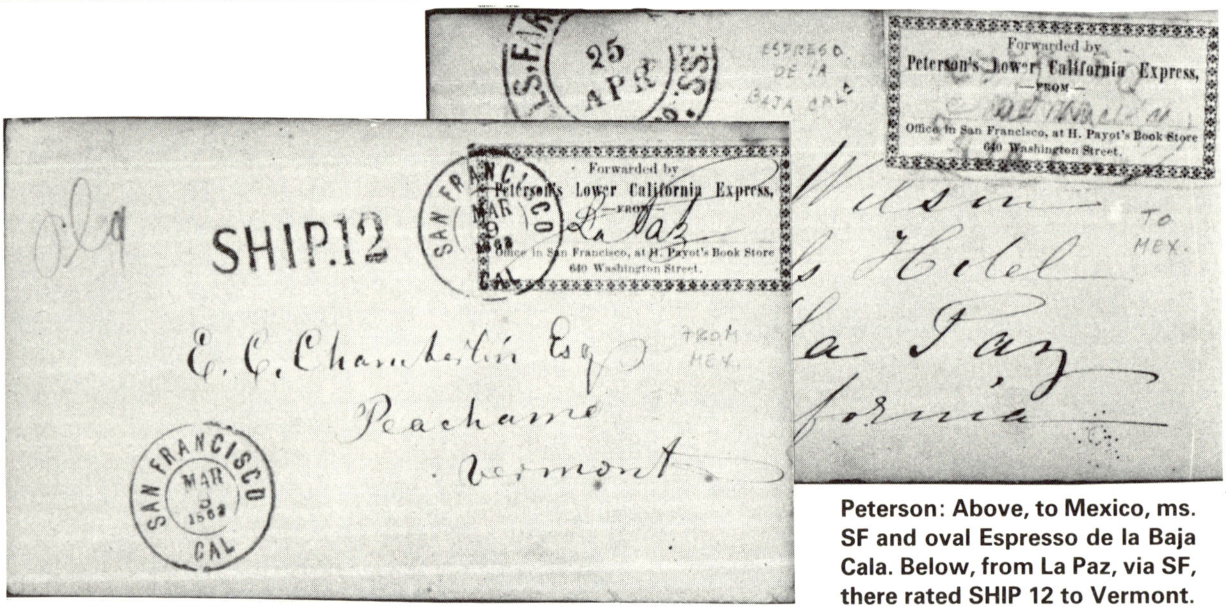

Peterson: Above, to Mexico, ms. SF and oval Espresso de la Baja Cala. Below, from La Paz, via SF, there rated SHIP 12 to Vermont.

Phillips Express

Dates: 1860-61.

Route: Nevada City to Dutch Flat, Alpha, and Washington.

Only a manuscript marking is known.

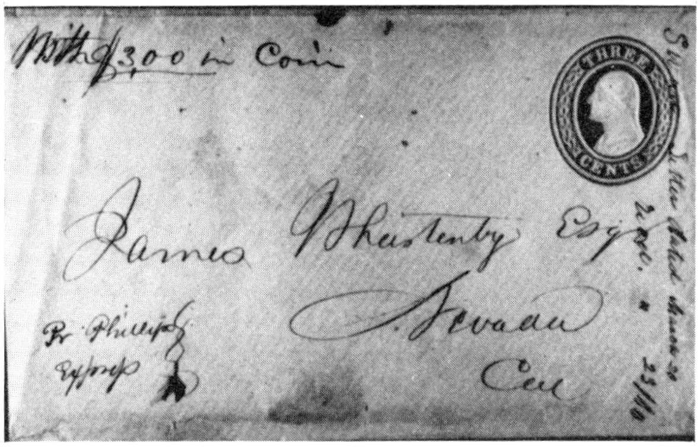

Philip & Gregory's Express

Dates: 1864-15 April 1866.

Route: Nevada City to Moore's Flat and Eureka South.

Pioneer Express

Dates: 1858-59.

Route: From Georgetown via Bottle Hill to the small camps on the North Fork of the American River.

Only one genuine cover is known; it is in the Wells Fargo History Room.

Pioneer Stage Co.

Dates: 1857-64.

Route: From Placerville to Virginia City (Nevada) via Strawberry Valley (California) and Genoa and Carson City in Nevada.

Pioneer Stage corner card postmarked Placerville.

Prindle's Express

Dates: 1860-66.

Route: From Yreka to the Scott and Klamath River mines.

Connected with Wells Fargo at Yreka.

Raveley's Express

Dates: 1854-56.

Route: Weaverville, Canon City, and the Trinity River mining camps.

Connected with Cram, Rodgers Express at Weaverville.

Reinkings Coast Express

Date: 1861.

Route: From Fort Ross to Tomales and Smith's Ranch, along the coast north of San Francisco.

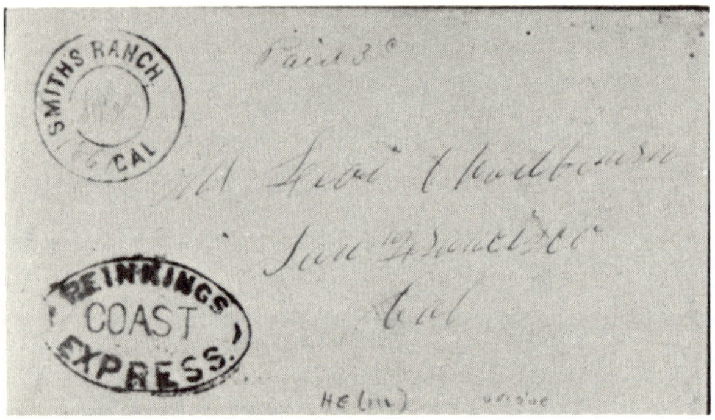

Reynolds & Co.'s Express

Dates: January 1850-fall 1851.

Route: From Sacramento and Stockton to the Southern Mines.

Connected with Todd and Adams expresses.

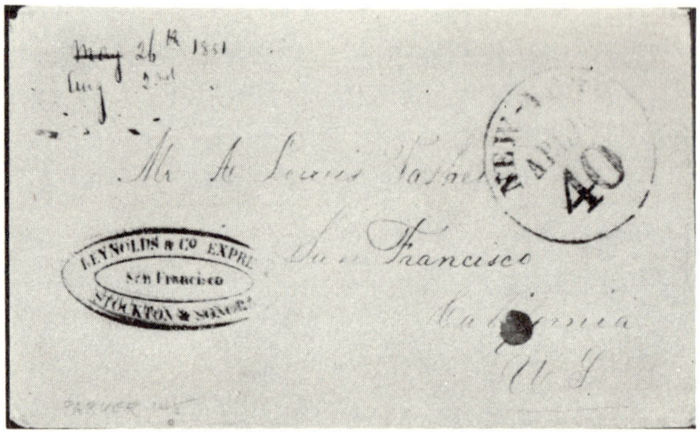

Reynolds double ovals Stockton and Sonora.

Reynolds, Todd & Co.'s Express

Dates: Fall 1851-April 1852.

Route: San Francisco, Sacramento and

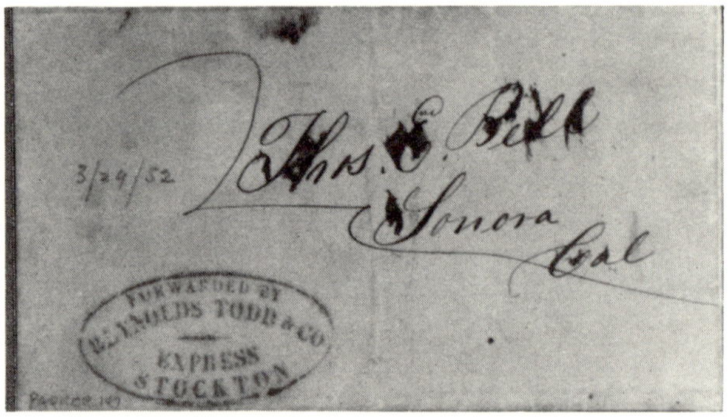

Reynolds handstamps on 1850-51 covers: ovals and rimless oval, both from Maine to Stockton; shield on cover from R. I. to SF; rimless circle on cover from Mass.

Stockton to the Southern Mines.

Rhodes & Lusk's Express
Rhodes & Co.'s Express
Rhodes & Whitney's Express

Dates: 1852-57.

Route: From Sacramento to Marysville, Tehama, Weaverville, Shasta, Yreka, and into southern Oregon.

Connected with Wells Fargo at Sacramento.

In 1852 **Rhodes & Lusk's Express** was formed to make daily runs from Marysville to southern Oregon, also a semi-weekly service to San Francisco.

In 1853 Lusk withdrew and James Rhodes continued for three years as **Rhodes & Co.'s Express.**

In 1856 Lusk rejoined Rhodes and the company was called **Rhodes & Whitney's Express.** At present there is no information concerning the Whitney named. This company lasted about one year. Counterfeits of the Rhodes & Lusk handstamp exist.

Ritner's Salmon River Express

Dates: 1855-58.

Route: From Shasta to the Salmon River Mines.

The red handstamp is almost indecipherable.

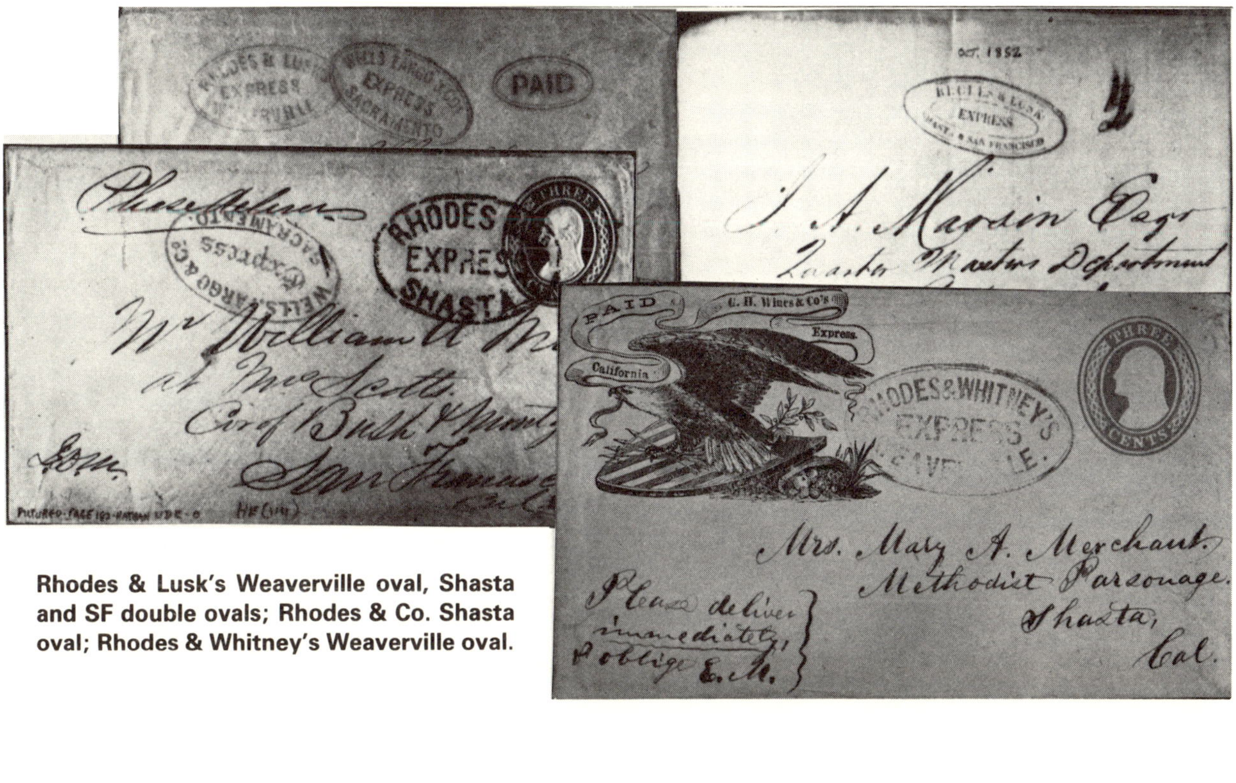

Rhodes & Lusk's Weaverville oval, Shasta and SF double ovals; Rhodes & Co. Shasta oval; Rhodes & Whitney's Weaverville oval.

Robinson & Co.'s Express

Date: Sometime between 1861 and 1865.

This is somewhat of a mystery express. Wiltsee lists it operating from Sacramento to Murphy's and San Francisco, but no year date is given. Nathan is unable to add any information. Konwiser and Mason date the Grizzly Bear frank as 1864. A cover bearing the frank was sold in the Barkhausen auction in 1955. The frank is illustrated in Nathan.

W. H. Robinson Express Agent

Dates: 1852-53.

Route: Sacramento to Murphy's.

Since Murphy's is in the eastern part of Calaveras County, W. H. Robinson's route, like that of Robinson & Co.'s Express, must have taken him through many mining camps in the richest area of the Mother Lode. (One cannot help wondering about the relationship between these two Robinsons who operated on the same route 12 years apart.)

Rowe & Co.'s Express

Dates: 1856-58.

Route: From Shasta to Weaverville and on to Big Bar and Canyon Creek on the Trinity River, a distance of about 80 miles.

Connected with Blake's express in Weaverville.

F. Rumrill & Co.'s Express
F. Rumrill & Co.'s Northern Express

F. Rumrill Downieville oval; straightline Northern Express; two types of Northern Express ovals.

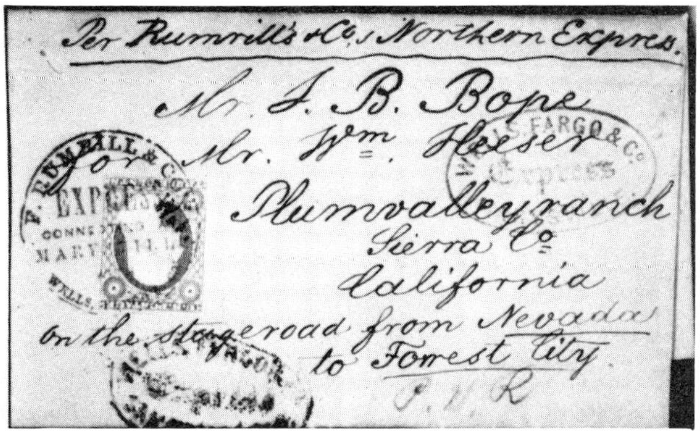

Dates: Late 1852 or early 1853-spring 1855.

Route: Up the North Fork of the Yuba River from Marysville to Fosters Bar, Camptonville, Goodyear Bar, and Downieville. From Downieville he left the North Fork and went north to St. Louis and Gibsonville.

Connected with Wells Fargo at Marysville. He used at least 10 different handstamps.

Rutherford & Co.'s Express

Dates: 1865-67.

Route: From Marysville to mining camps along the Feather River.

Connected with Wells Fargo at Oroville. This express was formed after the dissolution of Wheeler, Rutherford & Co.'s Express. The scroll-type printed frank illustrated has had the name "Wheeler" deleted.

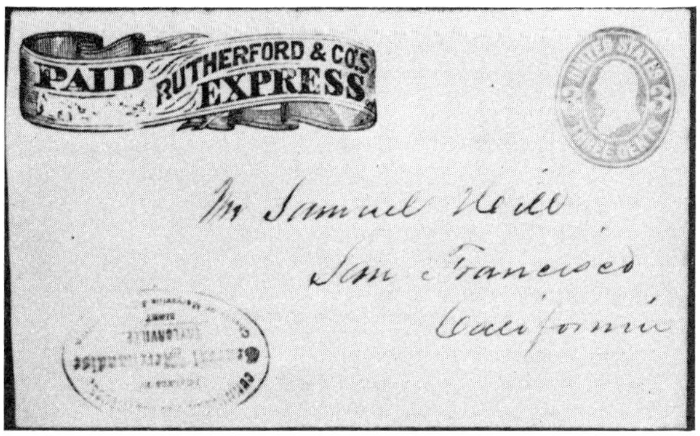

Sanders & Dixson's Express

Date: 1860.

Route: Virginia City (Nevada), Genoa (Nevada) to Placerville (California).

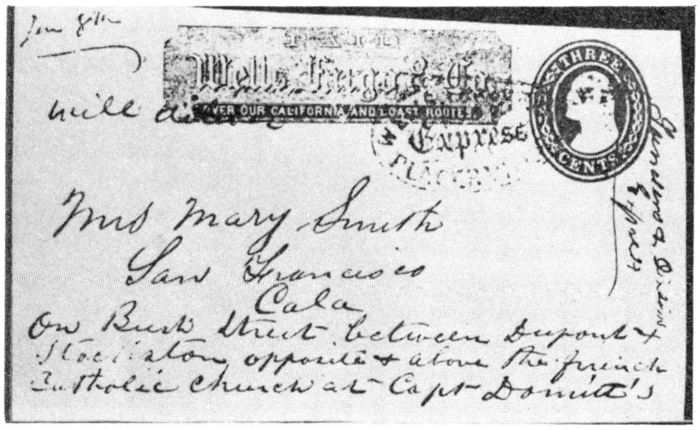

Connected with Wells Fargo at Placerville. Manuscript franks on two covers are known.

Saunders & Co.'s Placerville & Carson Valley Express

Date: 1858.

Route: From Placerville to Genoa (Nevada).

Connected with Wells Fargo at Placerville.

Schoch's Copper City Express

Dates: 1863-65.

Route: From Shasta to Buckeye, Churntown, Pittsburgh, and Copper City.

Two printed franks are known.

W. E. Singer ovals; Singer & Morrow oval and franks; Singer, Meek & Co. frank.

Service & Roberts Yuba Express

Dates: 1853-54.

Route: From Marysville to the mines along the Yuba River.

W. E. Singer & Co.'s Feather River Express

Singer & Morrow's Feather River Express

Singer, Meek & Co.'s Express

Dates: 1850-57.

Route: From Marysville to mining camps in the Feather River area including Forbestown, Bidwells Bar, Ophir, Buck's Ranch, and Spanish Ranch.

In seven years W. E. Singer had five different expresses, all serving the same area. While the route did not change, the partnerships did. From 1850 to 1852 it was **Singer & Fargo's Express;** from 1852 to 1855 it was **Singer, Dean & Co.'s Express.** For several months in 1855 it was **W. E. Singer & Co.'s Feather River Express.** From late 1855 to 1857 it became **Singer & Morrow's Feather River Express,** and in late 1857 it was **Singer, Meek & Co.'s Express.** Soon afterward Singer and Meek sold out.

Only the last three companies are known to have used franks. At least three varieties of the printed corner card and two handstamps exist.

Snow's Express

Dates: 1854-55.

Route: Sonora and surrounding camps.

An 1855 cover with black double oval handstamp has been reported.

Spargur's Honey Lake Express

Dates: 1860-64.

Route: From Oroville to Quincy to the Honey Lake area in Lassen County.

Stelle's Express

Dates: 1854-55.

Route: Between Stockton and the Mariposa mining camps.

A blue oval handstamp of this express is reported.

Stewart & Jones' Express

Date: 1865.

Route: From Dutch Flat to Excelsior and Meadow Lake.

H. R. Stiles' Express

Date: 1855.

Route: Daily from Nevada City via Red Dog, You Bet, across the Bear River to Gold Run, Secret Town, and across the North Fork of the American River to Iowa Hill; about a 25-mile route.

Stolp's Express

Date: 1851.

Route: From Nevada City to Red Dog and

Little York.

J. A. Stone's Express

Dates: 1864-3 February 1865.

Route: From Nevada City to Red Dog, Little York, and Dutch Flat.

Connected with Wells Fargo at Dutch Flat.

Stringer & Co. Express

Date: 1856.

Route: From Dutch Flat into the Mountain Springs area of Placer County.

Its frank is known only in manuscript form.

Susanville Express

Nothing is known of this company except for its large adhesive label on a paste-up addressed to C. P. Braynard, Red Buff, Tehama County. *(See p. 230.)*

Swart & Co.'s Express

Date: 1854.

Routes: From San Francisco to Oakland where the line branched, running east to San Pablo and Martinez and south to Alameda,

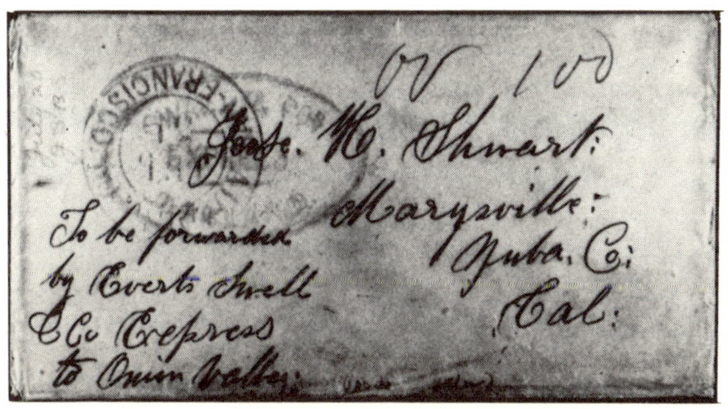

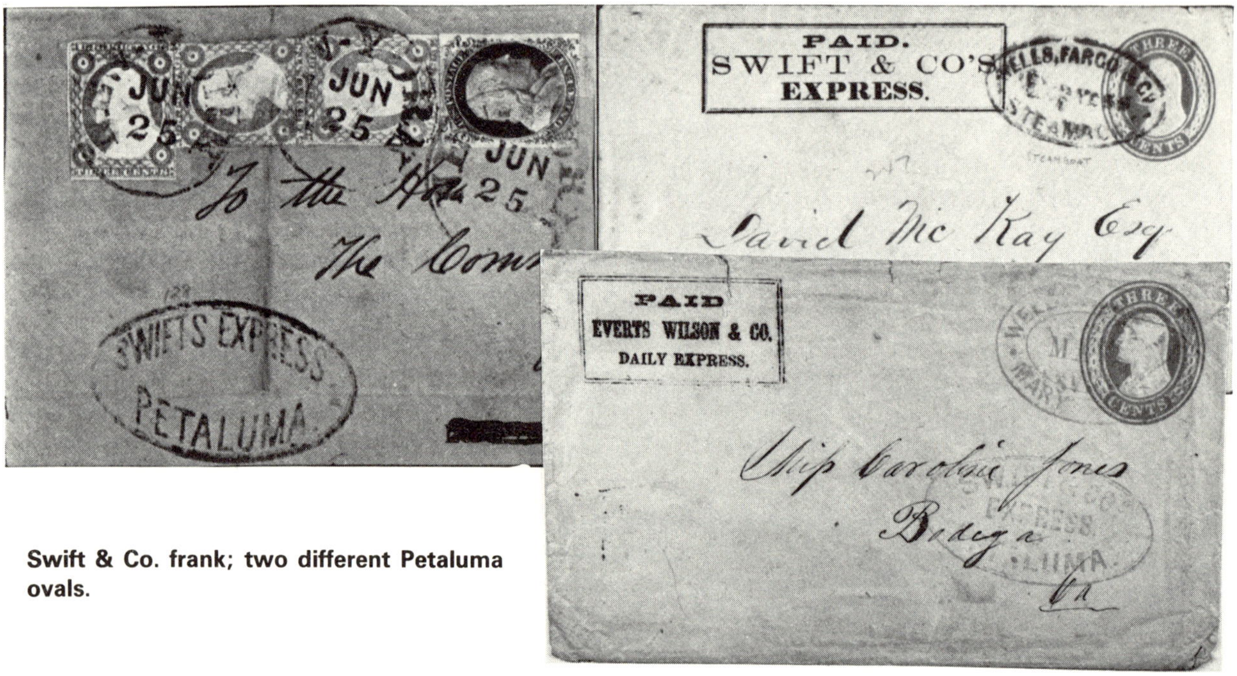

Swift & Co. frank; two different Petaluma ovals.

Union City (Centerville), and San Jose.

Connected with Adams & Co. Express.

Swift & Co.'s Express

Dates: 1854-59.

Route: From San Francisco to Bodega, Petaluma, and Santa Rosa.

Swift sometimes used the Wells Fargo steamboat service between Petaluma and San Francisco, as well as Wells Fargo at Santa Rosa. *(See p. 231.)*

Grant T. Taggart's Shasta & Weaverville Express

Dates: Spring 1866-1875.

Route: Between Shasta and Weaverville, and later from Shasta to Yreka and Fort Jones.

Taylor's Express

Dates: June 1851-spring 1852.

Route: From Shasta to Sacramento.

Connected with Gregory's Express. Taylor's covers are known only by manuscript marking.

J. A. Thompson Expresses

Dates: 1856-61.

In 1856 J. A. Thompson was operating **Thompson & Co.'s Express** from Petaluma to Santa Rosa and north. In the same year he began operating the **Carson Valley Express** from Mokelumne Hill via the Big Trees Road to Genoa in the Carson Valley of Nevada. From 1859 to 1861 he operated this express from Placerville via Strawberry Valley to

Thompson & Co. frank and Healdsburg oval.

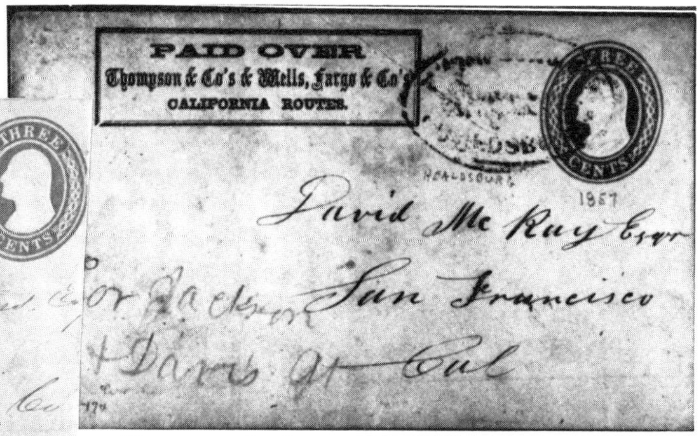

Thompson's Carson Valley oval on cover postmarked Carson Valley — one of two proven carried by "Snowshoe" Thompson.

Genoa. Thompson's exploits in carrying mail over the Sierras on skis are told on p. 153.

Tibbetts & Co.'s Excelsior Express

Dates: 1864-65.

Route: Nevada City through the Excelsior District to Meadow Lake.

In 1865 William Tibbetts took as a partner a Mr. Organ and the firm continued as Organ & Tibbetts Excelsior Express until 1868.

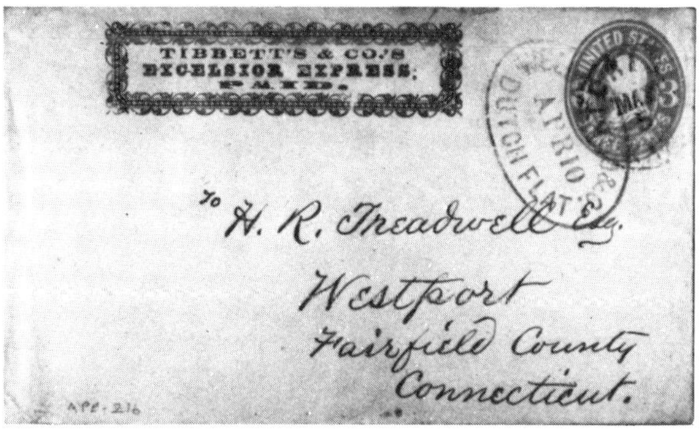

Tickner Expresses

The Tickner family could not resist the express business. Three brothers and a sister all operated expresses in the Siskiyou and Klamath Mountains between 1860 and 1880.

H. C. Tickner began it all with a mail express from Yreka to mining camps on the Klamath River until 1862. In 1866, from his headquarters in Yreka, he was running to the mining camps of the Scott River Valley.

In 1871 his brothers, W. A. and F. L. Tickner, began operating an express from Yreka to Fort Jones and Rough and Ready (Etna). In 1874 the sister, Lavinia Tickner, began an express in the same general area of Yreka, Scott's Bar, Fort Jones, and Indian Creek, all in the Scott Valley on tributaries of the Scott River.

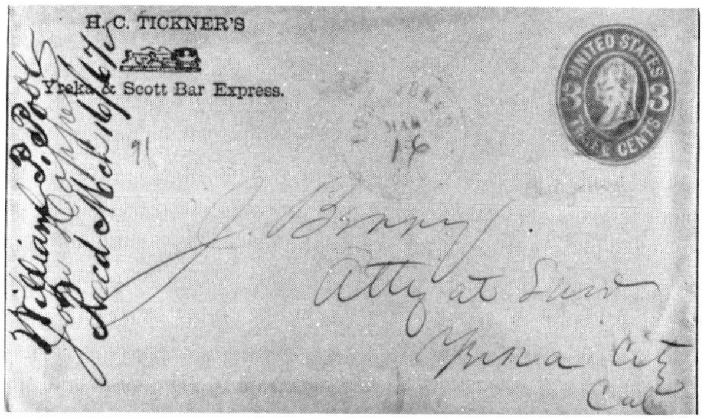

The Tickners are reputed to have used various handstamps for their expresses. The H. C. Tickner handstamp and printed frank pictured here are the only ones known to the author.

Tinnen & Owens Weaverville & Shasta Express

Dates: October 1867-May 1868.

Route: Weaverville to Shasta.

They had a printed frank and a circular handstamp, both of which are pictured on the cover. Only three Tinnen & Owens covers are now known to exist.

Todd & Bryan's Express

Alexander H. Todd was one of the original expressmen in California. Beginning in 1849 as a one-man express he quickly found that he could not handle alone his volume of business. He took in first Benjamin L. Bryan and then E. W. Colt. They organized a company known as Todd & Bryan's Express. Their route was from San Francisco to Stockton and out to the Southern Mines. On 13 July 1850 Bryan left the company.

Todd & Co.'s Express
Todd & Co.'s Oregon & California Express

Dates: 13 July 1850-October 1851.

Alexander H. Todd and E. W. Colt operated as **Todd & Co.'s Express** from San

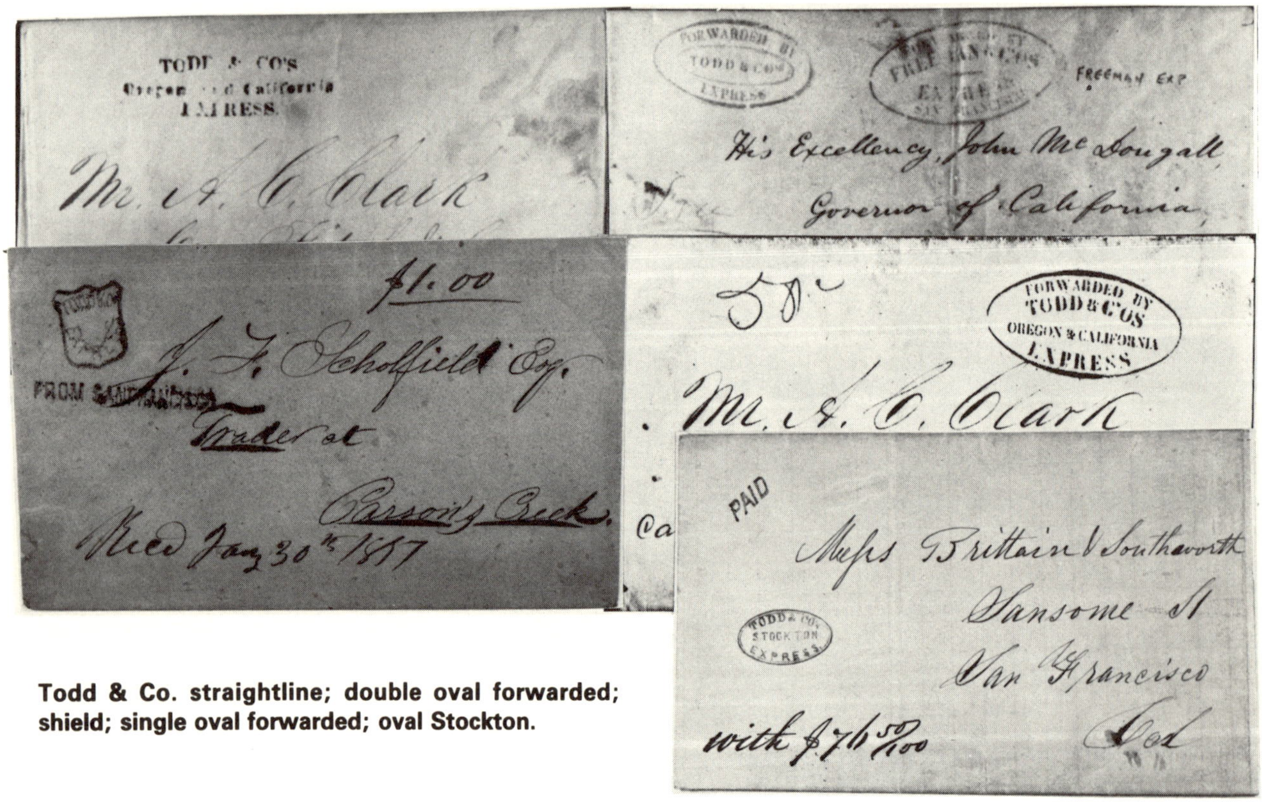

Todd & Co. straightline; double oval forwarded; shield; single oval forwarded; oval Stockton.

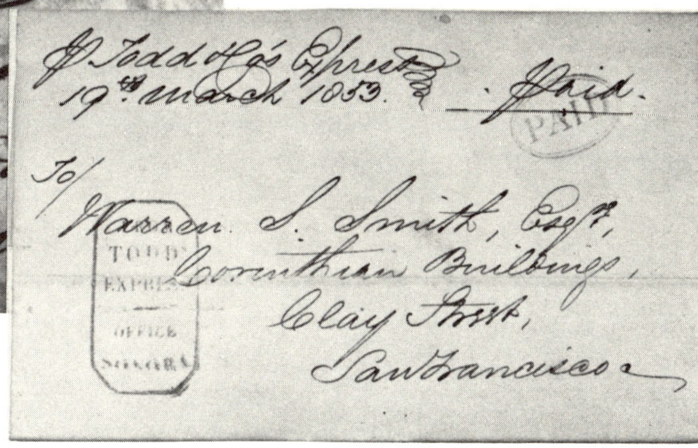

Todd's: SF oval; Sonora tombstone.

Francisco to Stockton and the Southern Mines. For a short time they extended a line to Oregon under the name of **Todd & Co.'s Oregon & California Express,** which they operated until October 1851. Six different handstamps are known for these companies.

Todd's Express

Dates: 27 April 1852-1 September 1853.

Route: San Francisco to Sacramento, Stockton, and the Southern Mines.

C. A. Todd, no relation to Alexander H. Todd, owned this express.

T. F. Tracy's Carson Valley Express

Date: June 1857 (short-lived).

Route: From Placerville via Carson Valley to Genoa (Nevada).

Only one handstamped cover is known at present.

Tracy & Spear's Express

Date: 1858.

Route: From Placerville to Georgetown via Kelsey and Spanish Flat.

This firm, owned by Theodore F. Tracy and J. J. Spear, Jr., lasted only about three months. Its only known franks are a "T & S" initialed manuscript frank, and a complete "Tracy & Spears" manuscript franked cover in the Wells Fargo History Room.

Truman & Co.'s Express
Truman & Chapman's Express

Dates: Early 1864-1868.

Route: From San Francisco to San Jose, Watsonville, and Santa Cruz.

According to the Truman Diaries in the Pomona City Library, James Christopher

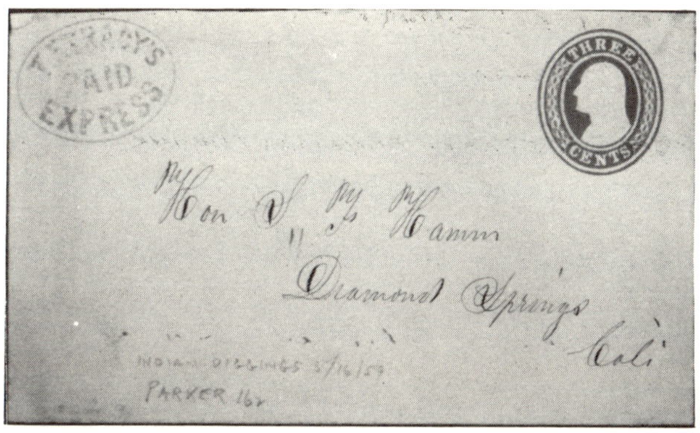

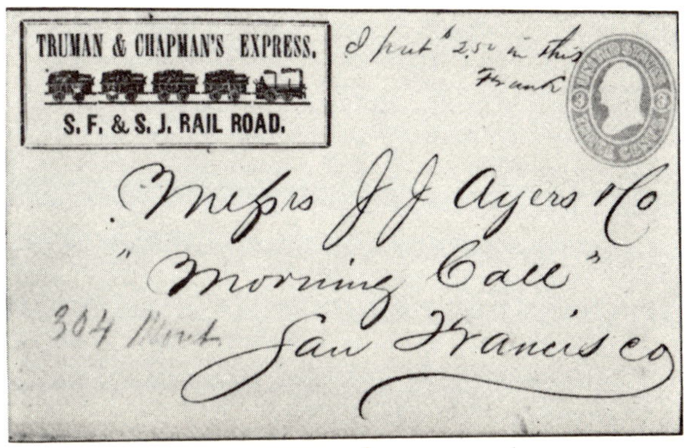

Truman & Chapman frank.

Truman & Co. franks and oval.

Truman arrived in California in November 1863. He began **Truman & Co.'s Express** in early 1864, and on 16 February 1864 he took in a partner named Chapman, forming **Truman & Chapman's Express.** In 1865 Chapman withdrew and Truman continued alone until 1868.

Tucker's Siskiyou Express

Date: 1858 (short-lived).

Route: From Shasta to mining camps in Siskiyou County.

Turner's Pony Express

Date: 1863.

Route: From Jacobsville (Nevada) to all parts of the Reese River Mines, then across the Sierras into California to Uniontown and Austin, which are near Weberville in the Placerville area between the American and Consumnes Rivers.

Vera's Express

Dates: 1857-58.

Route: Between Marysville and Oroville.

Wand & Davis Express

Date: 1864 or 1865.

There is an unresolved dispute about this express. Wiltsee claims that it was a transcontinental express operating overland from California to New York in 1865. Hertz cites the Berthold Papers to the effect that it was an express operating "in and around San Francisco." Other collectors maintain that it was an express that operated from the Southern Mines.

In support of Wiltsee's claim is the bottom line of the Wand & Davis Express frank: PAID THROUGH.

Warren's Mormon Island Express

Dates: 1849-50.

Route: From Mormon Island to Sacramento.

This letter-and-treasure-carrying express, owned by James L. Warren, was an adjunct to Warren & Company's Merchandise Store on Mormon Island.

It is known that this pioneer expressman used a handstamp, although no cover is known to the author.

Washburn's Express

Dates: 1856-57.

Route: Headquartered in Camptonville, this express ran to Marysville and the mining camps along the forks of the Yuba River.

Wellington Express

Dates: About 1858; and 1865-66.

D. Wellington operated an express from Nevada City to Moore's Flat and Eureka South about 1858. He used a blue oval handstamp WELLINGTON EXPRESS.

In 1865 and 1866 Wellington, with S. P. Dorsey, operated another express from Aurora at the northern end of the Owens Valley. In 1864, as a result of a boundary shift, Aurora became part of Nevada and was renamed Esmeralda. According to Nathan, the firm ran from Aurora to Kearsage, Nevada. Unfortunately no post office was ever recorded for Kearsage in either Nevada or California. However, there was a town of Kearsarge in Inyo County (California) in the Owens Valley about 120 miles south of Aurora. It is quite possible that the Wellington and Dorsey express ran from Aurora to Kear-

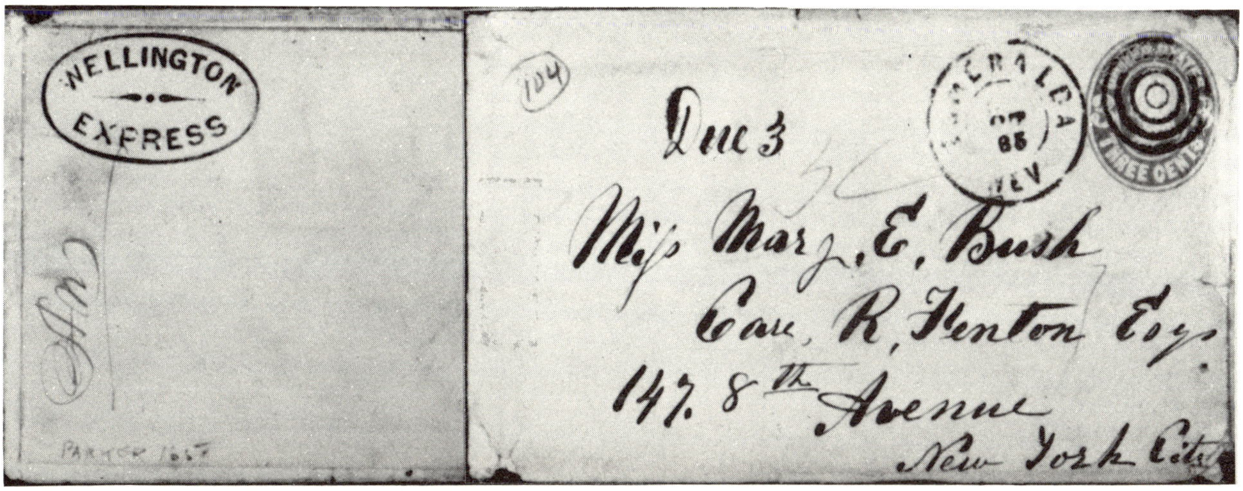

Wellington Express paste-up.

L. H. Wells', Wells & Co.'s franks.

sarge, California, rather than to a Kearsage, Nevada. Regardless of the exact route, **Wellington, Dorsey & Co.** used the same handstamp as the **Wellington Express** of 1858.

Illustrated here are both sides of a paste-up.

L. H. Wells' Moore's Flat & Eureka Express
Wells & Co.'s Moore's Flat & Eureka Express

Dates: 1864-65.

Route: From Moore's Flat to Eureka South, Nevada City, and Emigrant Gap.

Wells' North San Juan & Humbug Express

Dates: 1864-67.

Route: North San Juan to Cherokee to Humbug (North Bloomfield).

L. H. Wells Express

Date: 1867.

Route: French Corral, North San Juan and North Bloomfield.

This company used the same frank as Wells' North San Juan & Humbug Express.

J. P. Wharton's Express

Dates: Late 1862-1864.

Route: LaPorte, Gibsonville, Jamison City, along Hopkin's Creek and Poorman's Creek.

Connected with Whiting & Co. at Gibsonville.

Wharton franks, one to Germany at 15¢ Bremen-Hamburg rate.

Wheeler's Express
Wheeler, Rutherford & Co.'s Express

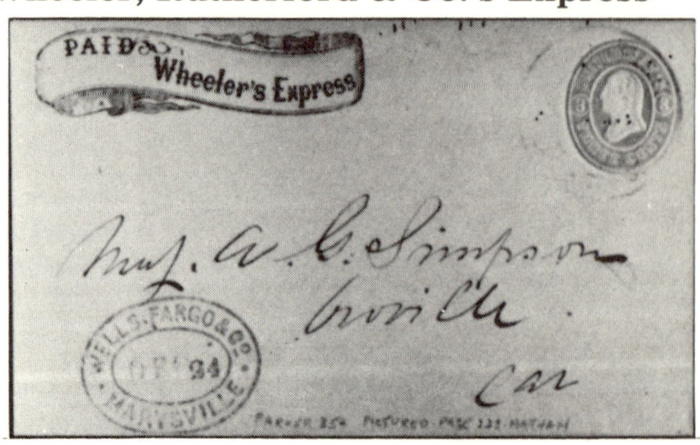

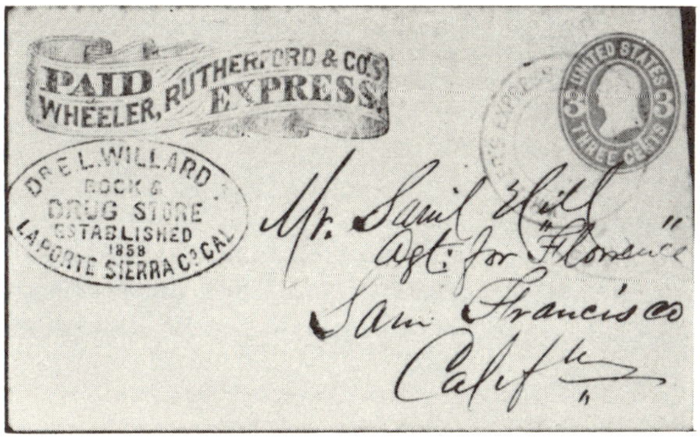

Dates: December 1863-1865.

Route: From Marysville to the Feather River and Gibsonville camps.

In December 1863 Sam Wheeler and his brother formed **Wheeler's Express.** On 27 February 1864 they changed the name to Wheeler Brothers Mountain Express. No frank is known for this company. Soon thereafter R. H. Rutherford joined the company which was re-named **Wheeler, Rutherford & Co.'s Express** and which lasted until sometime in 1865.

Whiting & Co.'s Feather River Express (Whiting & Co.)

Dates: 1857-68.

For a decade Whiting & Co.'s Feather River Express was the leading operator in Northern Plumas County and the Magalia Ridge in Butte County, with Quincy as its headquarters. During the long life of this company they used at least nine different printed franks and handstamps as well as an orange adhesive label.

Whiting & Co. circle and oval; three frank types; label.

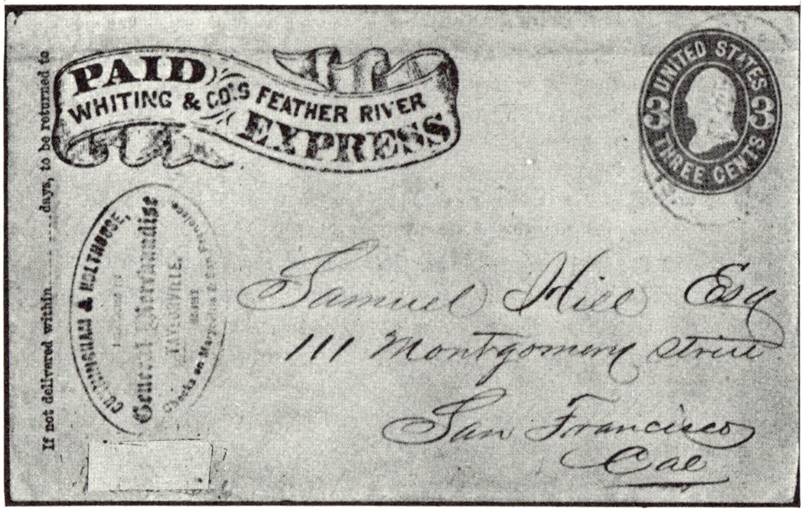

Whiting & Co.'s Feather River Express ribbon frank on envelope from Taylorville to SF, docketed 1866.

Whiting & Co. Feather River Express on 3¢ envelope to Germany, also 28¢ in stamps, 1864; Gibsonville postmark and ms. "Paid 31¢." By PCM (prepaid rate was 28¢ at this date).

Double circle Wingate's Oakland Express to Sacramento on Pacific Union Express envelope.

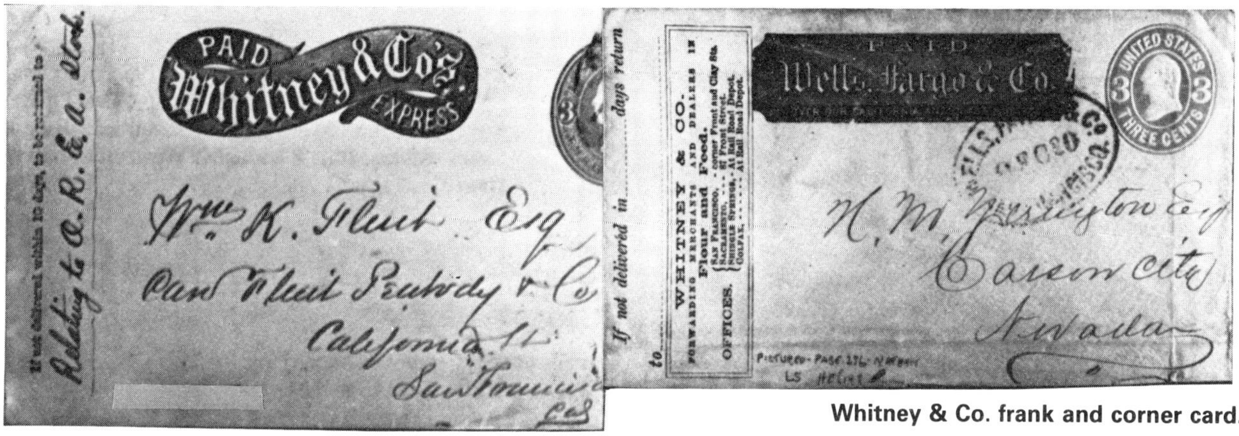

Whitney & Co. frank and corner card.

Whitney & Co.

Date: 1855.

Route: San Francisco, Sacramento, Carson City (Nevada).

Though primarily a forwarding company for flour and feed, Whitney may have carried mail. Their ribbon frank was not used until after their purchase of Bamber & Co. in 1875.

Whitney's Nevada Express

Date: 1854.

Route: From Nevada City to mining camps on the Yuba River, including French Corral.

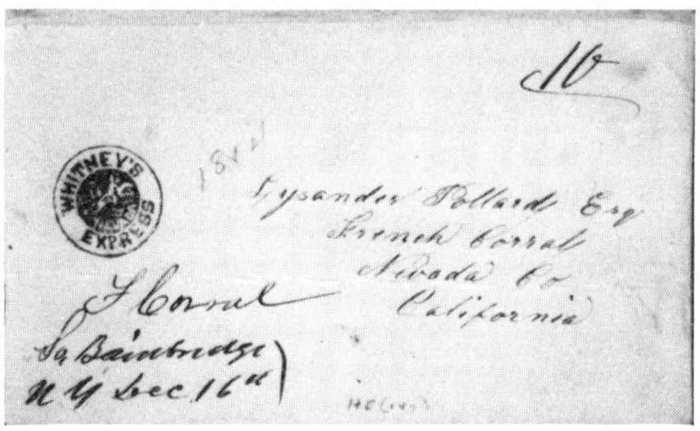

Wiggins' Express

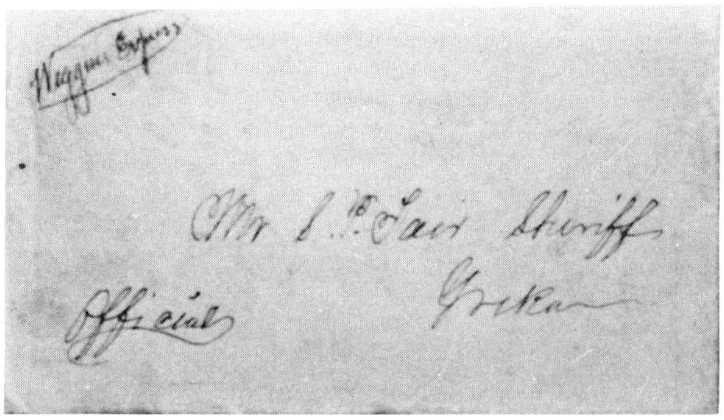

Dates: 1857-58.

Route: From Yreka to Scott's Bar.

The company's only known frank is in manuscript.

Wilson's Stage

No information is available concerning this firm. On a Wells Fargo & Co. franked 3¢ entire, addressed to Carson City, Nevada, there appears in manuscript "By Wilson's Stage."

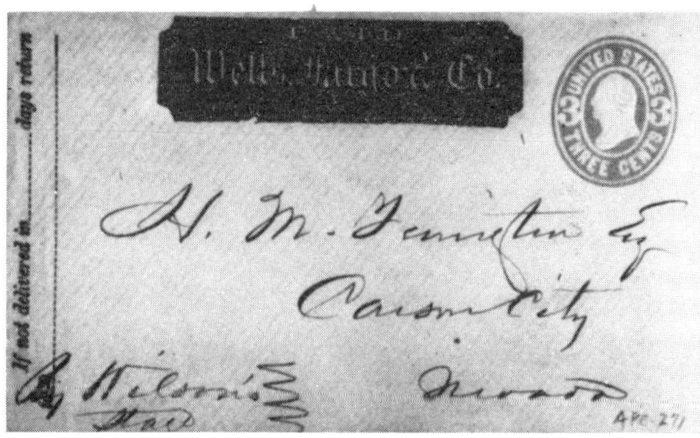

G. H. Wines & Co.'s Express

Dates: 1855-57.

Route: This company was widespread, operating throughout California and Oregon as well as serving Mexico, Nicaragua, and the Atlantic states.

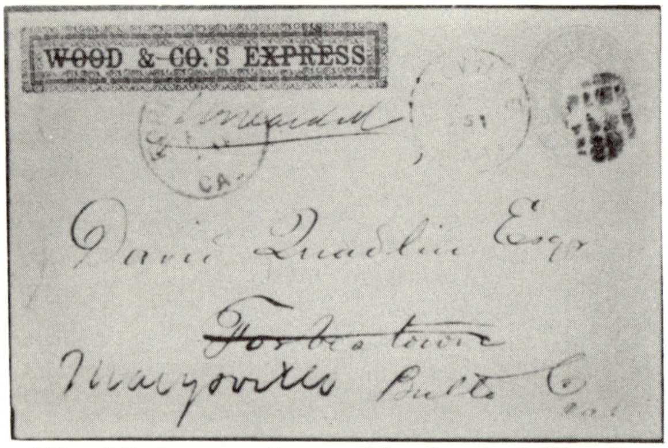

Wood & Co.: two types of franks; A. J. Wood's frank.

G. H. Wines & Co. frank.

Wingate's Express

Dates: 1868-early 1870s.

Route: Between San Francisco and Oakland.

For mail south he used the service of Truman & Co.; for mail north he used Pacific Union Express. Co. *(See p. 226.)*

Wood & Co.'s Express
A. J. Wood's Oroville, Susanville and Taylorville Express

Dates: 1864-May 1868.

Route: Between Oroville, Chico, Taylorville and Susanville.

Wylie & Co. Esmeralda & Monoville Express Co.

Date: Early 1860s.

Route: Between Esmeralda (Nevada) and Monoville (California).

Its only known frank is a manuscript on a Wells Fargo cover.

Zack's Express

[One-man express]

Dates: 1865-66.

Route: Between Downieville and LaPorte over the Gibsonville Ridge.

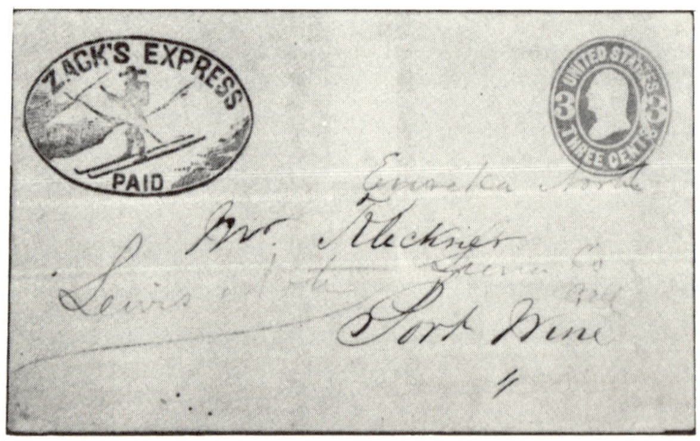

Since there was no stagecoach road across the difficult terrain, Granville Zacharian traveled on snowshoes in winter and on horseback the rest of the year. Zack's frank and handstamp are illustrated in Nathan. The 1857 date at the bottom of the third handstamp may be a printer's error; however, the authenticity of all Zack handstamps has been questioned by some historians.

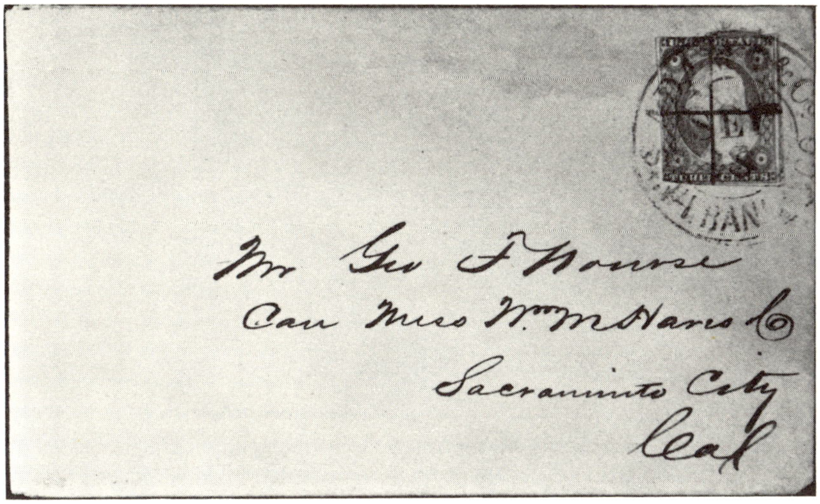

ADAMS & Co./SAN FRANCISCO double circle tying precancelled 3¢ 1851.

ADAMS & Co./NEVADA oval to SF, with precancelled 3¢ 1851.

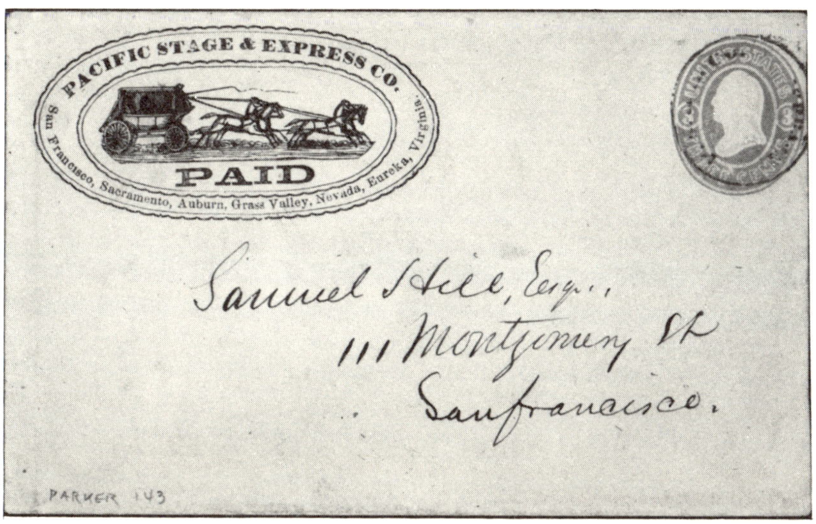

Pacific Stage & Express elaborate frank on 3¢ envelope cancelled P.S. & E. Co. circle.

ALTA EXPRESS label in black and orange on reverse of handmade stampless envelope.

SUSANVILLE EXPRESS label in black on green paper.

F. W. Blake shield, Weaverville, on 3¢ Nesbitt to Yreka.

Swift & Co. fancy shield frank on 3¢ Nesbitt to SF with WF STEAMBOAT oval.

Frank of Crawford's Middle Fork Express and oval handstamp on envelope to Sacramento. Also WF Greenwood oval.

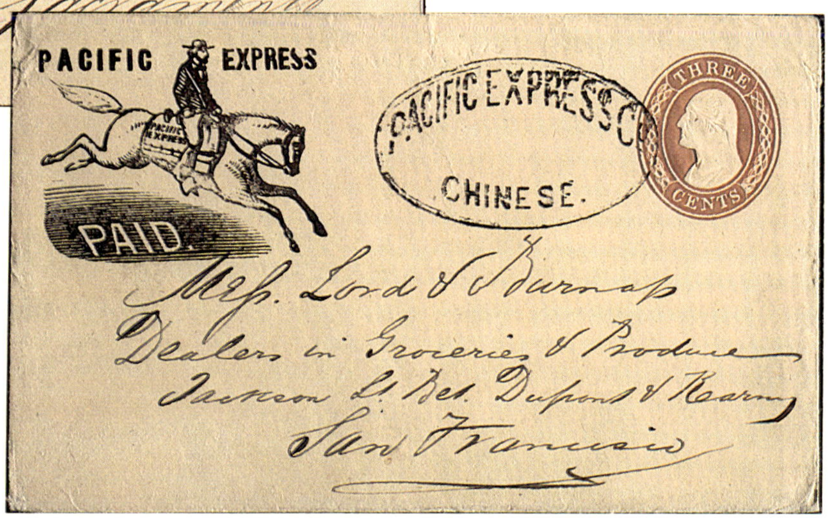

Fancy Pacific Express frank and oval CHINESE on envelope to SF.

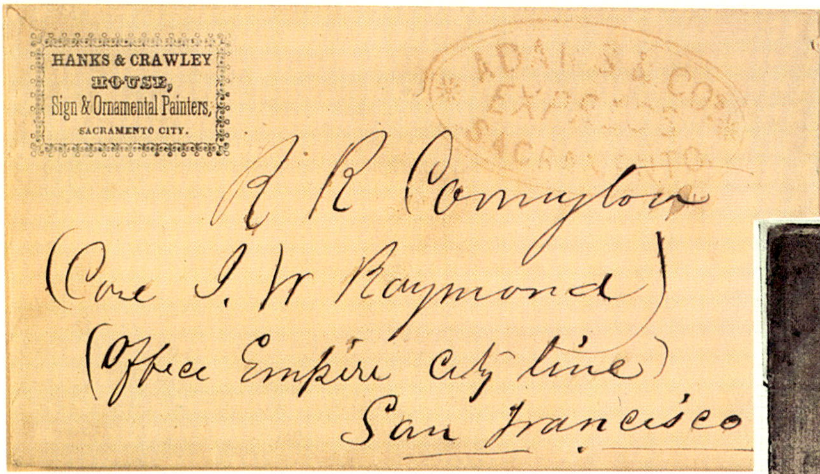

Adams & Co. Sacramento oval to SF, on envelope with painter's corner card.

S. Hervey Express oval on May 1859 copy of *Hutching's California Magazine.* Only example recorded.

"Prospecting Miners" design on unpaid cover from Marysville to Maine.

Goodyears Bar to Sacramento with 3¢ 1851. Rare design "The Grizley."

Flag and cannon patriotic used from Long Bar to SF with 3¢ 1857.

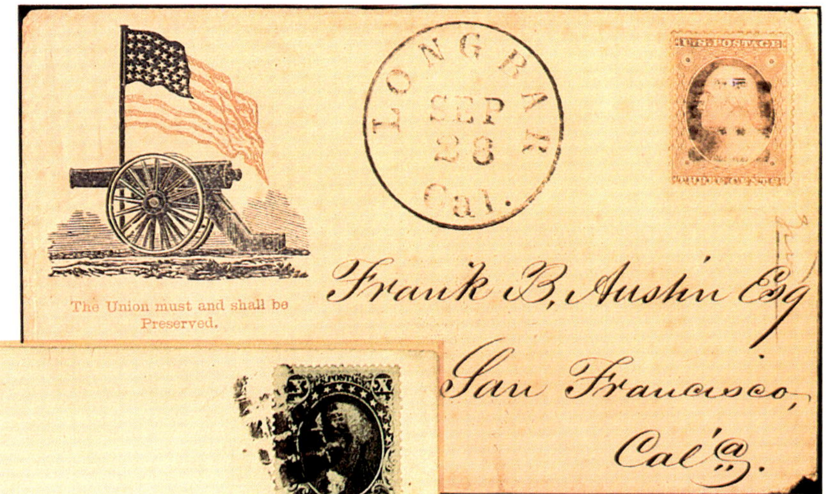

Marysville to Salem, Mass., with 10¢ 1857 on caricature patriotic.

Flag design printed by Hossack & Crawford, Sacramento, on unpaid cover to Maine.

Los Angeles to Canada, 15¢ rate made up by 1857 issue, on flag and cannon design envelope. Chicago exchange marking.

Hat design corner card on envelope from SF to N.H., with pair of 3¢ 1851. Oval Via Nicaragua (Sullivan) handstamp.

Allover advertising envelope of SF stationer to Yuba City at circular rate, paid by pair of 1861 1¢.

Sonora to Tenn., paid by 10¢ 1855, on envelope with corner card of Sonora furniture manufacturer, 1856.

SF cog cancels on pair of 1861 3¢ to Mass. Allover advertising envelope of H. M. Newhall & Co.

Downieville to Canada, April 28, 1862, "Paid 15" by a combination of 1861 issue stamps, all cancelled by Downieville star.

Forbestown to N.H., with pair of 1856 5¢ paying postage.

Honolulu to SF. Carried by STMR. CITY OF NORFOLK. Rated SHIP/6 at SF.

Bordeaux, France, 4 OCT 58 to SF. Double (over 15 grams) outward ship fee paid by 40c and 80c French stamps. By private ship *Bangkok* to SF Mar. 21, 1859, there rated SHIP/6.

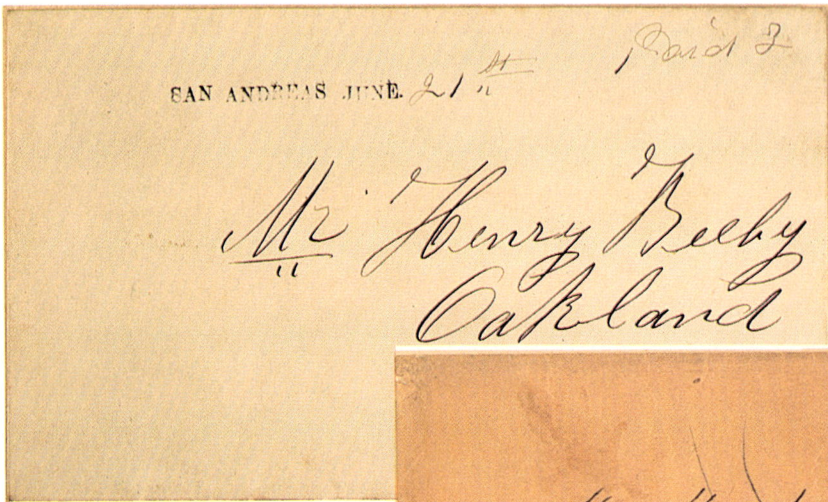

SAN ANDREAS straightline on cover to Oakland "Paid 3."

SONORA, CALIFORNIA/FEB, 11. 1852 tying 1851 3¢ pair on cover to Baltimore.

Columbia Dec 12/52 to Maine, with two 3¢ 1851; "Due 10" because overweight. Straightline postmark with ms. state.

COLUMBIA,/CAL (Jan. 12, 1854) to Syracuse. Another style of Columbia straightline on unpaid letter.

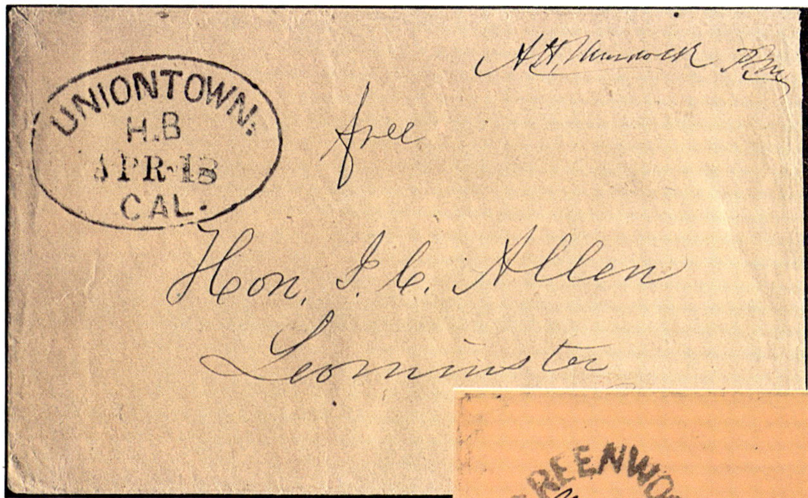

Oval UNIONTOWN: H.B. on cover to Mass. franked by postmaster.

Rimless oval GREENWOOD/CAL on paid stampless to Ohio.

Oval NORTH SAN JUAN (part double) struck twice on 10¢ envelope to Ill.

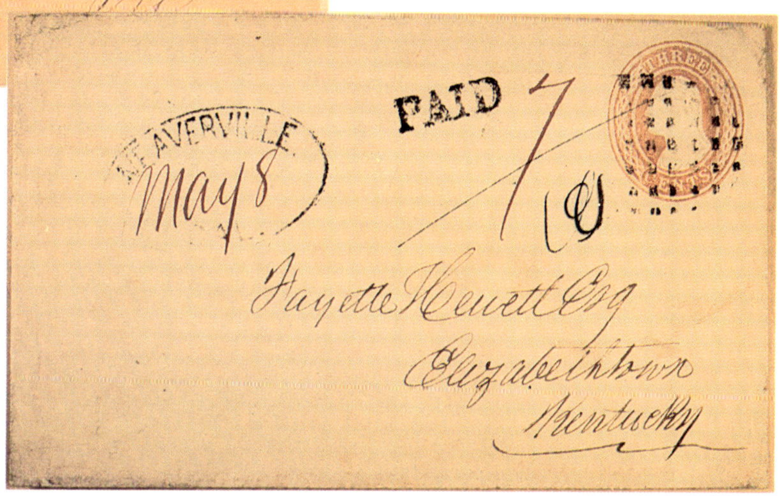

Weaverville oval to Kentucky on 3¢ envelope, balance "PAID 7" in cash.

Stockton to Canada, unpaid. Non-standard circular postmark and 15. Buffalo exchange marking UD. STATES.

Murphy's to Conn., double letter with 10¢ 1857 on 10¢ Nesbitt. Unusual circle postmark.

Irregular circle of Foster's Bar and oval PAID to N.H., with 1851 3¢ pair.

Sawyers Bar Cal Jany 24/61 to Mass. The transcontinental rate made up by ten copies of the 1857 1¢.

Chapter 7

The Overland Mails

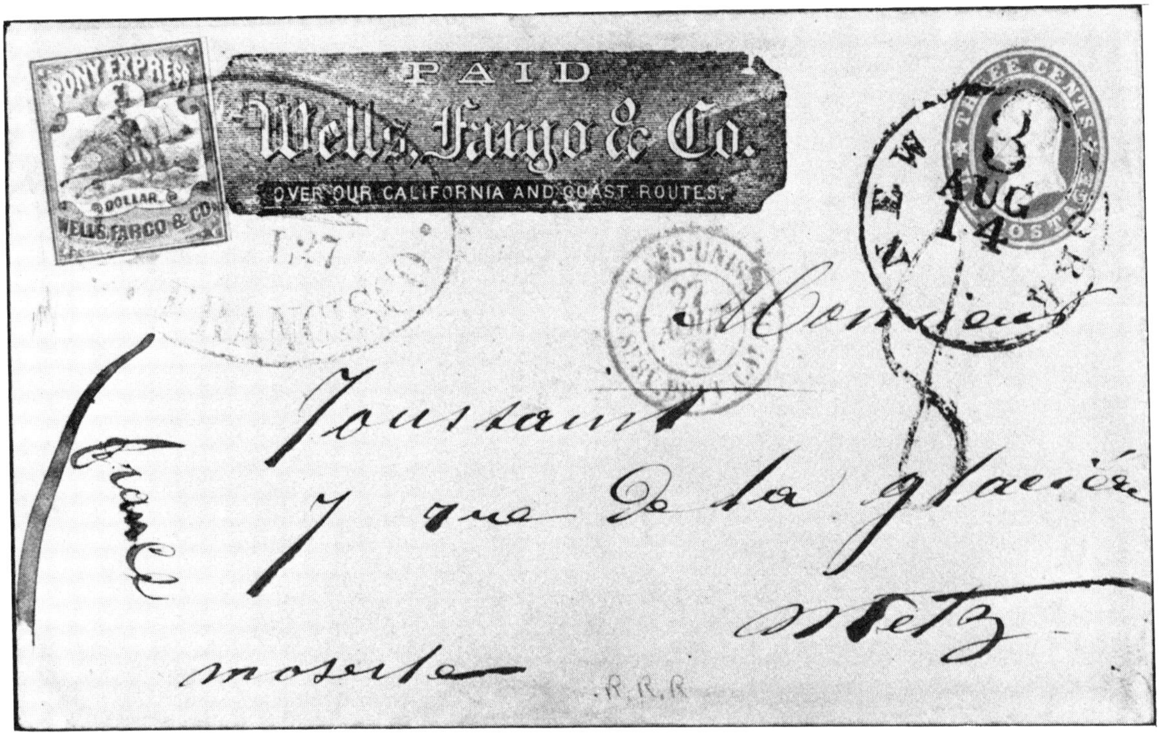

SF, July 27, 1861, to France by Pony Express. Oval SF running pony and $1 WF stamp to pay express rate. Rate to France was 15¢ per ¼ oz. (part payment not recognized); thus rated 8 decimes (15¢) due in France.

Triple rate cover from Benicia in overland mail (Butterfield route) to Mass. in January of 1861.

The Overland Mails

Since 1849 Californians had been vociferously expressing dissatisfaction with their "mail isolation." Even though they were on the remote western edge of a nation, they saw no reason that they should not be postally united with their eastern counterparts by a transcontinental mail service. They knew that only Congress could satisfy their complaints with postal service generally, and with the slow sea delivery of mail in particular.

California wanted an overland mail service. Her first senators, Frémont and Gwin, were pressing Congress to authorize the funding of a transcontinental railroad. Although a network of railroads existed east of the Mississippi, Congress had misgivings about the feasibility of a railroad across the vast plains, deserts, and mountains of the West. It was not until 1853 that Congress even would instruct the Army to survey possible routes for a future railroad. In the meantime, however, Congress was willing to provide at least the beginning of an overland mail delivery.

On 1 July 1850, Samuel Woodson began a contract service to carry mail monthly from Independence, Missouri, to Salt Lake City. He was granted a four year contract at $14,000 per year, and his route was to follow the Oregon Trail. It was a route 1,150 miles long, and 400 miles had to be covered by pack animals due to the mountainous terrain. Woodson simply could not meet his schedule even in the summer, and in the winter snows it became worse. By 1852 Brigham Young, governor of the Utah Territory, was complaining that the service was almost valueless. Woodson, however, kept trying and the Postmaster General was patiently understanding until the four-year contract expired in 1854 and Woodson gave up, and was succeeded by W.F.M. Magraw as contractor.

Woodward and Chorpenning

California senators still were dissatisfied; they wanted a complete service to California. To appease them, in early 1851 Congress had authorized a mail route from Sacramento to Salt Lake City, there to connect with Woodson's route east. The successful bidders were Absolom Woodward and George Chorpenning. Over a 900 mile route (later reduced to 768 miles) they were to use pack-mules to provide a monthly service for $14,000 per year. They were to carry mail from Sacramento, Placerville, Carson Valley, across the Humboldt Sink to Salt Lake City. The first mail in May took 36 days to reach Salt Lake City. From then on, the situation grew worse. Woodward was killed by Indians in November and Chorpenning had to struggle on alone. The December and January

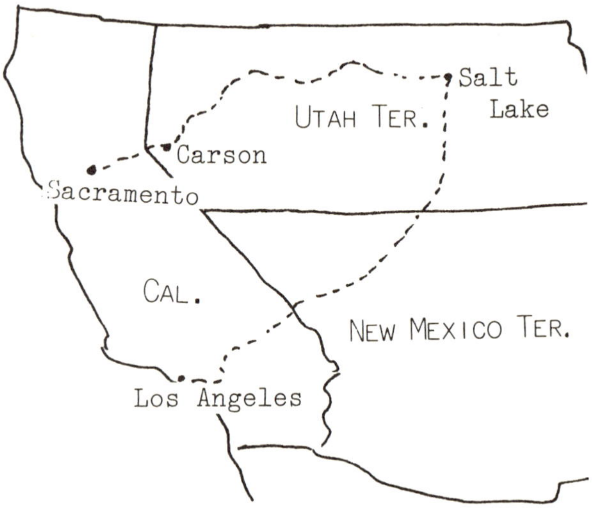

Chorpenning routes.

mails could not get through the snows of the Sierras. The February mail was carried on foot for the final 200 miles after the pack animals had frozen to death. No one was happy with such poor service, even though in following winters Chorpenning changed his route, going by boat to San Pedro and then by packmules over the old Mormon Trail to Salt Lake.

In 1852 and 1853 there was continuous debate in Congress on how to improve the unsatisfactory situation. In 1854 and 1855 the route controversy continued, especially when Congress began considering a consolidation into a single mail route which would combine mail and passenger service by stagecoach. In May 1856 California Senator John B. Weller sharpened the issue by presenting Congress with a petition signed by 75,000 Californians, demanding that Congress commit funds for an improved mail overland route to be used by stagecoaches.

Seventy-five thousand votes could not be ignored, especially in a Presidential election year, but the "sticky issue" of the exact route remained. Portents of the future Civil War were becoming apparent. Both Northern and Southern senators were aware that a transcontinental railway someday probably would follow whatever stagecoach route was chosen.

The Jackass Route

In the meantime, Congress had received a petition from the citizens of El Paso County, Texas, for a stage line to be built to San Diego. On 18 August 1856 Congress authorized a mail contract for a stage line to be constructed from San Antonio, Texas (since this city was easily accessible from the East via New Orleans) to San Diego, and left the exact route to be chosen by Postmaster General Aaron Brown, a pro-slavery Tennessean. Brown established the route to run from San Antonio to El Paso, then to Las Cruces in New Mexico, on to Tucson, Arizona, thence along the Gila River to Fort Yuma, to terminate in San Diego. It was a route across 1,475 miles of desert and mountain wilderness, serving only a few sparsely settled communities. Though called a stagecoach line, the first 180 miles out of San Diego required that all passengers ride on muleback. The remaining miles were over such rough terrain that mules had to replace horses in pulling the coaches.

John Birch, one of the most capable expressmen in the West, came out of retirement to receive a four year contract for semimonthly service on this route at $149,000 per year, with 30 days allowed for each trip, to begin operation on 1 July 1857. Shortly after Birch began setting up relay stations and buying equipment, he drowned in the sinking of the steamer *Central America* off the Florida coast on September 12th. George Giddings and R. E. Doyle took over Birch's contract. The service was increased to weekly in October 1858 and the mail subsidy was increased to $196,000. Shortly after, the El Paso-Fort Yuma section was dropped, as it was duplicated by the Butterfield line. The San Diego-Fort Yuma and El Paso-San Antonio portions remained but with diminished use. In 1859 a new pragmatic Postmaster General informed Congress that the annual cost of the route was $196,448 and receipts were only $601, at a cost of $326 per letter carried. The San Diego-Fort Yuma section

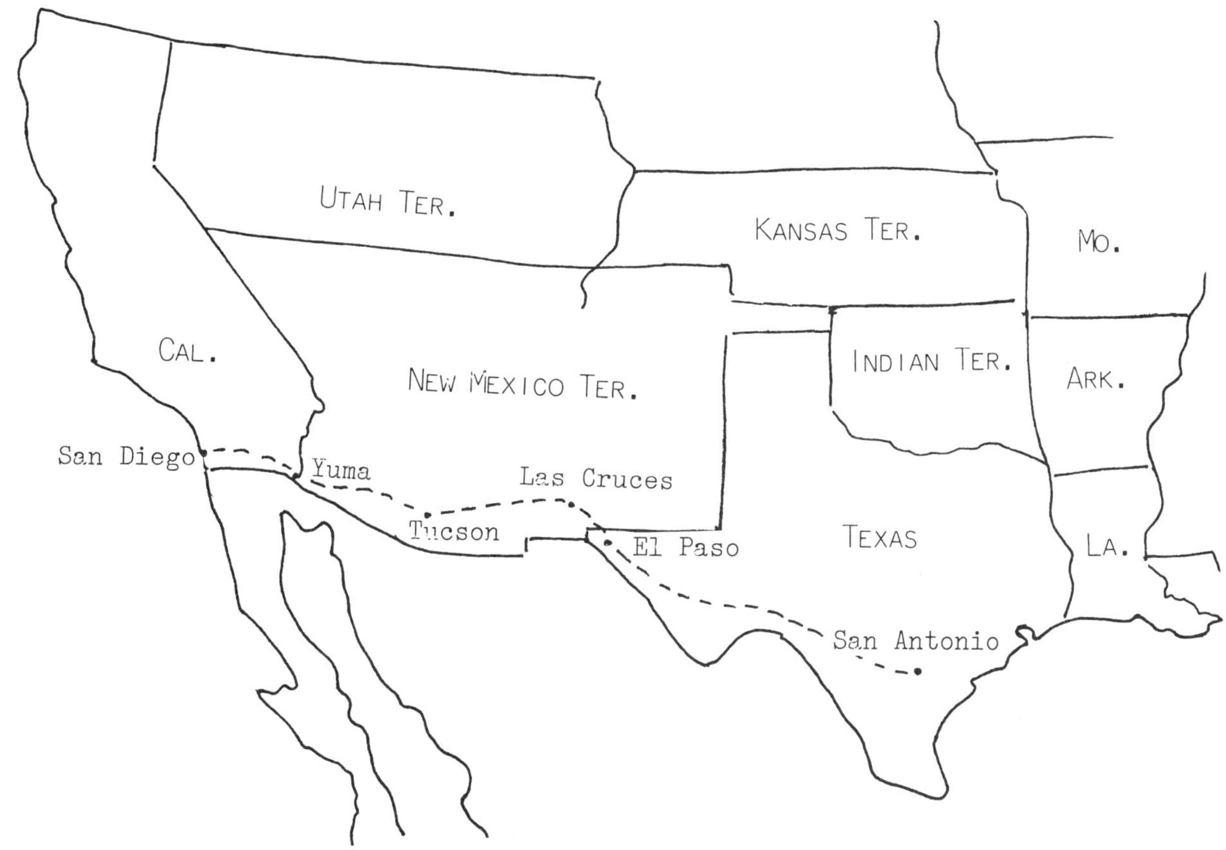

The San Antonio to San Diego, or "Jackass" route.

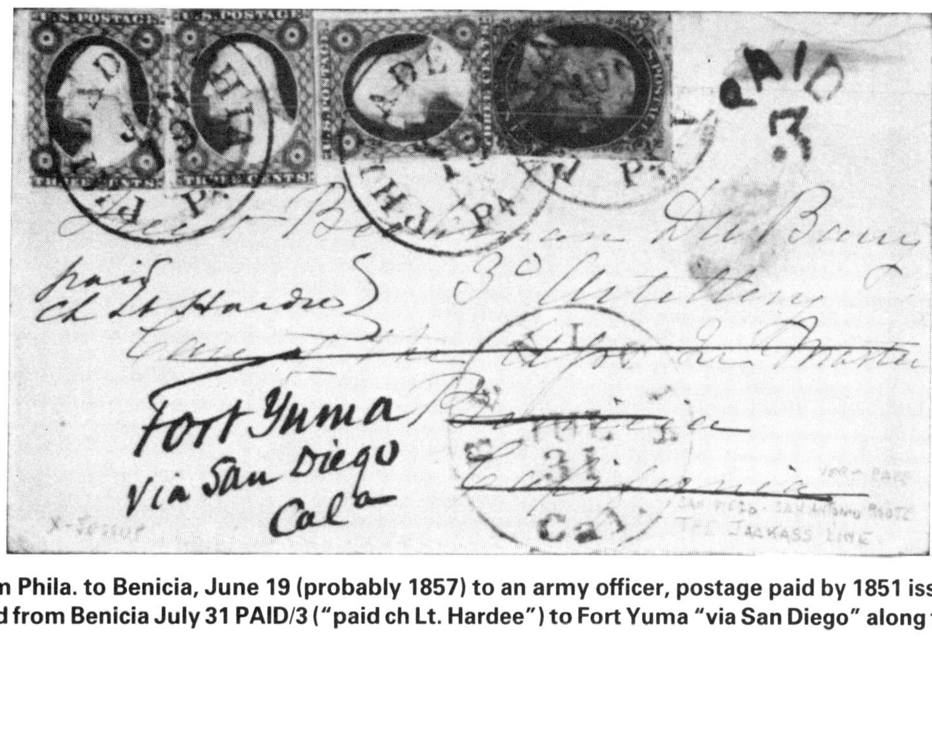

Cover from Phila. to Benicia, June 19 (probably 1857) to an army officer, postage paid by 1851 issue stamps. Forwarded from Benicia July 31 PAID/3 ("paid ch Lt. Hardee") to Fort Yuma "via San Diego" along the Jackass route.

Cover from Dutch Flat carried on Stockton-Kansas City route. One of two letters carried eastward the entire distance, according to *PMG Report*.

was dropped in 1860, the remaining portion surviving until the Civil War. The San Diego-San Antonio route had been nicknamed the "Jackass Route" in honor of the mules which had furnished most of the transportation.

Almost unknown to most collectors, but worth mentioning, is the fact that for a few months, from 1 October 1858 to 30 June 1859, there was a mail subsidy contract granted to carry mail from Stockton, California, to Kansas City via Albuquerque. The bulk of the mail on this route was between Albuquerque and Kansas City, Missouri. Only three letters were carried the entire distance: one westward and two eastward, one of which has survived. After spending $80,000 and receiving only $1,255 in receipts, the Post Office Department canceled the contract after nine months of operation.

Meanwhile the clamor for an efficient postal service to California by land continued. After a year's observation, Congress agreed that both the government and California deserved better than this. Thus, in September 1858 Congress finally authorized issuing a mail and passenger contract for semiweekly mail deliveries by stagecoach from the Mississippi River to San Francisco. This was an implementation of the Post Office Appropriations Bill of 3 March 1857, which had stated that "the Postmaster General be authorized to contract for the conveyance of the entire letter mail from such point on the Mississippi River as the contractors may select, to San Francisco . . . at a cost not exceeding $600,000 per annum for semi-weekly service." The same Act gave the Postmaster General the absolute right to choose the contractor, which, in effect, gave him control over the route to be utilized, as long as it ran from the Mississippi River to San Francisco.

When Congress in 1857 approved this new mail contract across the western continent, the Northern Congressmen assumed that the route would follow the old Oregon Trail and the main immigrant route to California. The assumption was based on the fact that this was the shortest, most direct route, about 1,800 miles long, although parts of it were impassable in winter. The Southern Congressmen and the southern-born Postmaster General, however, remembered the enabling clause which gave Postmaster General Aaron Brown the choice of contractors. Although the contractors were to select the route, Brown decided it would start at two

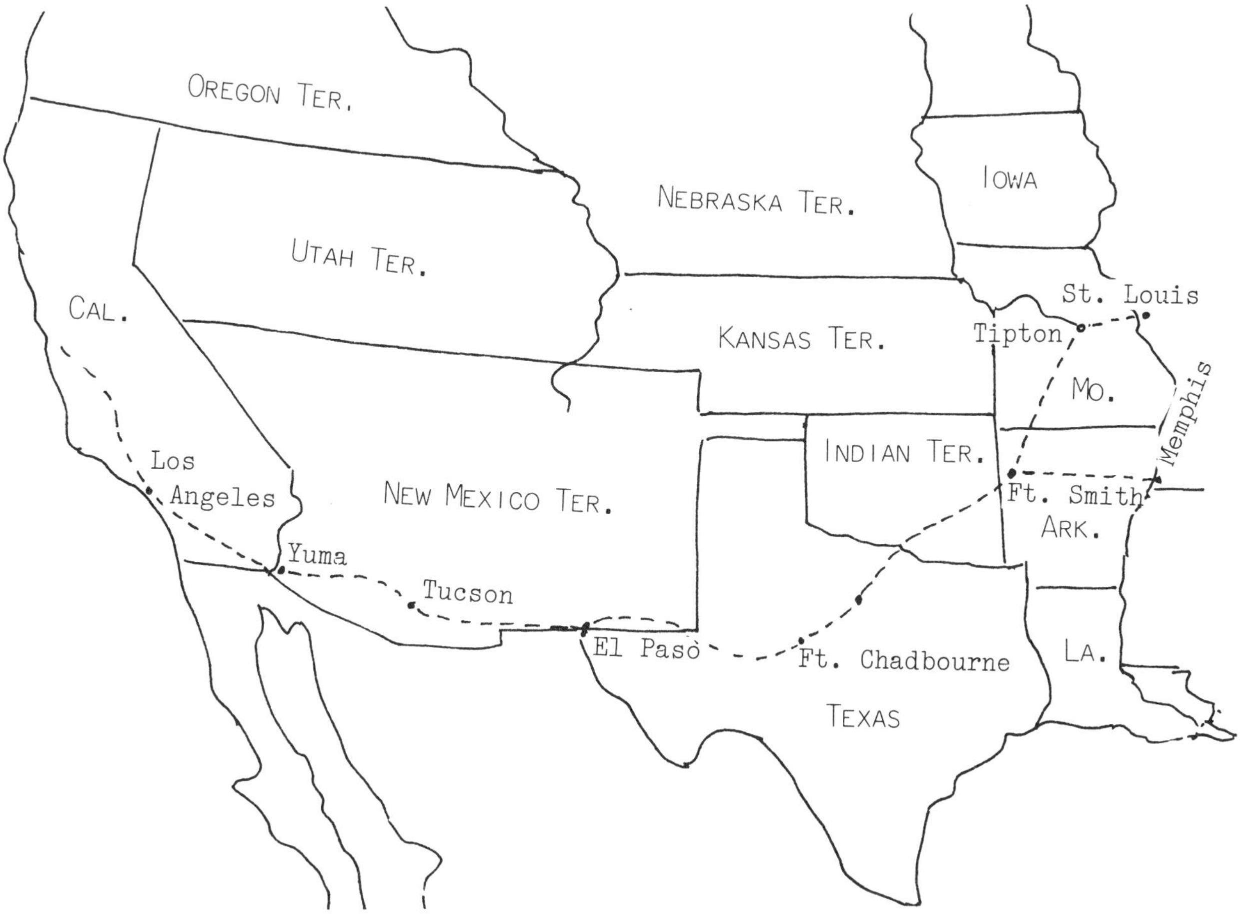

Butterfield overland mail route.

points on the Mississippi River, at St. Louis and at Memphis, Tennessee, his home town; these routes would join at Fort Smith, Arkansas (Brown's original choice was Little Rock), then curve in a southern arc to El Paso and Fort Yuma, to Los Angeles and north to San Francisco. California, along with many in the North and East, was outraged at the choice of a route 2,795 miles long, which they promptly dubbed the "Oxbow Route."

The Overland Mail Company

The successful bidder for this southern route was John Butterfield and a group of associates. A personal friend of President Buchanan, Butterfield also was an experienced stage line operator in New York State. Earlier he had helped Wells and Fargo form the American Express Company in 1850. In a few months Butterfield had spent $1,000,000 in surveying the route, building relay stations, improving roads, bridging streams and contracting with ferries, buying 1,800 horses and mules, purchasing 250 coaches and celerity wagons, and hiring experienced drivers. He was ready to operate by September 1858. Since he was required to make the trip in 30 days, his wagons and coaches rolled day and night, stopping only for meals and team changes.

Butterfield and the Overland Mail Company were not without financial problems. The tremendous cost of outfitting the line was accentuated when it turned out that passenger patronage was limited because the coaches carried only 10 or 12. The original

A variety of Overland envelopes: Woodside ms. postmark to Hanover, N.H., with 10¢ 1857; Napa City paid stampless to Conn., 1860; SF to N.H., 1859, with 10¢ 1857; Vallejo FREE to Washington; Petaluma to Illinois, an unusual combination of 3¢ imperforate and 3¢, 1¢ perforated; a remarkable 1860 cover from SF to Tenn., paid all the way (under 3,000 miles) by a 3¢ 1857.

Covers showing different four and six horse coaches: SF to Boston, six horse Hutchings & Rosenfield design; Nevada City to Wis., J. Hamlin design; Creek ms. postmark on four horse design to Maine; elaborate four horse envelope from SF June 1859, the 15¢ rate to New Brunswick overpaid 1¢.

mail subsidy of $600,000 did not meet the operational expenses. The problem was compounded when in March 1859 Congress failed to pass a Postal Appropriations Bill, leaving the holders of mail subsidies unpaid. By the end of April the Post Office Department owed the Overland Mail Company $300,000. At this time Wells, Fargo & Co. made high interest loans to the Butterfield line.

Historians long have argued over the exact financial relationship between Wells Fargo and the Overland Mail Company. (The destruction of Wells Fargo records in the San Francisco earthquake and fire in 1906 have made it difficult to resolve the controversy.) Nevertheless, California advertisements for the Overland Mail Company from 1858 on show Louis McLane as the agent of the Overland Mail — and McLane was the general manager of Wells Fargo. Also, the three largest stockholders of the Overland were directors of Wells Fargo. The general consensus of reputable historians is that by July 1861 Wells Fargo controlled the Overland Mail Company, though it may not have owned it.

Although Californians and Easterners

Handstamped, printed, and ms. Overland endorsements on covers carried on the Butterfield route: oval marking from Camptonville to N.Y. state, 1860; small straightline, SF to R.I., 1860; ms. on quintuple letter SF to Washington, 1859, with five 10¢ 1857; curved ribbon type from Santa Clara to Ill.; ms. (in great detail) Auburn to Vt., 1860; large straightline on 10¢ envelope from Stockton to Mass.

WF envelope with "OVERLAND VIA Los Angeles" route indication. A way letter received by Butterfield driver from a WF express agent along route. Not postmarked until reaching St. Louis.

Rare "via Visalia" corner card, but used on intrastate letter to SF. Not conveyed on Butterfield route.

Columbia, Cal., to Maine, by Butterfield route. Rate paid by 10¢ 1857.

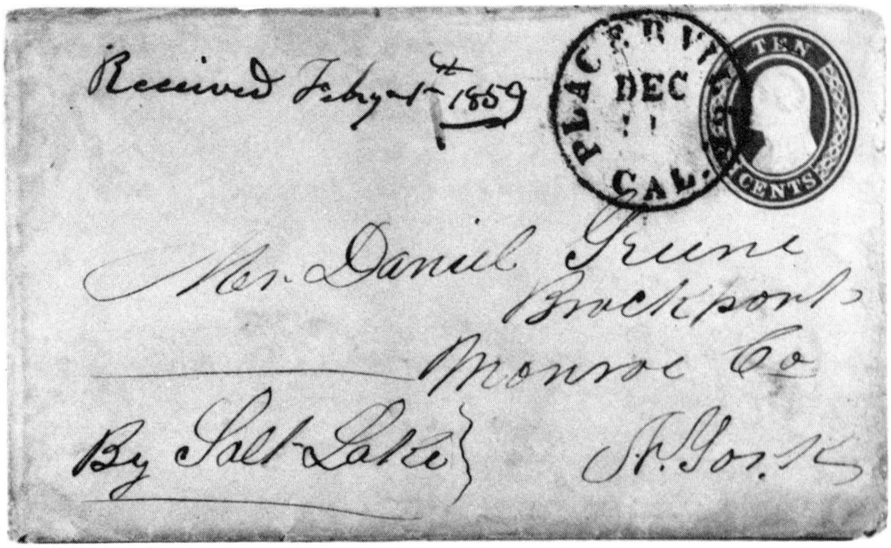

Placerville to N.Y. state on Chorpenning-Hockaday route, 1858. Seven weeks in transit. Carried on foot across the Sierras.

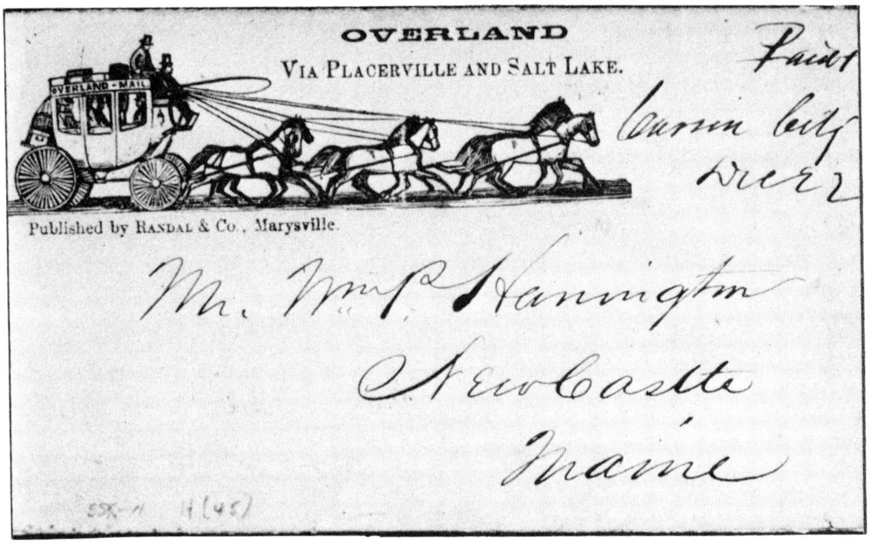

"Overland Via Placerville and Salt Lake" envelope of Chorpenning route. "Paid 10" in cash. Ms. "Carson City."

Timbuctoo to Maine (probably 1860) over Chorpenning route. Forwarded to Belfast, Me., rated "Due 4," including 1¢ way charge for carriage between Vinalhaven which had no post office and Carver's Harbor to be remailed.

Forbestown to Washington FREE on post office business. Since the year is unknown, it is not certain whether this travelled via Placerville and Salt Lake during period of Chorpenning contract or later in Central Overland period.

alike had criticized the lengthy route, nevertheless within two years the Overland Mail Company was delivering more mail than the subsidized Atlantic and Pacific steamship lines which had been in operation since 1849. This, also, despite the fact that postmasters automatically sent all mail by steamer via the Isthmus unless the sender specifically designated "via Overland." The Butterfield Overland Mail operation still stands as one of the great triumphs of transportation.

In spite of the success of the southern Overland Mail, there were still contract lines operating over the central route between Placerville and St. Joseph. George Chorpenning still had a contract from Placerville to Salt Lake City, which in 1858 had been placed on a semimonthly schedule with 20 days allowed for each one-way trip, and with a mail subsidy of $34,000 per year. John M. Hockaday & Company since early spring of 1858 had completed the eastern part of the route from Salt Lake to St. Joseph on a weekly schedule with a 22-day limit for each trip, and a subsidy of $190,000 per year. In 1859 the Postmaster General complained that the actual cost of this overland route was $320,000 annually with receipts of only $5,412.

But this service, averaging six weeks, was too slow to please the Californians. In addition, they primarily were mail lines rather than passenger lines. Yet their route was almost 1,000 miles shorter than the Butterfield road. The time was ripe for someone with an improved organization to prove the utility of the central route. And someone did!

C.O.C. & P.P.E.

The Central Overland California and Pikes Peak Express Company was the creation of three men who previously had made their fortunes in the freighting business, primarily carrying supplies for the Army. They were William H. Russell, William B. Waddell, and Alexander Majors. Three more dissimilar men could hardly be imagined. William Russell was an incorrigible "wheeler and dealer," impulsive, imaginative, dynamic, and reckless. William Waddell was a stolid, phlegmatic, deliberative, cost-conscious and penny-pinching office manager. Alexander Majors was a sober, deeply religious, experienced freighter, more at home with a wagon train than at a desk. Russell dominated the other two, but all three were bound together by a fierce loyalty.

In the 1850s these three men in a loose partnership controlled most of the large scale

Forerunners and competitors of C.O.C. & P.P.E.: Leav'h City & Pikes Peak on 3¢ envelope to Ohio; Hinckley & Co., Feb. 1861, into mails at St. Joseph, to Mass. Crude straightline Denver Central, etc., and Jones & Russell, Denver City (two dates) to NY. Entered mails at St. Joe May 10, 1860. Jones & Russell, Denver City, on 1859 map cover to N.Y. state.

freighting on the western frontier. By 1858 they owned 3,500 freight wagons, 40,000 oxen, and employed almost 5,000 men. Their entry into the stageline business was almost by accident. A gold strike had been made in Colorado and a gold rush ensued. William Russell, who in 1858 had helped found the city of Denver, decided to establish a stage line from Leavenworth, Kansas, to Denver. He offered Waddell and Majors partnerships, but they refused. It was not their kind of business. Russell went ahead anyway, and in February 1859 he formed the Leavenworth and Pikes Peak Express, with John S. Jones as a partner.

Soon afterward, in May 1859, Russell and Jones expanded by buying out Hockaday's line running from Missouri to Utah, the eastern portion of the central route. Neither venture was financially successful. Soon Russell was unable to meet the payments on $525,532 in debts and promissory notes, mostly owed to his freighting partners. He appealed to Waddell and Majors for help, and almost in despair they decided that they had to get into the staging operation in order to salvage their investments.

Thus on 28 October 1859 Waddell, Majors, and Russell took over the Leavenworth and Pikes Peak line and the Jones & Russell Express, and in February 1860 they organized the Central Overland California and Pikes Peak Express (C.O.C. & P.P.E.). Russell became president and Waddell and Majors supervised administration and operations. They began by absorbing the Leavenworth and Pikes Peak Express and the Jones & Russell Express into the new organization. Now committed to staging, they decided to enter the field vigorously in competition with the Butterfield Overland Mail. In 1860 the Chorpenning mail contract between Placerville and Salt Lake was annulled and the route awarded to Russell, with a mail contract for $33,000 per year to operate semimonthly over the former Chorpenning route from Placerville to Salt Lake. This was less than one quarter the actual cost of operation, but it was a beginning for a company dreaming of a potential $1 million mail subsidy.

Hinckley's Express which had begun operating in 1860 between Colorado and Fort Kearney, Nebraska, via Denver, primarily serving the Colorado mining camps and using the facilities of other stage lines to carry its treasure, was a potential rival. C.O.C. & P.P.E. contracted with Hinckley's company to become the exclusive carrier of its mail and gold boxes beyond Hinckley's usual area of operations. Then in May 1861, C.O.C. & P.P.E. bought out Hinckley's entire company.

By June 1860 Congress was debating postal bills which would pay $600,000 for service over this central route. Russell proposed to the Post Office Department that the C.O.C. & P.P.E. would carry the entire mail from California to Missouri, but the subsidy must be $900,000 per annum for a triweekly service. Congress adjourned without taking any action. Then on 29 August 1860 the Post Office dealt a devastating blow to the C.O.C. & P.P.E. It awarded the Western Stage Company a mail contract for weekly service from Missouri via Omaha and Fort Kearney to Denver. This shattered the monopoly of the C.O.C. & P.P.E. on the central route.

Though their stage line was operating over the entire central route, the C.O.C. & P.P.E. was in serious financial difficulties. They had been unable to secure any substantial mail subsidies, and their expenses were double their income. Their freighting business, once their financial foundation, began losing Army contracts. The War Department refused to acknowledge some of their claims for unpaid services, on the excuse that Congress had failed to appropriate sufficient funds. Soon C.O.C. & P.P.E. was losing more than $1,000 per day, and Russell, Waddell and Majors no longer had their freighting profits to fall back on. Their only hope was to secure a large mail subsidy. To do this, they

C.O.C. & P.P.E. oval, St. Joseph, on 3¢ star die to Utah, endorsed "Free/W. H. Russell."

C.O.C. & P.P.E., Denver City, to NY, 1860, and there forwarded with 3¢ 1857. Yellow label at left.

Cover with 3¢ 1857 delivered by Hinckley's Express to C.O.C. & P.P.E., who carried it to St. Joe where it entered mails July 15, 1860.

Central Overland and Pony Express route.

had to dramatize the superiority of the central route.

The Pony Express

It was then that they initiated the heroic and romantic Pony Express in the belief it would unquestionably prove the superiority of the central route over Butterfield's southern Oxbow route. They envisioned receiving a $1 million mail subsidy if they were successful. "The Pony Express" has caught the imagination of a people nostalgic for a legendary past. It has become a symbol, like the frontier marshal taming a town, or the booted, spurred and six-gunned cowboy and gunfighter. It is almost an epic as it tells of young men, like Bob Haslan and Buffalo Bill Cody, riding without relief 108 miles and 322 miles when they found the next rider dead or injured, outracing Indians, through dust of the sear-

Running Pony, San Francisco, in blue, Nov. 28, 1860, to NY, with 10¢ 1857. Black C.O.C. & P.P.E. St. Joseph oval.

Running Pony San Francisco oval Dec. 5, 1860, to NY; also C.O.C. & P.P.E. oval; on reverse Dec. 19 strike of encircled oval (earliest date known).

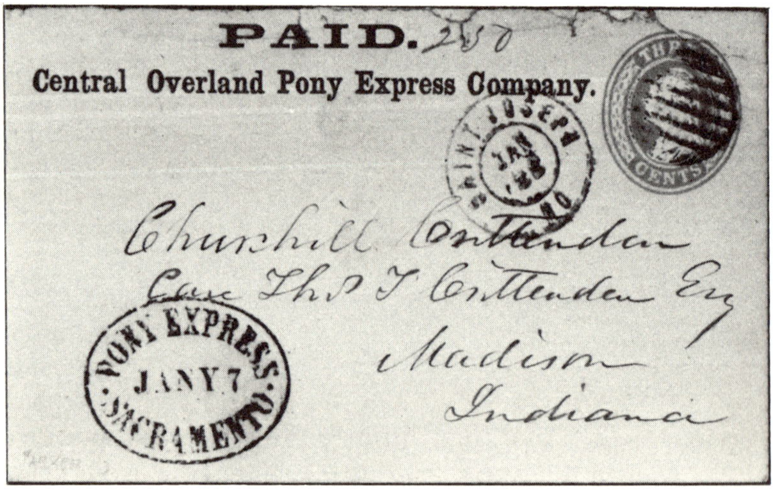

Sacramento oval Jan. 7, 1861, to Indiana. PAID 2.50/Central Overland Pony Express Company. The rate indicates a ¼ oz. letter.

ing desert, through sleet and snow of the mountain ranges, as they spanned the western half of the continent in 10 days, each rider galloping about a 50-mile relay, stopping only to change horses every 10 to 15 miles. It is a saga which has been told and retold, and it

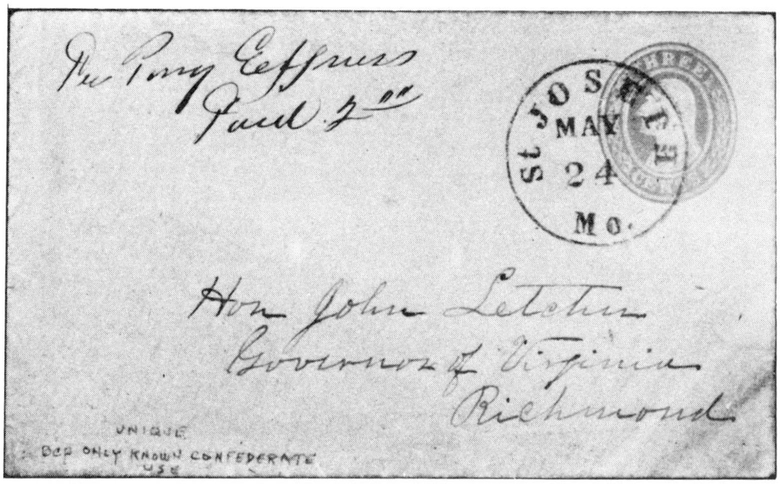

St. Joseph, May 24, 1861. "Per Pony Express/Paid 2.00." Used to Virginia, then in Confederacy.

California Pony Express/New York oval to SF, May 7, 1861, on Wells Fargo envelope "½ ounce PAID."

need not be repeated here.

But it was a serious business, not an heroic romance, with Russell, Waddell, and Majors. In February 1860 Russell was advertising in Leavenworth, Salt Lake City, and Sacramento for hundreds of mares. He was sending out wagon trains of materials to build relay stations at intervals of 10 to 15 miles, until there were 153 way stations and relay posts along a 1,966 mile route. By 3 April 1860 the first riders were ready to leave San Francisco and St. Joseph. Eighty riders using 500 horses could each carry 35 pounds of mail by night and day until 10 days later welcoming crowds would hail their arrivals at the terminal towns. It was pure drama. Never had mail moved with such speed across the continent. During the one and a half years it operated, the Pony Express carried 34,753 pieces of mail. Only its speed justified to letter writers the cost of its use: $5 per ½ oz. until August 1860 when ¼ oz. letters were accepted at the half rate of $2.50. On 1 April 1861 the rate became $2 per ½ oz. through 30 June. Though the charges were high, the expense of operation was incredibly higher. And, unfortunately, too many of those who acclaimed the heroics continued to send their ordinary mail by older and cheaper routes. The Pony Express carried an average of only 35 letters per trip in the early days.

From the very beginning the Pony Express was an economic liability, operated by a

Scarce green C.O.C. & P.P.E. St. Joseph encircled oval, Jan. 20, 1861, to SF.

company always on the edge of bankruptcy, but sustained by the hope of securing a large mail subsidy. Then in 1861 another series of disasters began. First, Russell became involved in a political and criminal scandal. For several years he had persuaded Secretary of War John B. Floyd to grant him "acceptances" for freighting services not yet performed, and then had used these illegal documents as security for loans. The Army's dilatory practice in paying its bills might have justified this shady scheme, but beginning in 1860 Russell had crossed the line into criminality. Goddard Bailey, a relative of Secretary Floyd and a clerk in the Department of Interior, had access to the negotiable bonds belonging to the Indian Trust Fund. Over a period of months Bailey gave $870,000 of these bonds to Russell, who used them as security for loans. In exchange, Bailey took unsecured notes from Russell, who signed as President of C.O.C. & P.P.E. Russell promised to return the bonds before 4 March 1861, the inaugural date for a new President and change of Administration.

But, after the election of Abraham Lincoln in November 1860, Bailey became frightened and wrote a letter of confession to the Secretary of the Department of Interior, who in turn immediately informed President Buchanan of the embezzlement. Buchanan was determined to leave office honorably, so on 23 December 1860 Russell and Bailey were arrested. The next day the House of Representatives appointed a committee to investigate, and hearings began on 27 December. Congress suspected that War Secretary FLoyd, who refused to testify and resigned on 29 December, had accepted bribes. Russell testified that he had given a C.O.C. & P.P.E. note for each bond and therefore was not guilty of fraud. Nevertheless, on 29 January 1861, a D.C. grand jury indicted Russell, Bailey, and Floyd. Russell pled immunity on the basis of "double jeopardy," citing his examination by a Congressional committee. On this technicality the grand jury reluctantly excused Russell from prosecution. Bailey simply "jumped bail" and disappeared. The case against Floyd was dismissed for lack of evidence of an intention to defraud the government, and Floyd returned to Virginia, ultimately to join the Confederacy. Though not convicted, Russell was a "tainted" man. That Majors and Waddell probably were unaware of Russell's methods used to finance the C.O.C. & P.P.E. is evident by their subsequent actions. Majors assigned all his assets to the company's creditors and, penniless, left the company. Waddell also signed over most of his assets, but remained with the company at its Leavenworth headquarters.

This series of events was followed by a

San Francisco Running Pony, June 26, 1861, to NY. Triple pony express rate paid by $2 red and $4 green Wells Fargo stamps. Two U.S. 10¢ (probably over the embossed 10¢) to make up the triple U.S. rate are missing.

final blow. The Civil War began, and with the secession of Texas on 1 February 1861, the Overland Mail Company had become a "jack-pot" for the Confederacy, which immediately began confiscation of the Overland Mail Company's rolling stock, horses, and provisions, and soon the Butterfield line was no longer in existence. (Butterfield had resigned in 1860 in an argument with his directors, and W. B. Dinsmore had replaced him as president, but common usage still referred to the Overland Mail Company as the Butterfield line.) With the Overland Mail Company no longer operational, the only California overland communication with the Union was the C.O.C. & P.P.E. on the central route, and it seemed reasonable that they should take over the Overland Mail Company's huge postal sub-

Pony Express San Francisco to Salem, Me., with Wells Fargo $2 for double rate.

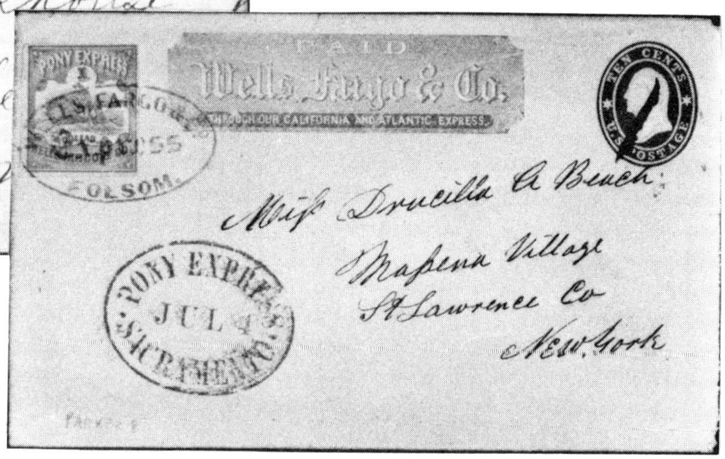

Sacramento oval and $1 red tied by WF Folsom on star die to N.Y. state.

sidy.

Nevertheless, the Overland Mail Company still had friends in Congress which immediately proposed that the Overland Mail Company's operation be transferred to the central route and that it establish a daily mail service east to west, in spite of protests by Senator Gwin of California that this was not fair to the C.O.C. & P.P.E.

Congress simply could not give a $1 million mail contract to a company headed by a man so recently embroiled in flagrant scandal. But it provided a loophole: it gave the mail contract to the Overland Mail Company, but allowed it to subcontract such services as needed. Since the Butterfield line had lost most of its equipment, it had no choice but to cooperate with the C.O.C. & P.P.E., but it was a hard-nosed negotiator, for it still retained the mail subsidy. The equipment salvaged from the southern route was sold to C.O.C. & P.P.E., and the St. Joseph to Salt Lake City and the Denver routes sublet to them. Russell was compelled to resign on 26 April 1861, and Bela M. Hughes, a cousin of Ben Holladay, became president of C.O.C. & P.P.E. The Central Overland could operate from Salt Lake City to the eastern terminal. Wells Fargo, which held liens against the Overland Mail Company, would control operations in the western 35 percent of the route, including that of the Pony Express.

The Pony Express was still running and Wells Fargo had no intention of allowing it to

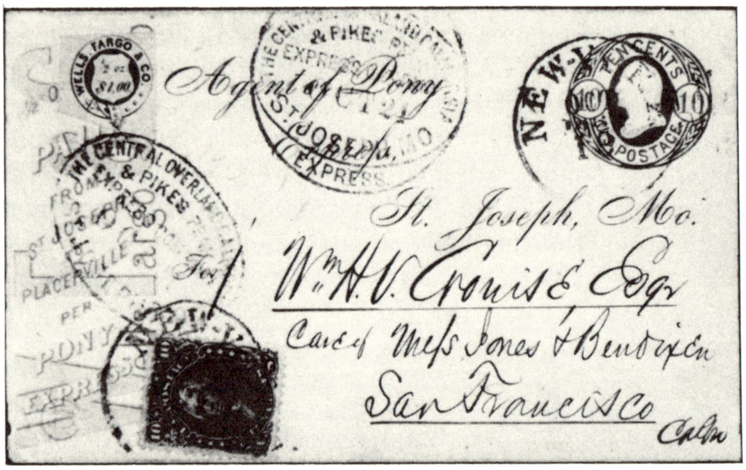

NY to SF, Oct. 19, 1861. Double letter with $1 garter stamp.

Double letter from Boston with $1 garter stamp and 10¢ August on 10¢ envelope.

Oval Pony Express Sacramento Sep. 15 (1861) to NY with $1 red WF stamp. Atchison, Kan., double circle Sep 27 cancels 10¢ star die.

die. They continued its operation and created adhesive stamps (a $2 red and a $4 green) to be used on the Pony Express mail originating in California. On 1 July 1861 the rate was reduced to $1, and Wells Fargo issued a new series of adhesives: $1 red, $2 green. These were used only on eastbound mail. For westbound mail, they used an entirely different frank, and a "garter" adhesive stamp to indicate an overweight letter. The completion of the transcontinental telegraph line on 24 October 1861 ended the Pony Express. The last

Two covers with Overland endorsements carried on Central Overland route after Pony Express period. SF to Boston with 3¢ 1861 (1862 or later); Linden to Canada, March 4, 1868, short paid with two blackjacks, rated 10¢ due.

A choo-choo cover used on the Central Overland route from SF to Mass., 1862 or later. The legend "Via Los Angeles" in the corner card has been thoroughly crossed out.

Pony Express letter was received in San Francisco on 20 November 1861. While the Pony Express was a financial failure, it did demonstrate the practicability of the central route — and marked the path for the first transcontinental railroad.

Ben Holladay

In July 1861 Ben Holladay had loaned the C.O.C. & P.P.E. $208,000, secured by a deed of trust to the company's property as collateral. Earlier he had signed a $100,000 performance bond for Russell. In December when the company could not pay the interest on the notes, Holladay demanded that the company be sold as bankrupt. On 22 March 1862 he bought the company's assets for $100,000. This purchase gave him control of the eastern half of the central overland route.

Ben Holladay was a wealthy, ambitious, and skillful businessman. Already known as the "Stagecoach King" in the east, he controlled a vast financial empire, which included 16 steamers plying the Pacific coast and the China trade. After picking up the ruins of the C.O.C. & P.P.E., for the next five years Holladay had a virtual monopoly on the eastern half of the central overland route. Leaving the Salt Lake to California run, now controlled by Wells Fargo, in possession of the Overland Mail Company, Holladay established new routes north and west through the Boise Valley silver mines and on into Oregon, and north into Montana. Under the name of the Holladay Overland Mail and Express Company, Holladay used his wealth to undermine his competitors. When gold strikes occurred in Montana and Idaho in 1862 and 1863, three expresses began serving the 400 mile run between the camps and Salt Lake City. They had no mail contracts, but were making a profit by charging $1 for each letter and $150 per passenger. The largest of these expresses was owned by J. P. Oliver.

Holladay visited the Postmaster General and emerged with a $13,271 per annum mail contract to serve Virginia City and Bannack City in Montana. At the same time he secured a $186,000 per annum mail subsidy to open a line between Salt Lake City and Walla Walla, Washington, via Boise Valley. He drove his competitors out of business by cutting rates, which he promptly raised again when they were gone. He consistently charged his passengers high rates, succeeding because his "was the only game in town." By 1866 Holladay employed 15,000 men, owned 20,000 vehicles and 15,000 draft animals, and he had $365,000 in mail contracts. On 1 November

Cover with Holladay Overland Mail and Express Company frank, Omaha to Washington, D.C.

A. J. Oliver Virginia City hand-made oval. "Cod 25."

A. J. Oliver Bannock and Salt Lake City oval on cover from St. Louis to Bannack City.

1866 he suddenly sold out to Wells Fargo.

As mentioned earlier, Holladay had observed the progress of the transcontinental railroad. He knew that on its completion feeder railroads would radiate in all directions. He foresaw the demise of the stage line glory, and he had no intention of being present at its funeral.

Strip of three of 5¢ buff from Stockton to Canada, 19 November 1861, on patriotic cover.

Chapter 8

California Illustrated Mail

Placervill May 22nd 1853
Dear Mama
As I have a few leisure moments this morning I propose to devote a share of them in writing to you hoping at the same time that there will be no hard feelings on your part for my being silent so long the fact of it

Lettersheet datelined Placerville May 22nd 1853.

California Illustrated Mail

Almost every tourist visiting another country mails home pictorial postcards. It is a simple way of sharing with the stay-at-homes the scenery and activities of another land. When the Gold Rush of 1849 began, California almost literally was a foreign country to the rest of the United States. For two decades the argonauts and immigrants to the Far West filled the eastbound mails with pictorial letter sheets and envelopes. These portrayed the landscape scenery, mining scenes, the miners' amusements and diversions, the current events of assassinations and executions, the disasters of fires and earthquakes, panoramic views of towns, and often comic-relief caricatures of a miner's life.

Letter Sheets

More than 60 publishers in California produced these pictorial letter sheets, and some of them also printed illustrated envelopes. Proprietors of stationery stores, booksellers, newspapers, and even express companies would commission artists to make woodcut blocks and lithograph stones, and then would print thousands of copies of these pictorial records which sold from 10¢ to 50¢ each. Almost always they were printed in black and white.

It is impossible to determine accurately how many different letter sheets and pictorial envelopes were printed. More than 1,000 letter sheets have been cataloged, if one includes the varieties of paper, and the reprints that are reworkings of the lithographs. Their popularity is shown by James M. Hutchings's claim to have sold over 90,000 copies of his famed "Miners Ten Commandments" sheet in one year. It was not uncommon for a newspaper advertisement to offer 10,000 assorted sheets in a single sale. Letter sheets may have been the most common means of communication used by Californians to their friends and families between 1849 and 1855. Nevertheless, they are rare today.

Letter sheets came in two types: (1) full page illustrations, on the back of which could be written a message, then the page folded and sealed, with space left on one panel for the address, thus creating a integral envelope; (2) half-page illustrations which were vignette letterheads, allowing more space for the written message. The half-page pictorials not only portrayed the subjects aforementioned, but also often were used by commercial firms as an advertising medium.

The majority of letter sheets and illustrated envelopes were published in San Francisco. The largest publishers were Hutchings & Rosenfield, Kurtze & Co.'s Wide World Office, Britton & Rey, and Charles Kimball the Noisy Carrier. Among other *known* San Fran-

The Miner's Ten Commandments.

Michigan City, published by Barber & Baker.

Burning of Nevada City, March 12, 1851. Publisher W. H. Baxter.

cisco printers, for many of the sheets are unsigned, were the following:

Anthony & Baker Co.
D. Appleton & Co.
Atwill & Co.
W. H. Baxter of Bower's Express
George H. Bell
S. F. Bogardus
Bonestill & Williston
T. C. Boyd
B. F. Butler
W. C. Butler
W. B. Cooke
Cooke & LeCount
Conner & Forrest
C. M. Fenn & Co.
Fisbourne & Gow
Gregory's Express
Frederick Hess
George B. Hitchcock
Marvin Hitchcock
J. M. Hutchings
Justh, Quirot & Co.
Kuchel & Dressel
Larrabee
Leland, McCombe & Co.
Wm. E. Loomis
R. & J. McLees
J. D. Page
Henry Payot
D. A. Pleckner
Pollard & Britton
Quirot & Co.
A. Rosenfield
Sterett & Butler
Sun Print
S. H. Wade

Printers in other California towns soon followed suit:

Sacramento: Barber & Baker
H. J. Biddleman
E. B. Davidson
Gardiner & Kirk
Hossack & Crawford
Lovegrove & Murray
Placer Times (newspaper)
Sacramento Union (newspaper)
Stockton: George A. Theall
L. C. Van Allen
La Porte: J. C. Lester
E. L. Willard
Marysville: R. A. Eddy
Randal & Co.
Columbia: A. A. Hunnewell
Towle & Leavitt
Downieville: Downieville Bookstore
Langton's Express
W. B. Monmonier
Placerville: A. W. Bee
El Dorado News (newspaper)
Hunter's Express
Nevada City: Forrest & Borden
J. E. Hamlin
A. W. Potter
H. R. Stiles

Lynching of Casey and Cora.

Adventures of Mr. Greenhorn. Published by Cooke & Lecount.

Commandments to California wives and accompanying envelope mailed at Jamestown.

Jackson: C. L. Parish
Sonora: G. S. Wells
Timbuctoo: Wessell's
Goodyears Bar: Wm. B. Monmonier
Oroville: Garnam & Lockwood
Folsom: Julius Jacobs
Georgetown: W. T. Gibbs
San Jose: D. H. Mason
Shasta City: George W. King
Rhodes & Lusk Express

Miners Envelopes

Among the first illustrated envelopes were the so-called "miners envelopes." These were covers depicting experiences in a miner's life, sometimes serious, sometimes comic. Since an envelope was limited in size, the picture was usually a single scene placed on the left side of the cover, although in rare instances the back of the envelope portrayed the scene. Unlike the letter sheets which illustrated hundreds of scenes, the miners' envelopes probably never exceeded 25 different types. All of the presently known designs are here illustrated.

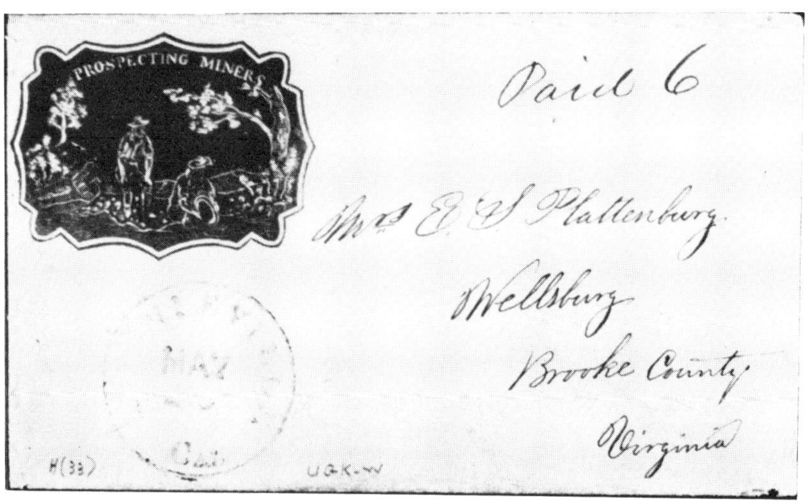

"Long Tom," two designs: Downieville to Maine with 10¢ 1855; Georgetown to SF with 3¢; "Prospecting Miners" on paid stampless to Virginia.

J.M. Hutchings designs: "The Expressman," ms. Jacksonville, Ore., on double letter to Ohio; "Prospectors," North Bloomfield to Maine; "Writing Home," Astoria, Ore., to Ohio, double with strip four orange brown; "Rocking the Cradle," Weaverville oval to Boston; "Mining Scene," Nevada City to R.I., paid stampless with "6" over "3" in postmark.

Designs published by Wide West, San Francisco: "In a Tight Place," SF to Vt.; "Ups and Downs," unpaid stampless to Georgia; "Up Hill Work," ms. Red Bluffs to Wis.; "Proceeds of the Day," Yreka unpaid stampless to Ohio; "The Traveling Miner," ms. Marysville in circle "paid" to Vt.

Comic envelopes from Noisy Carriers: Michigan Bluff to Maine with 10¢ 1857; Rough & Ready to NY state, with 10¢ 1855; ms. Tehama on 1855 paid stampless to Mich.; Napa City PAID 6¢ to Charleston, 1854.

Noisy Carriers: "Long Tom," Michigan Bluff paid stampless to Maine; comic designs: Rough & Ready to Stockton with 3¢ 1857; Weaverville to Virginia, 10¢ 1857. "Miners' Coat of Arms," (Barber & Baker), Sacramento to N.Y. state; reverse of F.C.M. Fenn envelope.

Civil War Patriotic Covers of California

Although California had been admitted to the Union as a "free state," the pro-South element was a strong minority. When secession began, many southern sympathizers left the state to join the Confederate forces. Those that remained behind were sufficiently vocal to inspire their opponents into displays of patriotic fervor.

Nine days after the 12 April 1861 attack on Fort Sumter, Senator Edward D. Baker took a leave of absence from the Senate and in New York called upon Californians "to rally 'round the flag." There were many people in the East who had returned from California after the initial Gold Rush. Enough of them responded to Baker's summons to arms that Baker was commissioned Colonel of the First California Infantry, and the unit was called to active duty. In October 1861 at the Battle of Balls Bluff, Baker was killed and the First California Infantry was shattered. The remnant of the regiment was absorbed into the 71st Pennsylvania Infantry, where it served for the rest of the war.

The National Conscription Act was never enforced in California, yet approximately 15,000 Californians volunteered for enlistment in the Union Army. The majority did not see active service in the fighting engagements of the eastern battlefields. The cost of transportation seemed to outweigh the Union Army's immediate need for men, and the Army Staff had other uses for the Californians. To insure that secessionist sympathy was kept under control, some California regiments were assigned to local garrison duty. Others were ordered to patrol the Overland Mail route from California to Salt Lake City to protect it from the fate of Butterfield's southern mail route. Other California troops policed the Indians of the Northwest.

One group, composed of about 2,000 men under command of Col. James Carleton, was known as the "Column from California." Beginning in April 1862, in small numbers so that they would not overtax the waterholes, these men were dispatched south from Oak Grove to "recapture our forts in Arizona and New Mexico." They proceeded to Fort Yuma, then on to Tucson and beyond. By late August these troops had reached the Rio Grande.

By late 1862 Massachusetts was having difficulty in filling its conscription quota. A group of young San Franciscans offered their services if Massachusetts would pay the costs of organization and transportation. The offer was accepted, and on 11 December 1862 the California contingent boarded the steamer *Golden Age* to sail to Boston. On arrival there, they were attached as the "California Company" to the 2nd Massachusetts Cavalry, and

Soldier's letter from a member of the Column from California at Fort Yuma with officer's certification. Entered mails at Los Angeles April 11, 1862, rated 20¢ due as double letter. Soldiers were permitted to send letters unpaid under this procedure.

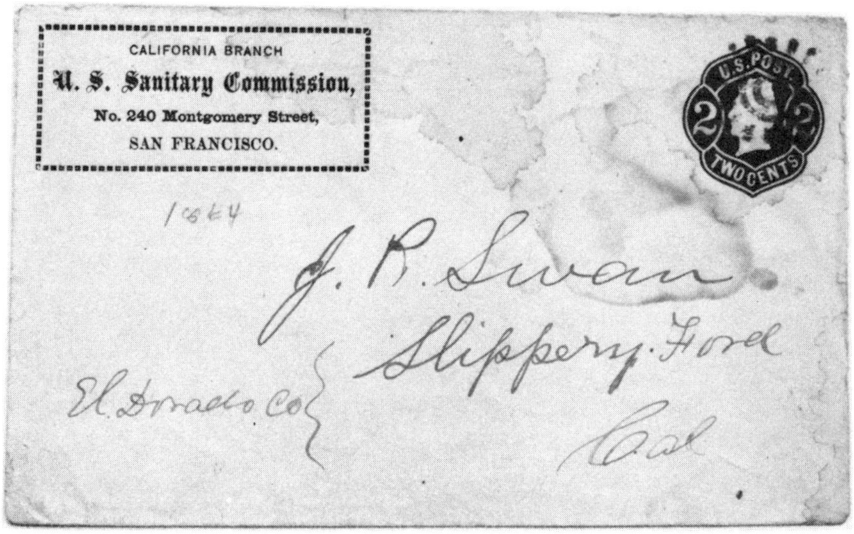

Corner card of U.S. Sanitary Commission office in San Francisco.

subsequently participated in more than 50 engagements. By the end of the war, the battlefield had become the final destination for possibly 500 Californians.

The greatest contribution California made to the Union was not military but economic. The $500 million in treasure that had been shipped east since 1849 had given the United States its first gold reserve. It had financed the industrial expansion of the North, which made the defeat of the Confederacy inevitable. The great Comstock discovery of 1861 added to the silver and gold that fueled the support of the Union armies in the field. When the Sanitary Commission (the Red Cross of the Union Army) was created for the relief of the sick and wounded in the war, California contributed $1 million, more than one-fourth of the entire amount given by the rest of the nation.

California designs: Portland, O.T., to Ill., 10¢ 1857; Woods Ferry to NY state, 1861 10¢.

Los Angeles to Canada. Rate remained 15¢ until Feb. 17, 1864.

Overall flag design used to Canada. Chicago Ill. PAID 15 exchange marking.

Stockton, May 1, 1862, to Switzerland. Prussian closed mail 35¢ rate paid by 1861 issue.

Flag design patriotics used from Todd's Valley, SF, Nevada City, and Georgetown. Last two California publishers.

Patriotic SF to Boston Sept. 18, 1861, with 1857 issue.

Patriotic Georgetown to Oswego, N.Y., Nov. 5, 1861, double rate paid by mixed 10¢ issues.

All of this is history, but it is a history of special interest to the postal historian, for the war spirit is seen on the California mail of the period. To show their patriotic support of the Union, Californians often placed their letters in patriotic propaganda envelopes. Even some of California's postmasters participated by creating special patriotic "killer" devices to cancel the postage stamps. Although about 11,000 different "patriotics" have been cataloged as used in the United States during the Civil War, their use within and from California was comparatively rare. These are highly valued by collectors, and some are illustrated here.

Sacramento shield on cover to NY; flag and UNION on cover postmarked Healdsburg.

Railroads & Choo-Choo Mail

Despite the progress made in overland and sea communication during the 1860s, Californians realized that the only successful alternative to the slow stagecoaches and maritime steamers was a railroad across the western half of the continent. The argonauts and immigrants to California had come from the East where towns were linked together by a network of railroads. They had accepted necessarily but reluctantly the transportation hardships of their new world. Nevertheless, they continually complained and demanded better. The steamer lines had lessened their isolation. The overland stage lines had resulted from their complaints about the inadequacies of the sea passage. They were convinced, however, that the end of their isolation and alienation from the rest of the nation depended upon iron rails.

From 1850 on, California's congressional representatives had dinned into Washington's ears the desires of their constituents. They knew that as early as 1835 Asa Whitney had laid before Congress a proposal for a railroad from Lake Michigan to the mouth of the Columbia River. A transcontinental railroad was not a new idea, nevertheless Congress could not bring itself to act. First, there were the physical obstacles: to construct a track across 2,000 miles, over two major mountain systems and through country largely uninhabited except by hostile Indian tribes. Then there were the political implications: the suspicions and antagonisms of the Northern and Southern factions about the route. Finally, there were the financial problems: a venture of this magnitude was beyond the capacity of private enterprise alone; government subsidies on an unprecedented scale would be required.

Yet the dream would not die. As the years passed and Congress debated endlessly, Californians began to build small railroads within the state. On 22 February 1855 the 22 miles between Sacramento and Folsom were

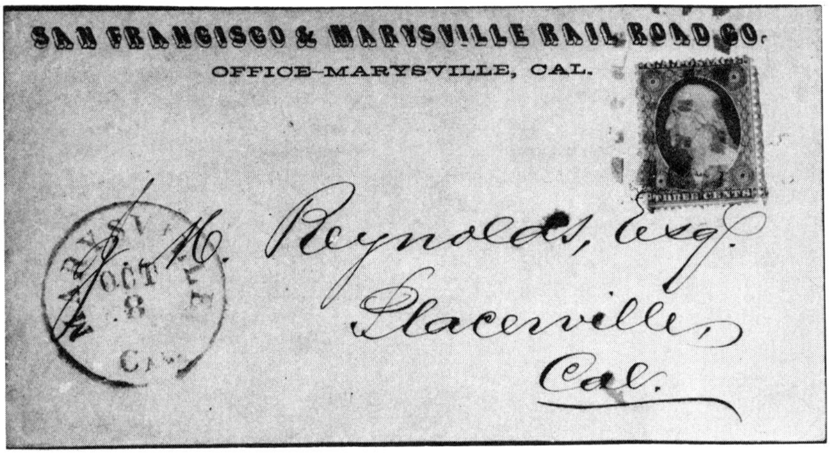

Corner card of San Francisco & Marysville Railroad.

spanned by California's first railroad, which later extended to Placerville. A short line railway was completed from Los Angeles to its port at San Pedro. Construction began on tracks from San Francisco to San Jose. Railroad surveys were made from Sacramento to Shasta. But these were small and local and did not touch the fundamental problem of an across-the-continent rail passage. They were, however, evidence of California's determination that railroads were the hallmark of future transportation and communication. The only hostile opposition came from the steamship lines, Wells Fargo, and the California Stage Company.

Theodore D. Judah, a brilliant construction engineer, became the spokesman of California's railroad proponents. Having been involved in the state's internal railway building, he became convinced that a track across the Sierra Nevada was feasible. The conviction became a fanatical obsession with him. He lobbied unsuccessfully for his idea in Washington, but returned to California convinced that the pressure of public opinion expressed in potential voting power could influence Congress.

In 1859 he persuaded the California State Legislature to call a Railroad Convention for the specific purpose of memorializing Congress in favor of a transcontinental railway over the central route. On 20 September 1859 the Pacific Railroad Convention met in the Assembly Hall at Kearney and Post Streets in San Francisco. More than 100 delegates were present, most from California but some from Nevada, Oregon, and Washington. On October 11 the Convention adopted a two-part Memorial: (1) a route survey should be authorized; (2) Congress should

Corner card of Pacific Railroad Convention, San Francisco.

promote construction by guaranteeing local bonds, making low interest loans, and approving land grants for public sale. Judah carried the Memorial to Washington, D.C.; Congress received the Memorial; Congress filed it without action.

Now the common folk of California began to express their opinion by the use of slogan-bearing letter envelopes, known to collectors today as "choo-choo" mail. As letters were sent east by steamers and by overland, their propaganda picture-cachets and serio-comic slogans were designed to catch attention, to keep a dream alive, and to remind the nation that the time had come to make the dream a reality.

Putt's Overland envelope, Sutter City to Salt Lake City, addressed to Brigham Young; two varieties of choo-choo envelopes, from Stockton, 1858, and SF, 1859.

Four different choo-choo designs: two by Sacramento printers from Sacramento to Phila. and Brooklyn; two SF, one to Ohio, one to Ill. at 3¢ rate, 1860.

In 1860 Judah made a physical survey which convinced him that from Sacramento through Dutch Flat an unbroken ridge extended upward toward Emigrant Gap and Carson Pass, on which could be laid the roadbed of a railroad. He took his survey to four of the state's wealthiest merchants, Collis P. Huntington, Leland Stanford, Charles Crocker, and Mark Hopkins. They were sympathetic to his proposal, but appalled at the potential cost. The outbreak of the Civil War, however, convinced them that Congress might now approve and subsidize such a venture. On 28 June 1861 they incorporated the Central Pacific Railroad to build 115 miles of track from Sacramento to the Nevada boundary. Needing an immediate $115,000 to begin construction, the "Big Four" posted 10 percent of the amount as a performance bond. (Each of the four subscribed for 150 shares of

Choo-choo covers from SF to Ohio and Mass., 1859; Folsom City to Mich., 1859; Georgetown to Iowa, 1862. All with 10¢ 1857, the last after demonetization.

SF to Prussia, May 14, 1869, in North German Union direct mail. Three copies of 3¢ 1861 on 3¢ envelope overpaying 10¢ rate. First run of Central Pacific from west to east.

the stock at $100 per share, and sold to other investors 830 more shares, most of which they later would buy back. Ultimately the Big Four would reap more than $200,000,000 from their cautious investment.) At 10 percent down, the Central Pacific began construction with only $15,800 in cash. Obviously, they soon were to be in serious financial straits.

Congress finally acted. Concerned with access to the gold and silver treasure of the West, and anticipating need for its military aid, Congress now became sympathetic to the desires of the West. The Pacific Railroad Act of 1862, later amended in 1864, was signed into law by President Lincoln. It agreed to support the Central Pacific Railroad building toward the east, and incorporated the Union Pacific to build westward. It would pay $16,000 for each mile of track laid across level land and $48,000 for each mountainous mile. In addition it granted to each railroad every alternate 10 square miles of land along a right-of-way 40 miles in width.

In spite of the Congressional Act, the Central Pacific was plagued by a constant money shortage. Construction costs far exceeded expectations. Payment was not made for track laid until it was completed. Profits from land sales could not be realized until the railroad was finished. The Big Four frantically borrowed money from their friends. They imported thousands of Chinese from Asia to provide cheap labor. They resorted to chicanery such as awarding all construction contracts to a "dummy" company which they solely owned, and to "persuading" state geologists and surveyors to testify that the Sierra Nevada mountains began eight miles east of Sacramento, thus allowing them $48,000 per mile for construction. Somehow they managed to avoid bankruptcy, conquered incredible physical obstacles, until on 10 May 1869 at Promontory Point, Utah, the golden spike was driven to join the bands of iron that united a nation. California's isolation had ended.

No one knows if the propaganda of the "choo-choo" mail had any real effect. The postal historian is satisfied to remember the platitude that "great rivers from small streams grow," and to seek eagerly for the few covers of this mail that were preserved by chance and fortune.

Historical, Political & Temperance Covers

"It pays to advertise" was an idea not limited to business interests. Supporters of moral crusades and of political candidates, as well as those proud of historical emblems, used illustrated envelopes to proclaim their interests. They cost the buyer only a few cents, they preached a message, and they made his letters more interesting to the recipient.

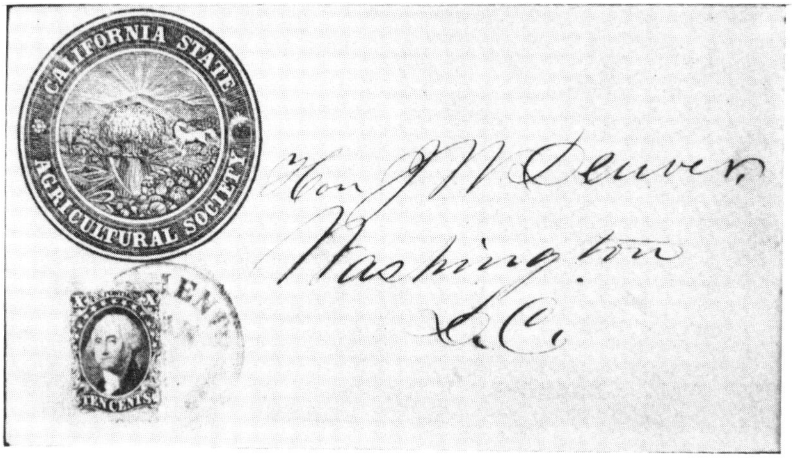

California State Agricultural Society seal used from Sacramento to Washington, D.C.

California Escutcheon on cover to R.I., carried to NY outside mails and there mailed by "Politeness of I.G. French."

Temperance envelope published by Barber & Baker, Sacramento. Weaverville oval postmark on 1855 paid stampless to Boston.

Advertising Covers

Enterprising merchants and business institutions, including the steamship lines, were quick to see that illustrated envelopes gave them free advertising. Usually the advertisement was a corner card or a printed frank. Sometimes it was an "overall" in which the entire envelope was an advertisement, on which the sender superimposed the address.

Gardner House, unpaid stampless, frank not accepted.

Bassette House, with 1855 10¢ to Conn.

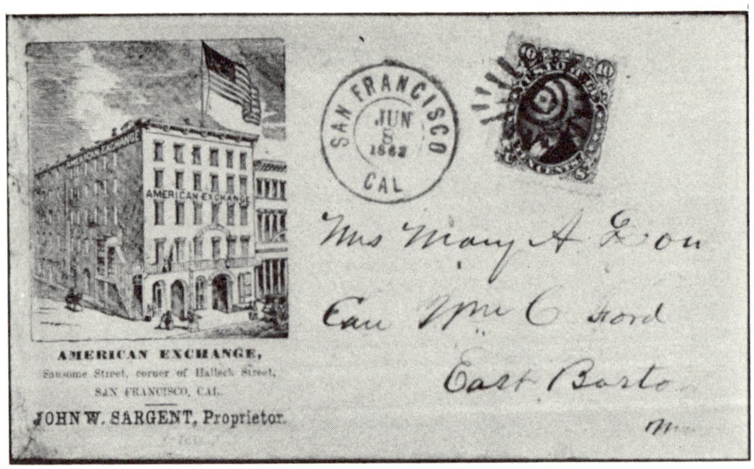

American Exchange, to Mass., 1862, with 1861 10¢.

Advertising corner cards of three SF hotels.

Covers advertising SF newspaper, to Auburn and forwarded PAID/3; shipping and commercial agent carried by WF Stockn. Messenger to Stockton; engraver from Sonora to N.J.; a school; a stationer.

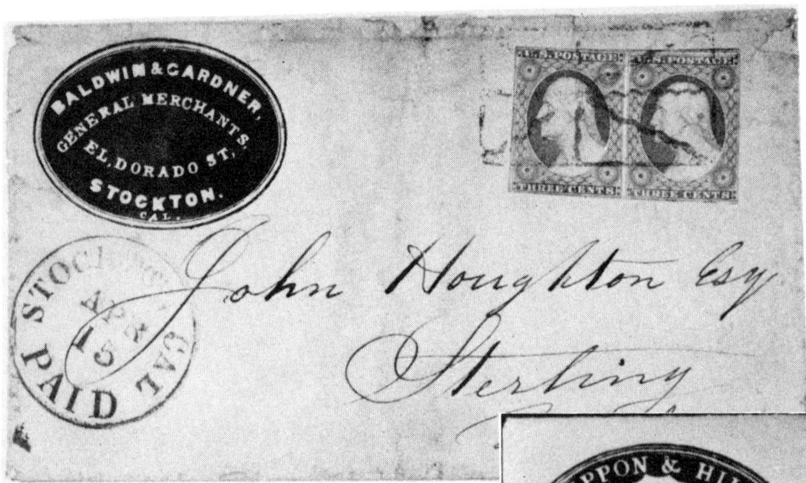

Corner card ad of Stockton merchant used to Mass. with pair of 3¢ 1851, "web" cancel. PAID in postmark.

Hardware and buggy ad on cover from Sacramento to Conn.

Advertising cover of San Luis Obispo wine dealer.

Allover advertising cover of Stockton bookseller.

Chapter 9

California Postal Markings

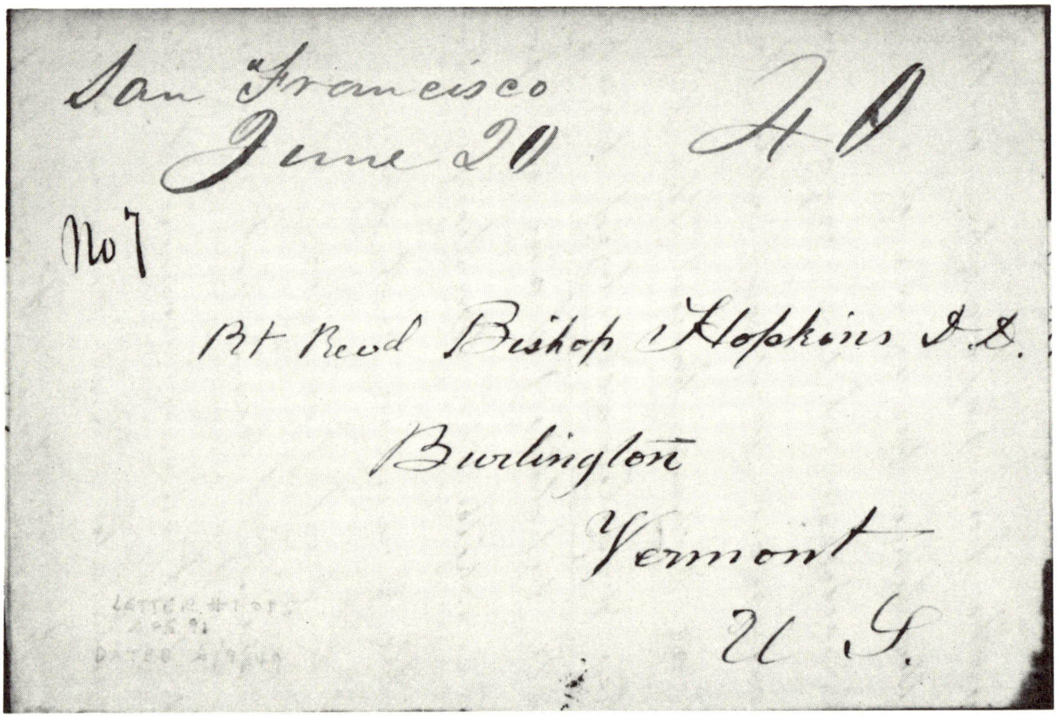

Two SF Letters carried east in the June 20, 1849, contract mail. The upper, datelined June 9, received a manuscript postmark dated June 20 (the sailing date) and was rated 40¢ to Vermont when it entered the SF post office. The lower letter, datelined June 19 and written by Peter Burnett, first governor of California, also received the manuscript "June 20" date. However, in the interim the straightline SF device had arrived; this is the first occasion of its use. At St. Louis the letter was forwarded to Liberty, Mo., for 5¢ additional.

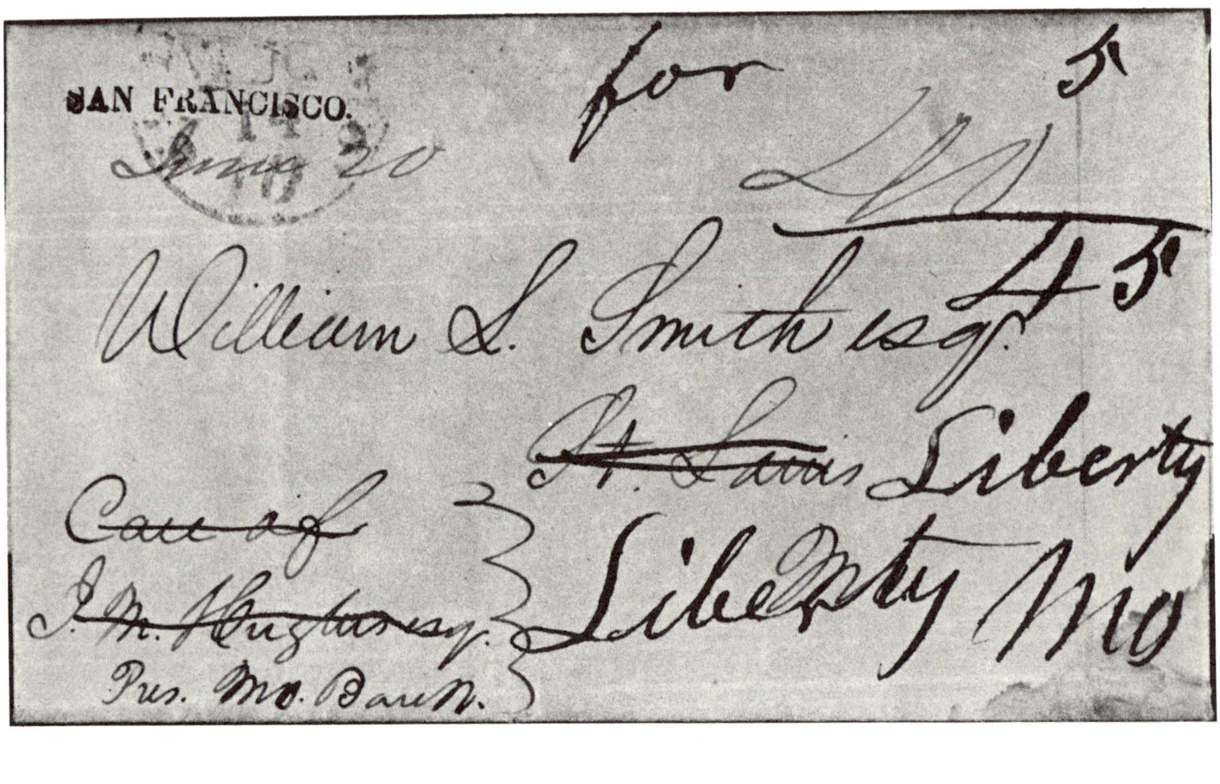

California Postal Markings

Prior to 28 February 1849 when William Van Voorhies arrived in San Francisco as the Special Agent of the U.S. Post Office Department for the purpose of establishing a postal system on the west coast, mail from California generally bore no official postal markings except those of the eastern post offices which were placed on ship letters arriving from the west coast. However, by the end of July 1851, 52 post offices had been officially confirmed within the state and letters from California now carried markings indicating postage rates and towns of origin, and in many cases the handstamps or manuscript markings of private agencies which supplemented mail distribution.

Letters of this period were rated for postage under the provisions of the Acts of 3 March 1847 and 14 August 1848, relating to the Post Office: 40¢ for a single letter coast to coast and 12½¢ for intra-California mail and letters between California and Oregon. The postage charge was indicated on the cover by a manuscript notation, by a handstamp, or by

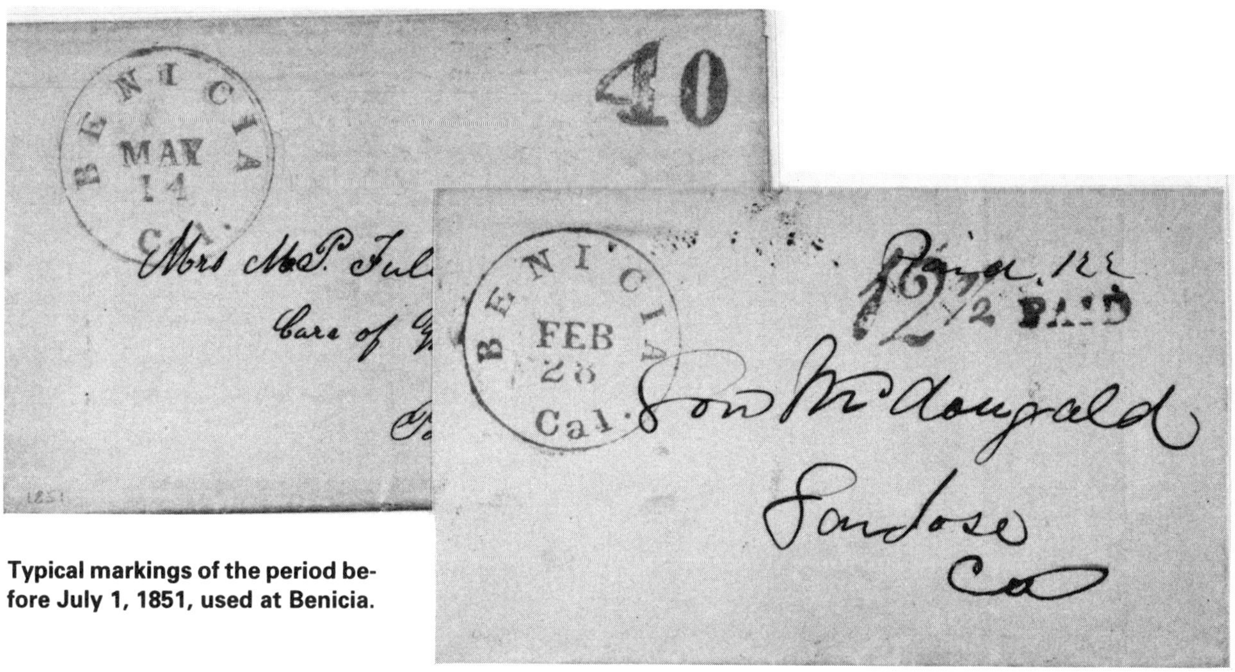

Typical markings of the period before July 1, 1851, used at Benicia.

Postmarks of some towns with offices established before July 1, 1851: Sacramento oval and circle with 12½ and 40 handstamped rates; Stockton circles with oval rate stamps PAID 12½ (last day of 12½¢ rate) and 40 (unpaid); SF circle and boxed 80 double rate; Monterey handstruck 25 for double rate, and 12½ on legal papers.

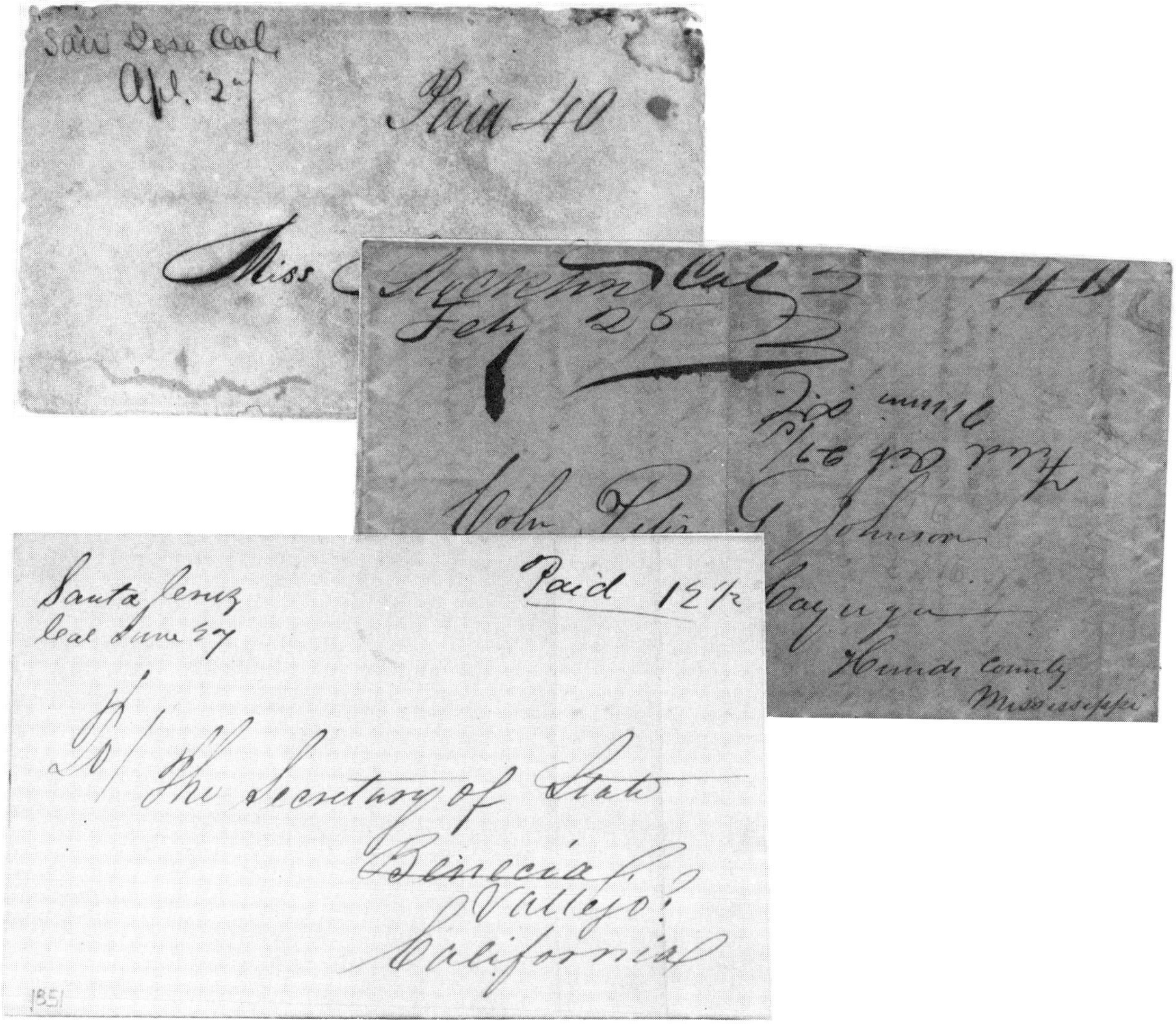

Ms. postmarks of early post offices: San Jose "Paid 40" to Texas; Stockton, Feby 25 (1850), unpaid to Mississippi; Santa Cruz, June 27, 1851, "Paid 12½" to Vallejo.

a figure within the postmark. At the post office where a letter entered the official postal system, the postmaster applied his townmark which incorporated the town and state names, and the month and day dates. Depending on the individual office, the device showed a variety of forms: a dated manuscript notation, a straightline handstamp, an oval handstamp or the more familiar circular postmark.

When the Act of 3 March 1851, effective 30 June 1851, was passed, only 21 post offices had been officially confirmed for all of California. Prior to 1 July 1851 these post offices had all used the 40¢ and 12½¢ rates and their multiples as postage rate markings. These 21 offices were:

- Benicia
- Culloma (Coloma)
- Fremont
- Haydenville
- Junction
- Los Angeles
- Mission San Jose
- Monterey
- Napa
- Nevada City
- Placerville

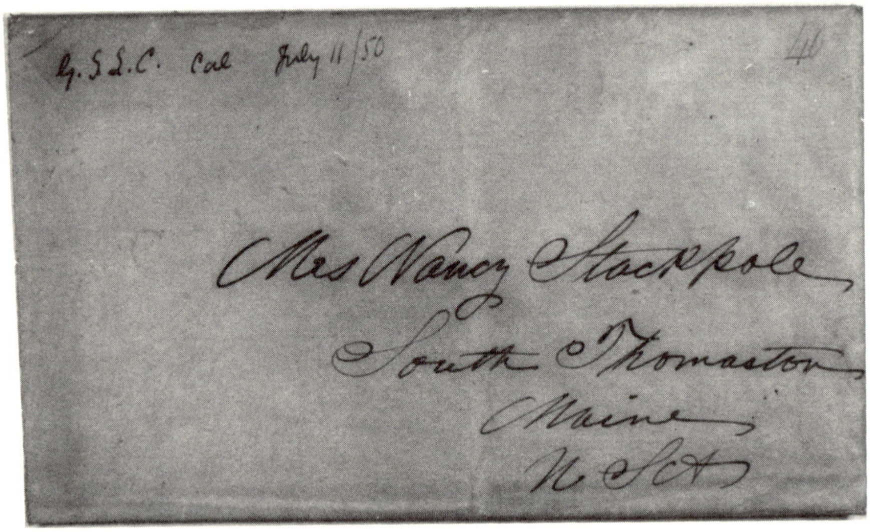

Ms. postmark "G.S.L.C. Cal July 11/50" — Greater Salt Lake City, when Utah was still included in California during quasi-territorial period (1 Feb. 1848-8 Sep. 1850).

Sacramento (City)
San Diego
San Francisco
San Jose
Santa Barbara
Santa Cruz
Sonoma
Stockton
Vernon
Weaverville

However, the postal historian and collector must note that during July 1851 31 more post offices were confirmed by Washington. Since there sometimes was a delay of up to six months between an appointment by

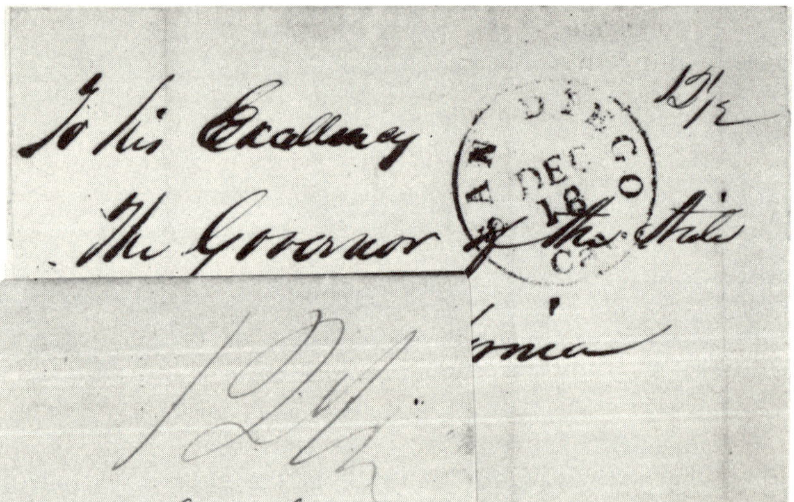

Early intrastate covers from San Diego and Culloma with ms. 12½.

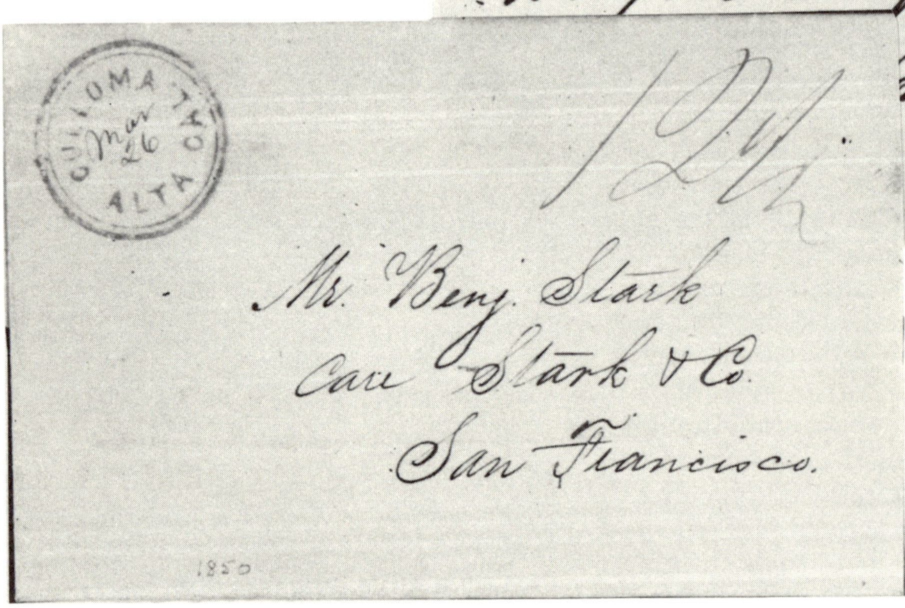

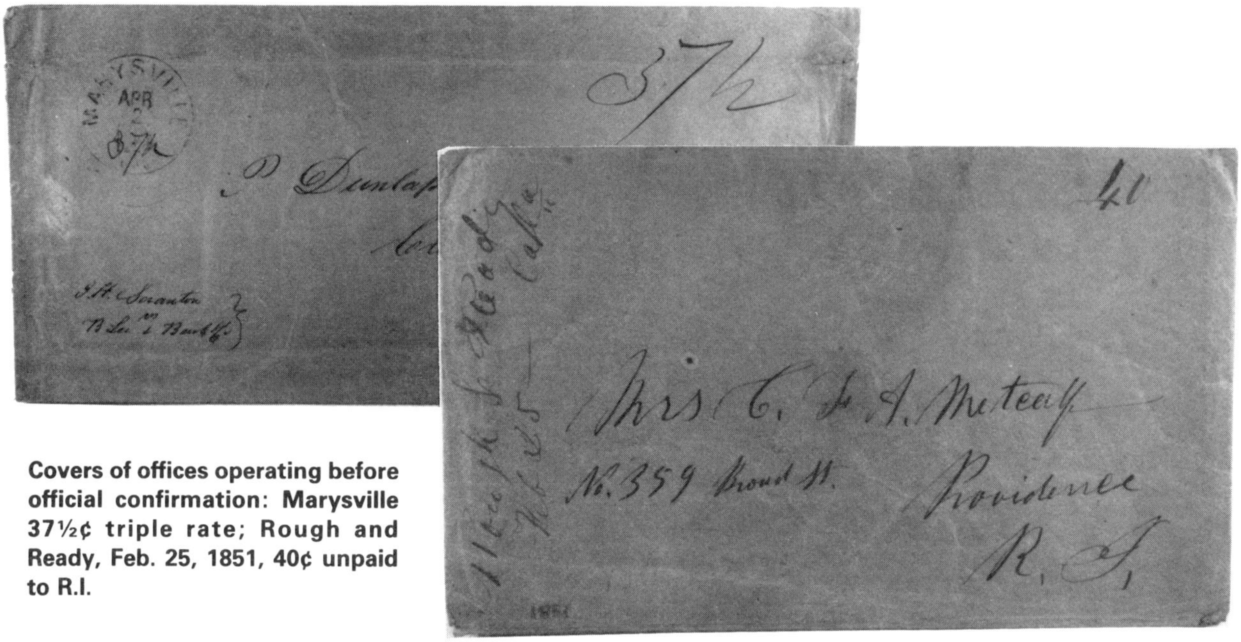

Covers of offices operating before official confirmation: Marysville 37½¢ triple rate; Rough and Ready, Feb. 25, 1851, 40¢ unpaid to R.I.

the California Postal Agent and official confirmation in Washington, D.C., many post offices actually functioned during this interim, so it is possible that covers marked at the 40¢ and 12½¢ rates may be found from some of the following offices. Those post offices where operation prior to 1 July 1851 has already been confirmed by covers showing 40¢ or 12½¢ rates are denoted by an asterisk.

Bidwells Bar*
Centerville
Double Springs
Empire
Graysonville
Green Springs
Hamilton
Horr's Ranch
Jackson
Knights Ferry
Livermore Ranch
Louisville*
Mariposa
Martinez
Marysville*

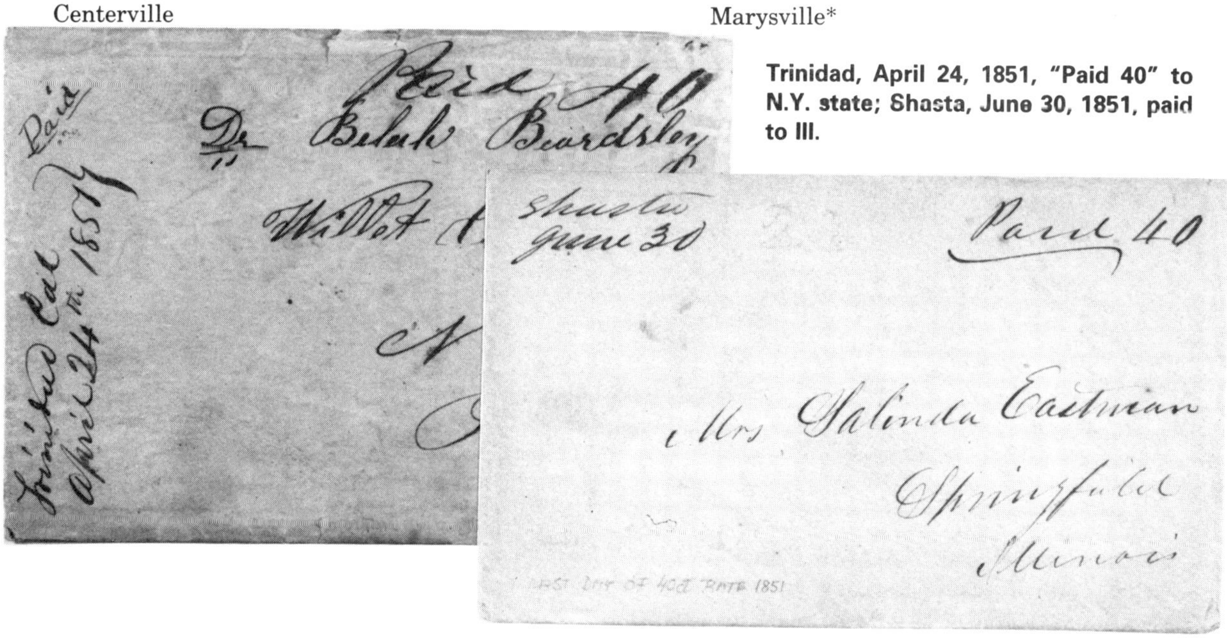

Trinidad, April 24, 1851, "Paid 40" to N.Y. state; Shasta, June 30, 1851, paid to Ill.

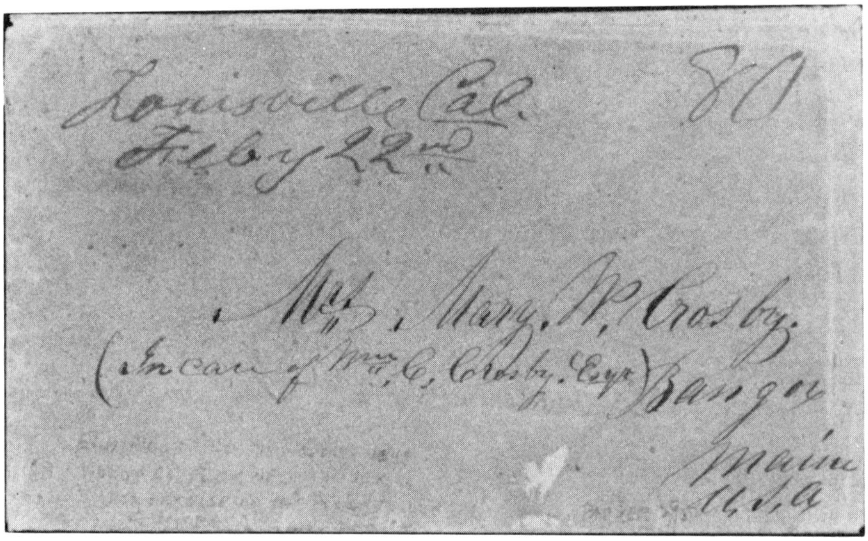

Louisville Cal./Feby 22nd ms. postmark on unpaid double letter to Maine.

Mokelumne Hill
Mountain Inn
Nicolaus
Oak Spring*
Rough and Ready*
San Joaquin
San Juan
San Luis Obispo
Santa Clara*
Shasta*
Sonora*
Trinidad*
Tuolumne City
Vallejo
Woods Diggings*
Yuba

Handstamped Townmarks

The Post Office Department did not begin supplying all postmasters with today's familiar dated handstamp until 1884. The *Postal Laws and Regulations (PL&R)* of 1847 had stated that handstamps would be furnished to post offices which produced $300 or more per annum in revenue. In 1852 a subsequent regulation provided that postmark instruments would be available from firms with government contracts; that the $300 revenue requirement remained for metal date-

Unusual double circles from Alvarado with 10¢ 1857 and Crescent City on 10¢ Nesbitt, both to Mass.

Non-standard circles from Plum Valley, Yreka, Martinez, Culloma, Big Bar, and Bear Valley.

stamps, but that post offices with revenues of $200 were eligible to receive wooden handstamps. Again in 1855 a new regulation raised the revenue required for receiving a metal handstamp to $500, but lowered that for a wooden one to $100. The postmaster could obtain the devices from government suppliers or other authorized sources and charge the cost to his operating expense allowance.

Most of the standard circular townmarks were of a common style: a single circle with a diameter of 29 to 33 mm., containing the town name, the month and day, and the state abbreviation. There were numerous variations of this basic type, some quite elaborate. The use of a year date was not general until 1855 except for some California straightline postmarks.

Many of the California post offices generated insufficient postage revenue to be eli-

Advertisement for postmark instrument.

Irregular circle postmarks of Murphy's, Mokelumne Hill (both intrastate), and Rough and Ready, unpaid to Me.

Oversize circles from Michigan Bluff, Gibsonville, and Michigan Bar; Collin style postmarks from Patterson, Bondville, and Rattlesnake.

Small plain circles from Maxwell's Creek, with 10¢ 1855 to NY; Gold Run, Callahan's Ranch, and Petaluma, all with the 3¢ 1861, the last on comic patriotic. Also Diamond Springs double circle, paid 6 to NY.

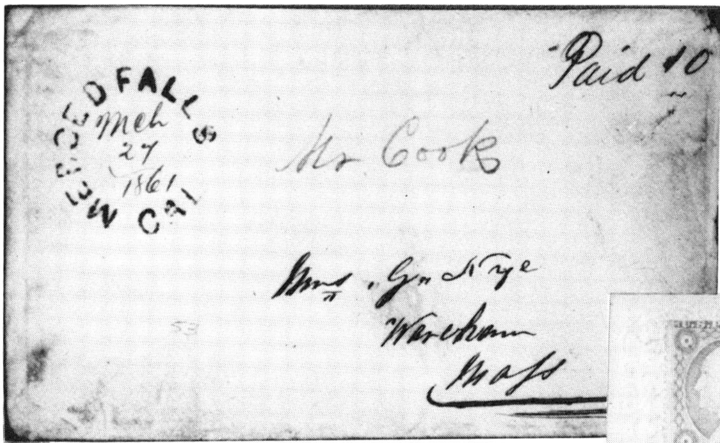

Rimless circle of Merced Falls on 1861 paid cover to Mass.

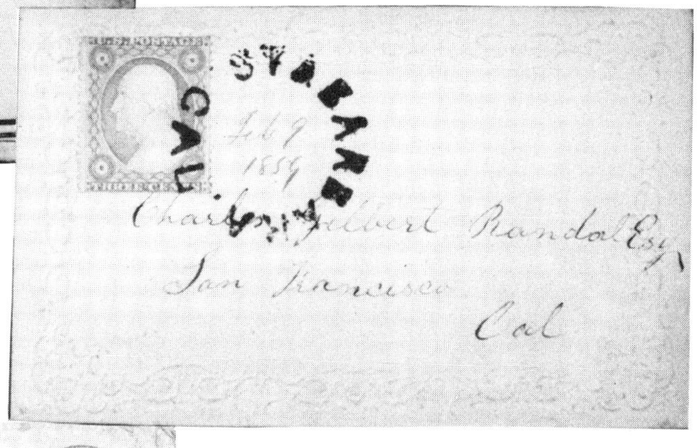

Rimless STA BARBARA on 1859 ladies' envelope with 3¢ 1857.

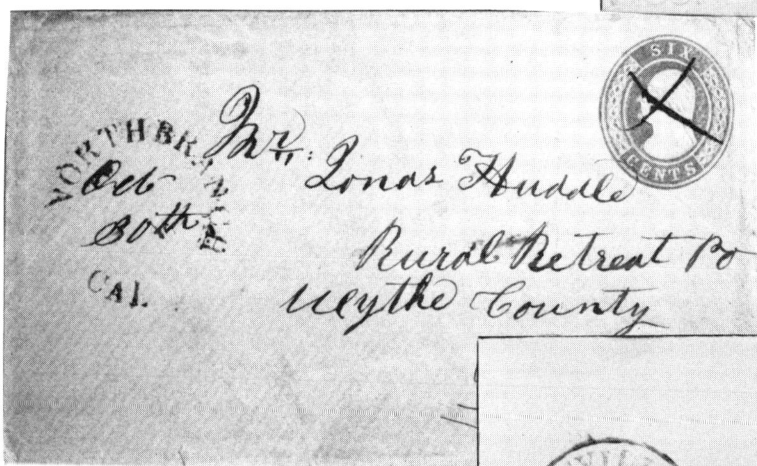

North Branch, no outer rim, on 6¢ Nesbitt to Va.

Marysville PAID/BY STAMPS and crossroads on 10¢ 1857.

gible for the issuance of a stock handstamp. While some postmasters purchased the stock styles from manufacturers who began advertising in local newspapers in 1851, others purchased individualistic devices. Still others were innovative and created their own handstamps from handcarved wood or cork. A Marysville postmaster purchased a unique device which read "Marysville Paid By Stamps."

Placerville postmark with 5, to Seattle Nov. 25, 1854. Whether this letter was actually carried through the Carson Valley by "Snowshoe" Thompson cannot be determined.

Placerville postmark incorporating 6 PAID on stampless letter to Ohio.

SHASTA/10 CAL. rimless circle indicating 10¢ due on folded letter to Vt.

Postmarks incorporating amounts due or paid.

From the introduction of circular dated townmarks, a number of California postmasters used a device which included not only the town name and month and day date but also a numeral indicating the postage rate. Later, even after payment by postage stamp was required, a few postmasters redundantly continued to handstamp covers with the ob-

Postmarks showing postage amounts: Culloma 80 double rate and curved PAID to Nantucket; Marysville 40 due to Mass.; Nevada City 3 PAID redundantly tying 3¢ 1851 to Yuba City; SF PAID/6 on 1853 letter to NY "Pr Str. Oregon."

solete devices.

Postmasters who did not qualify for free handstamps from the Post Office Department and who did not individually purchase their own supply simply wrote in ink the name of the post office and the date, thus applying a manuscript postmark. A few postmasters had a local printer make up a stamp from newspaper type — straightline postmarks — and used these until they purchased or were issued more standard devices. Again, fires were a common disaster and handstamps would be destroyed, or a replaced postmaster would take his with him, and the sequence would begin once more.

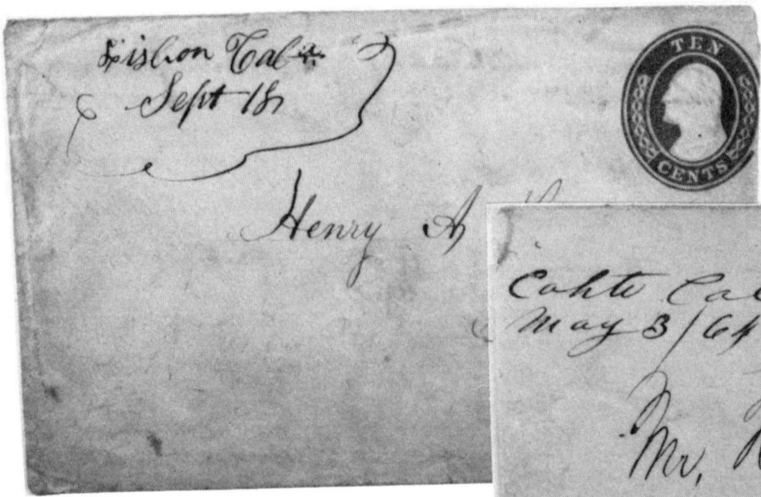

Ms. Lisbon on 10¢ Nesbitt to Conn.

Ms. Cahto to Mass. with 3¢ 1861, 1864.

Ms. Illinoistown on 6¢ Nesbitt to Pa.

Ms. Long Bar to Maine with 10¢ 1857, Oct. 18, 1858.

Moore's Flat to Pa. with margin copy of 10¢ 1855.

Greenwood ms. postmark on stampless PAID/6 to Me.

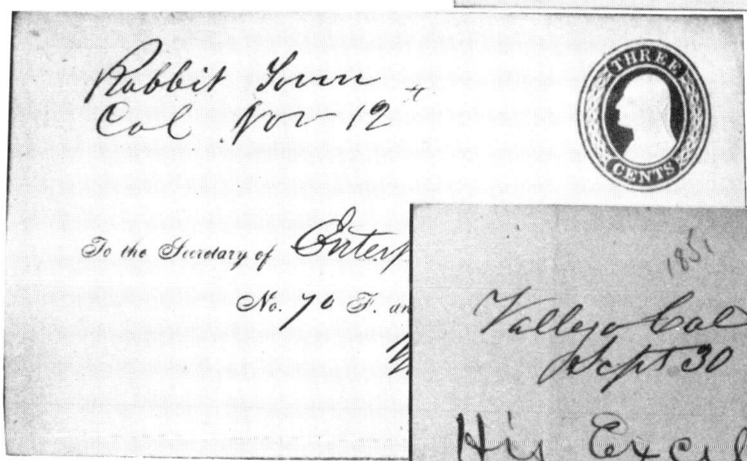

Rabbit Town on 3¢ Nesbitt to Yuba City.

Placerville double circle to Vallejo and ms. Vallejo forwarding postmark, 1851.

Manuscript Postmarks

During the first two decades of California's postal history, more than 400 post offices used manuscript postmarks. Some of the towns where they were used were short-lived and their postmarks are among those often referred to as "ghost town" postmarks. As the population of California increased and mining camps and villages became established towns, however, postmasters obtained more conventional handstamps and the manuscript postmarks disappeared, to become a

Straightline postmarks of Sacramento, with date and PAID/40 on 1848 cover to Iowa; Placerville to SF; two varieties of Sonora straightline: in one line on unpaid letter to SF, there advertised and forwarded to Md., for total of 16¢ due; in two lines with date, with pair 3¢ 1851 to Del.

collector's delight. (See Appendix 3 for a complete list of manuscript postmarks known to the author to have been used through 1869.) Shown here are some illustrative examples of their use on covers.

Straightline Postmarks

During the interval between use of the manuscript postmark and receipt of the more common circular instrument, approximately 27 California post offices may have used straightline handstamps. The straightlines were usually newspaper type fastened to a wooden block. They could be made up in moments in any newspaper or printer's office in town. It is perhaps surprising that so few were used, and disappointing also are the poor strikes of many of them on covers.

Town	*Date*	*Types*
Alameda*	1866	1
Angels/Cal	1853-54	1
Benicia Cal	1849	2
Carrollton*	1856	1
Columbia/Cal	1853-54	3
Eureka/Sierra Co.	1860	1

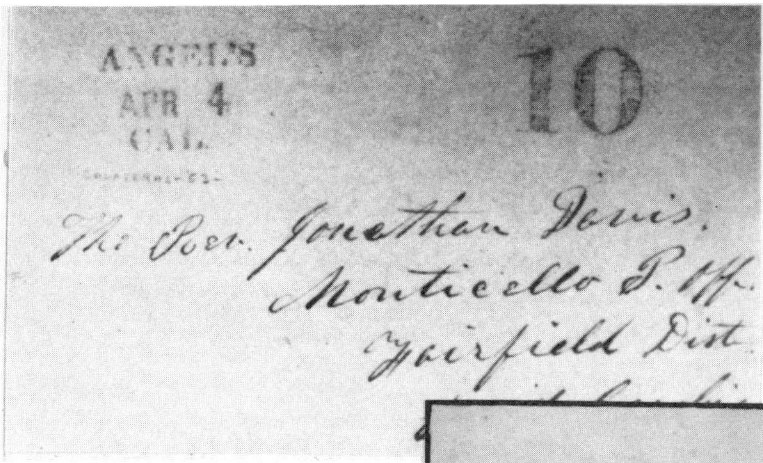

ANGEL'S/APR 4/CAL on unpaid cover to South Carolina.

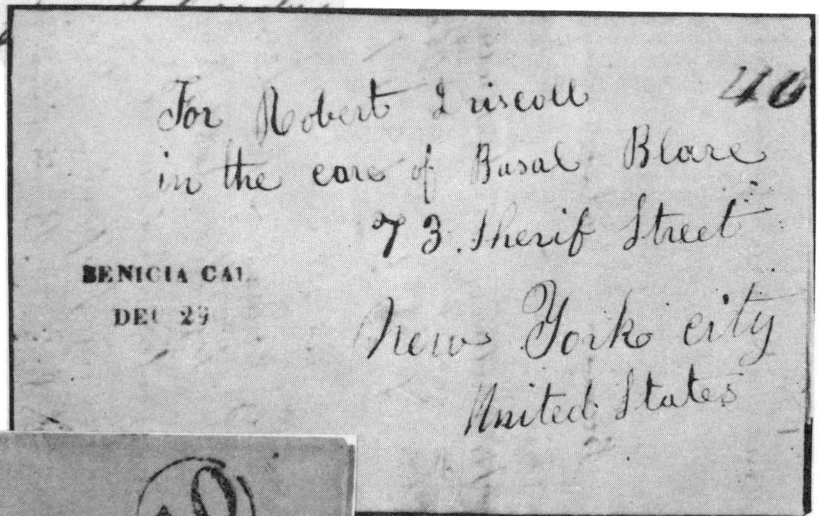

BENICIA CAL to NY, 40 due.

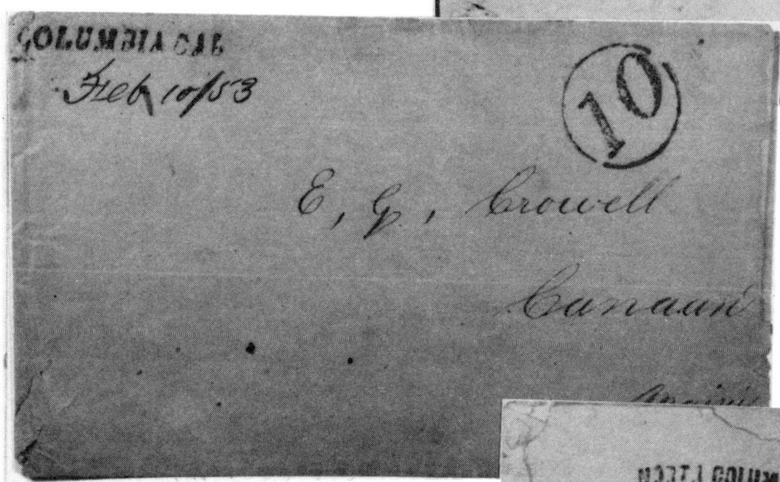

COLUMBIA CAL, Feb. 10/53 on cover to Me., rated 10 due.

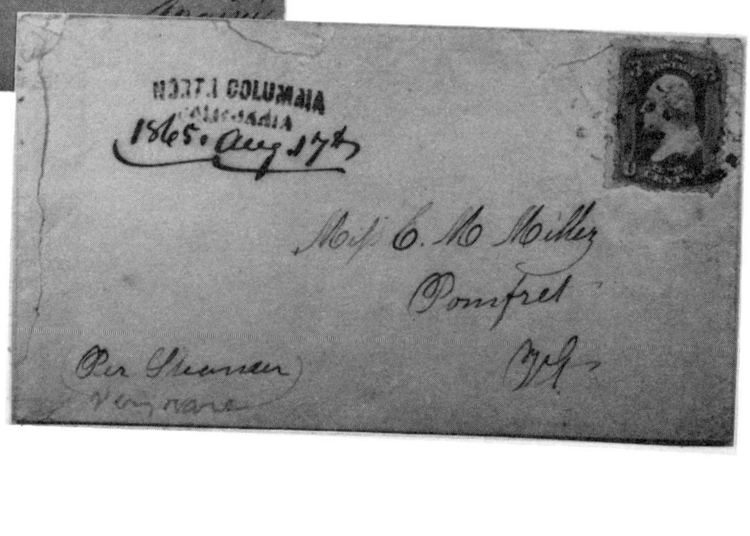

NORTH COLUMBIA/CALIFORNIA to Vt. on 1865 cover with 3¢ 1861.

Iowa Hill on telegraph company corner card to Sacramento, 1856; also cancelling stamps.

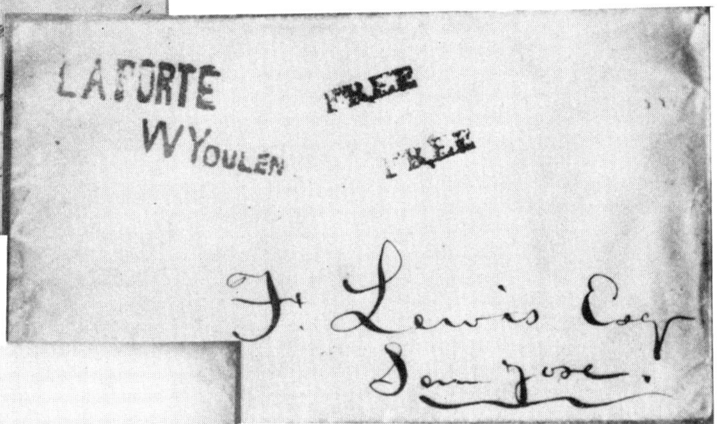

Laporte / W Youlen [pm] and FREE to San Jose.

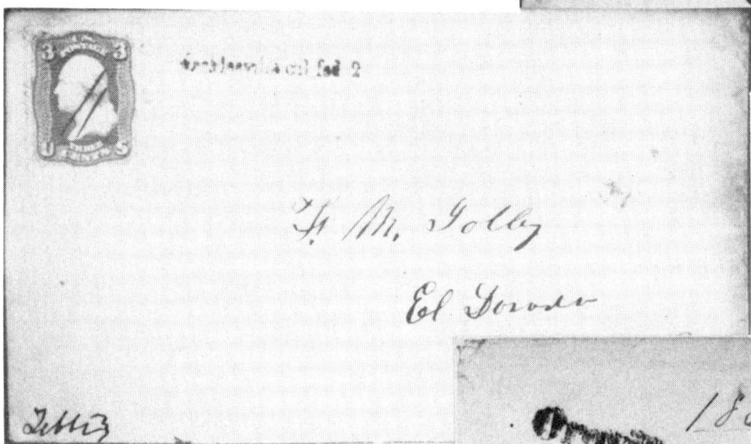

Markleeville cal in tiny lower case letters to El Dorado with 3¢ 1861.

Oroville/Cal May 31, 1856, on paid stampless to N.Y. state.

MOKELUMNE HILL CAL, on 1854 letter to N.H., PAID/6.

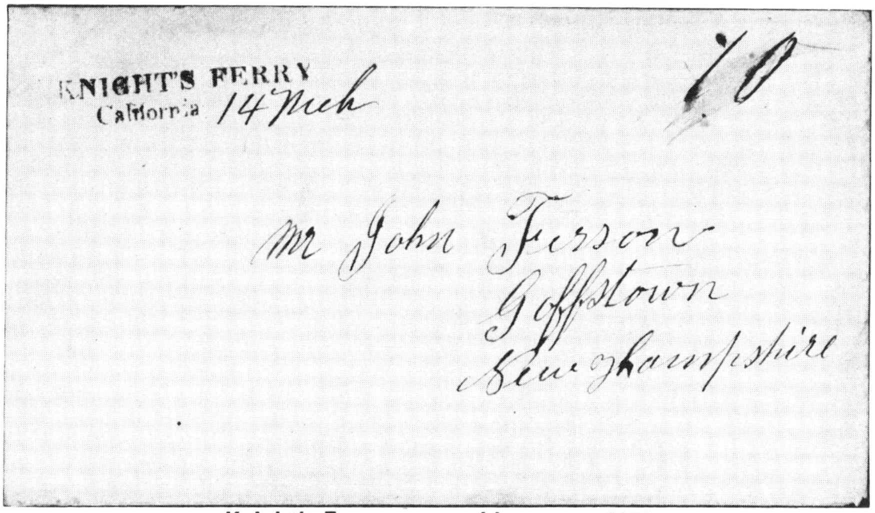

Knight's Ferry on unpaid cover to N.H.

Gibsonville	1855	1
Iowa Hill	1856	1
Knights Ferry	1851	1
La Porte	1857	2
Markleeville	1863	1
Marysville*	1851	1
Mokelumne Hill Cal	1854	2
Monterey, Cal	1850-51	1
Moores Flat*	1857	1
North Columbia	1865	1
Oroville, Cal*	1855	2
Placerville Cal*	1850	1
Sacramento	Aug 1849	2
San Andreas	1850	1
San Francisco	Jun-Aug 1849 Apr 5, 1865	2
San Leandro	1854	1
Shasta*	1851	2
Sonora	1851-52	4
Stockton*	1856	1
Todds Valley Cal	1856	1
Watsonville*	1853,55	2

Probably some straightlines were applied by express companies, not the post office. In the preceding list they are indicated by an asterisk (where two types are known, one

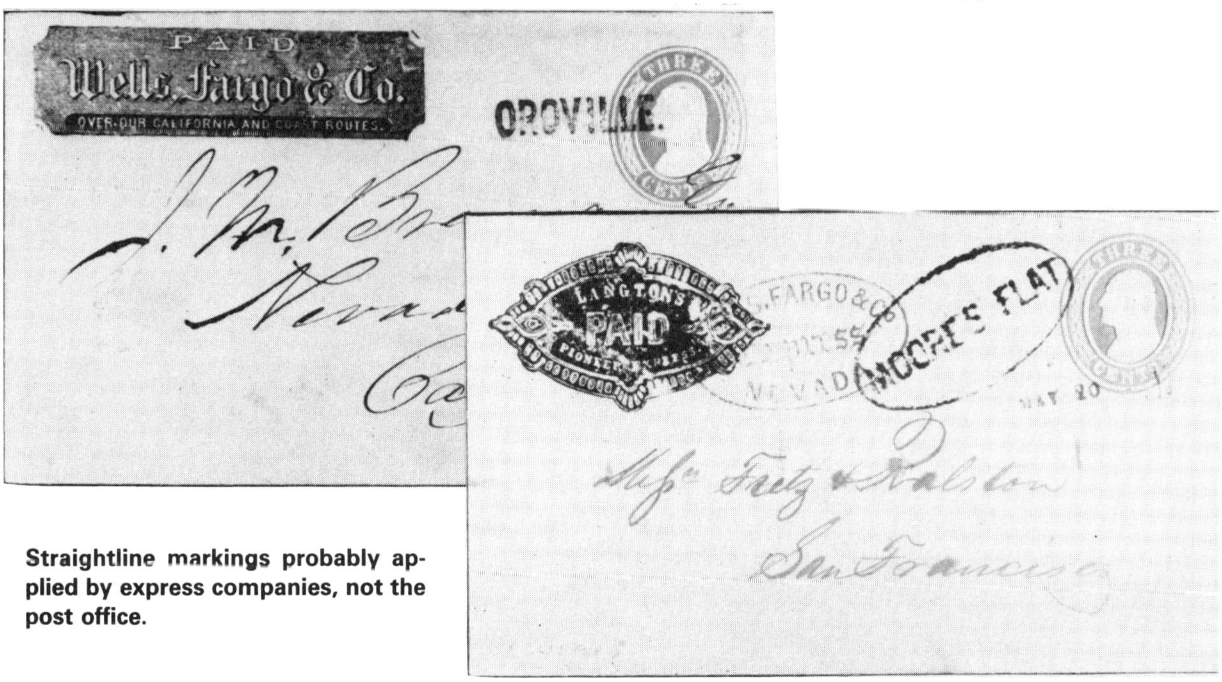

Straightline markings probably applied by express companies, not the post office.

is attributed to the post office, one to express use). Early postal historians accepted all of these straightlines as being post office townmarks, but recent research has raised valid questions about this conclusion. For example, the Carrollton straightline: a mining camp named Carrollton was established in 1855 on the American River, and Wells Fargo opened an office there in the same year. However, there is no evidence that a post office ever existed at Carrollton. A more reasonable conclusion is that this straightline was a handstamp applied by the Wells Fargo agent. The same reasoning can be applied to the following straightlines, all of which appear on Wells Fargo franked 3¢ 1853 Nesbitt entires: Marysville, Oroville, Shasta, Stockton, Placerville, and Alameda. All of these town names are in similar capital letters; few include the state abbreviation; none are dated. The same suspicion applies to the Watsonville straightline found on an American Express franked 3¢ entire, and to the North San Juan and the Moores Flat straightlines in an oval frame found on Langton Express franked 3¢ entires. The fact that these straightlines sometimes cancel the postage stamp does not resolve the question. Although a postal regulation forbade the express companies to apply their handstamps to postage stamps, this regulation was frequently ignored, especially on letters carried outside the mails. The problem is compounded in the case of Oroville, Placerville, and Alameda, for there are covers of these towns which bear valid postal straightlines.

Oval Postmarks

Scarcest of all are the pre-1869 oval postmarks. Some postmasters purchased or made them, perhaps for aesthetic reasons or because some manufacturer offered them at a reduced price. They had been fairly common in the east in the early 1800s and some had remained in use well into the 1840s. Their use in California, however, was comparatively short-lived. Succeeding postmasters substituted the more widely accepted circular types which had become standard. Only 16 oval postmarks are known at the present time to have been used in California prior to 1869:

Alleghany Cal Sierra County
Chico Cal
Clarksville Cal
Eureka Sierra Co.
Forbestown, Cal
Greenwood Cal
Hicksville Cal
MacDermott's Bridge Cal
Moore's Flat
Mormon Island Cala
North San Juan
Punta Arenas Cal
Sacramento Cal
Snelling's Ranche Cala
Uniontown H.B. Cal
Weaverville Cal

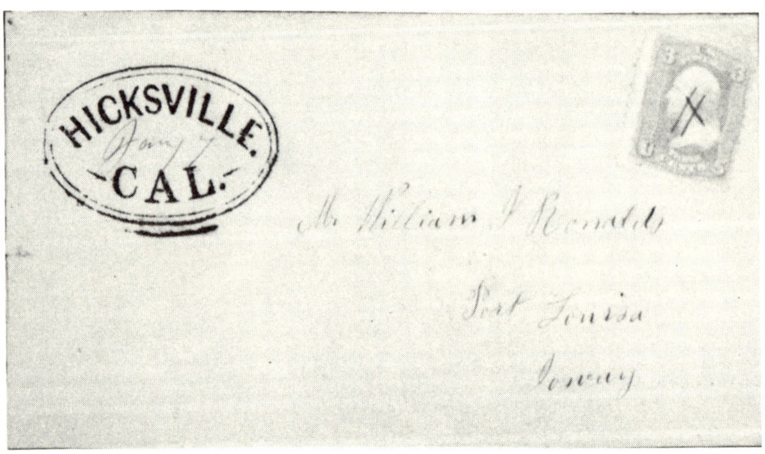

Hicksville double-lined oval on cover to Iowa with 3¢ 1861.

Alleghany Sierra County oval to Mass.; Clarksville, Apr 4/64 to Ill.; Chico double-lined oval on unpaid letter to Pa.; Eureka/Sierra Co. oval on Langton envelope; Forbestown rimless oval with margin pair of 3¢ 1851 to N.Y. state.

Mac Dermott's Bridge oval on 10¢ Nesbitt to N.H.; Mormon Island double oval to Mass. Sacramento oval postmark and PAID/40 to Iowa; Weaverville oval on unpaid letter to Oregon Ter.; Snelling's Ranche double oval to N.Y. state with 3¢ 1861.

Lake Bigler ms. postmark to Mass. with 10¢ 1861.

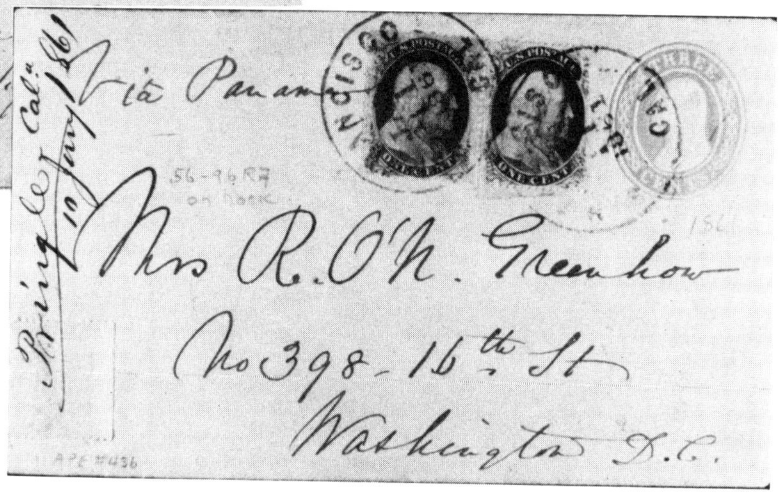

Pringle ms. on 1861 envelope to Washington; 5¢ added postage on back. Also SF postmark.

Phantom Townmarks

All postal historians and collectors eagerly seek covers of towns that no longer exist. Some of these are "phantom" towns, known only by manuscript notations — places for which no post office was ever authorized, and in some cases unknown even to the map makers. Perhaps a store owner or a stagestop proprietor received a letter for mailing, and decided to immortalize himself by marking the point of origin with a handwritten notation. At least 25 such covers are known, six of them from the 1849-69 period:

Corbets Ranch
Dagget's Run
How's Ranch
Job's Store
Lake Bigler
Pringle

Ghost Town Postmarks

During the period between 1849 and 1869, 236 of the post offices which had been established were permanently discontinued. Transient populations declined; new settlers (perhaps disliking the crudely descriptive names chosen by the rough miners) insisted on new names — some of the towns remain as lonely relics, crossroads now too small to justify a post office. For a brief period they were important enough to achieve the dignity of a postmark, then time and civilization passed them by. (See Appendix 4.)

All of these postmarks from now nonexistent towns are eagerly sought by collectors, but most highly prized are those on the true "ghost town" covers. These are from towns that have disappeared beneath the waters of man-made lakes, or have melted back into the earth, or exist only as weed-covered ruins, many known only as historical markers, all part of the debris of a dramatic past. Many were so short-lived that their only town identification is a manuscript postmark; others survived long enough to give us covers with handstamped postmarks. Covers from 62 of these 1849-69 ghost towns are known at present. (An asterisk signifies that a post

Postmarks of California ghost towns: Aurum City (vertical ms.) on paid stampless to Ill.; Fremont to Benicia at 12½¢ intrastate rate; Quartsburg circle on cover marked "Free"; Poverty Bar to N.Y. state with two 5¢ 1857; Reynolds Ferry on 1859 cover to Mok Hill with 3¢ 1857; Round Tent, 1859, on 10¢ Nesbitt to Me.; Sinks (Sink of Tehon) to SF; Virginia on 1862 cover to Me.

office under another name existed at the same site at a later date for a period of time. All have since gone out of existence.)

Acqueduct City (Amador)
Aurum City (El Dorado)
Bidwells Bar* (Butte)
Bottle Hill (El Dorado)
Butte City (Amador)
Cedarville (El Dorado)
Chuntown (Shasta)
Clay's Bar* (Calaveras)
Don Pedro's Bar (Tuolumne)
Elbow (Fresno)
Elderton (Shasta)
El Dorado Ranch* (El Dorado)
Eliza (Yuba)
Elk Horn (Contra Costa)
Eureka North (Sierra)
Ferry Point (Del Norte)
Fosters Bar* (Yuba)
Fremont (Yolo)
Fugitt* (San Joaquin)
Hamilton (Butte)
Illinois Mills (Tulare)
Jay Hawk (El Dorado)
Johnson's Ranch (Yuba)
Keeneysburgh (Tulare)
Kilna (Shasta)
Konigsberg* (Alpine)
Last Chance (Placer)
Locust Shade (San Joaquin)
Long Bar (Yuba)
Middletown (Shasta)
Mill Valley (Calaveras)
Minersville (Trinity)
Monoville (Mono)
Oro City* (Placer)
Ownesville (Mono)
Owsley's Bar (Yuba)
Pacific Home (Sonoma)
Petersburgh (Tulare)
Phillips Flat (Mariposa)
Pleasant Valley (Mariposa)
Plumas (Yuba)
Poverty Bar (Calaveras)
Quartzburg (Mariposa)
Rattlesnake Bar (Placer)
Red Dog (Nevada)
Reynolds Ferry (Calaveras)
Round Tent (Yuba)
Salsbury (Sacramento)
San Carlos*(Tulare)
Sarahville (Amador)
Sink (Los Angeles)
Soldiers Bridge (Lassen)
Split Rock (Mariposa)
Staples Ranch* (San Joaquin)
Stringtown (Butte)
Two Mile Bar (Calaveras)
Unionville (El Dorado)
Vernon (Sutter)
Viola (Sacramento)
Virginia (Placer)
Webster (Placer)
Willow Point (Yolo)

Napa City circle with two P's and after removal of first P.

Stamp Obliterators

From the first printing of postage stamps in 1847, the Post Office Department had insisted that all stamps placed on mail be defaced in some way to prevent their reuse by unethical persons. The majority of postmasters in small California post offices simply drew ink lines across the stamps, or sometimes used the townmark to obliterate the stamp until the latter practice was prohibited in 1860. Prior to 1869 postal regulations did not prescribe any standard "killer," so postmasters innovatively designed their own obliterating devices, often made from wood or cork, or purchased a metal obliterator that happened to be advertised in a newspaper or Post Office Department circular.

The most common obliterators were geometric designs: vertical or horizontal bars, framed and unframed dotted lines, square or oval or rectangular grids, and concentric circles or targets. Some were unusual: varieties of stars, rays radiating from a solid center, cogwheels, clam shells, spider-like figures, varieties of "crossroads," Masonic emblems, and patriotic designs. Various scholars, attempting to classify them, have created a variety of names for them. Frequently the same post office used several different designs, seemingly at the whim of the individual postmasters or postal clerks.

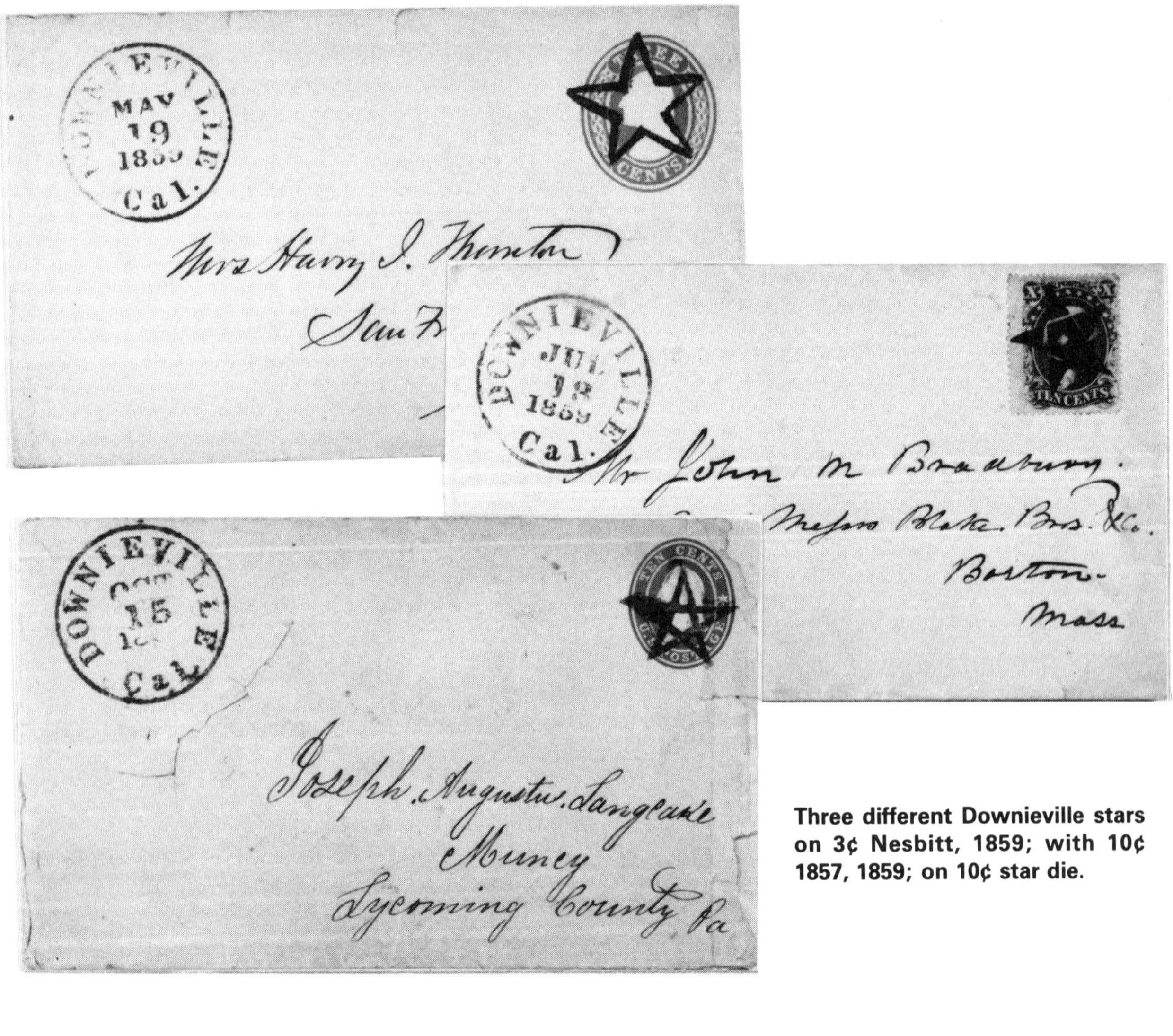

Three different Downieville stars on 3¢ Nesbitt, 1859; with 10¢ 1857, 1859; on 10¢ star die.

Distinctive cancels: Georgetown star and quartered cork with 1861 issue; Bidwell's Bar elongated grid on 10¢ Nesbitt; fancy cork oval from Mok Hill; Knight's Ferry circular grid; North Branch spider web on 10¢ Nesbitt; segmented cork from Rattlesnake; Timbuctoo quartered corks on 1862 cover.

A variety of Sacramento cancels: three types of quartered circles on 3¢ 1861; circled T on 3¢ envelope; petal cancel on 3¢ env.; leaf on 3¢ 1861; diagonal barred grid on 3¢ 1861; quartered cork and stylized leaf on 3¢ envelopes. Last three with ad corner cards.

Sonora shell cancel on pair 3¢ 1851 used to Mass., 1852, also straightline postmark.

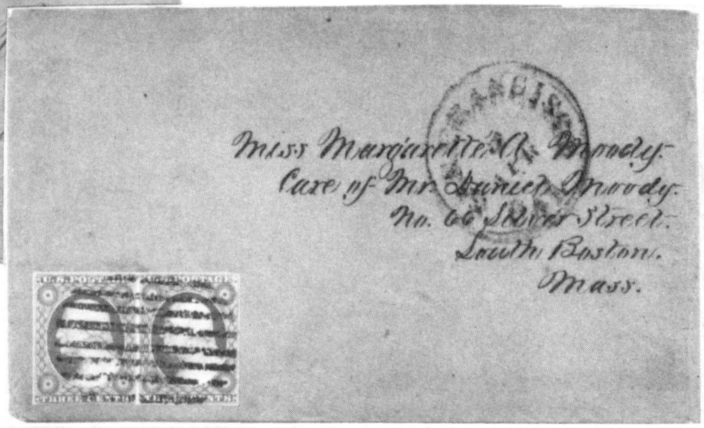

SF oversize grid (for two stamps) on pair 3¢ 1851 to Mass.

SF barred circle and truncated cork cancels to Mass. with 3¢ 1861; Vallejo pinwheel and Maltese type cross with 3¢ 1861; Sutter Creek grid of squares on double letter.

Rate numerals: Mok Hill PAID and 6 on 1854 cover to N.H.; SF PAID and 3 to Culloma; PAID 3 from Santa Barbara; Sacramento PAID and outlined 3.

Paid and Unpaid Markings

Prior to the general availability of postage stamps in California (19 September 1851), letters received a numeral marking, either handstamped or in manuscript, to indicate the correct postage rate. If the postage was prepaid in cash the cover was marked "Paid," again by handstamp or manuscript notation. The absence of the word "Paid" always meant that the postage was to be collected from the addressee.

The term "Due" was not generally adopted until after 1 April 1855, when prepayment was required. It was then used to indicate an amount which was due if the prepayment was not sufficient to cover the actual postage or if the letter had incurred an extra charge, such as a forwarding or advertising fee.

Nevertheless, prior to the Act of 3 March 1855 requiring mail to be prepaid by stamps

Different forms of "Paid 10": ms.; handstamped PAID and ms. 10; PAID/10; boxed PAID and 10; PAID in oval and 10 in oval; PAID/X.

Unusual 10 rate markings: X from Mok Hill; dotted circle from Sacramento; TEN from Auburn.

An attempt to use an envelope stamp cut from the envelope to make up the 10¢ rate to Ohio. Such practice was illegal, so the cover was rated "Due 3."

or money as of 1 April 1855 and with the further requirement effective on 1 January 1856 that all mail carry affixed postage stamps or be sent in stamped government envelopes, the preponderance of California mail was unpaid and did bear rating marks that indicated payment was required from the addressee. These requirements, of course, did not apply to letters carried outside the mails by the private express companies or by the local carriers of intracity mail.

If a totally unpaid letter was deposited in a post office after 1 April 1855, it was not forwarded as due mail, but was detained for postage. The postmaster notified the addressee by a post office form that a postage due letter was being held for him and would be held for a maximum of 30 days. If the addressee returned the required postage, the letter had the necessary stamps affixed and

An envelope mailed without postage to Watsonville, hence DETAINED FOR POSTAGE at SF on 19 Jan. The addressee was notified and sent a stamp or cash, a 3¢ stamp being applied on 28 Jan. and the letter sent on.

was sent on to the addressee. If no reply to the notice was received after 30 days, the letter was sent to the dead letter office. If, however, a letter was partially prepaid the required rate, it was sent on immediately to the addressee with a due marking indicating the balance to be collected on delivery.

Free Mail

Congressmen and certain governmental officials, including many postmasters, were granted the franking privilege. Such covers were marked "Free" in manuscript or by handstamp and were signed by the individual holding the privilege (together with the name of his office) when that individual was sending the letter. In most cases, such mail could also be sent free to an individual holding the franking privilege, in which case no authenticating signature was required. Covers prepaid with postage stamps are occasionally found which are also struck with a "Free" handstamp. This redundancy was accidental because of the carelessness of a postal clerk.

SF, two covers FREE on post office business; Sonora FREE with pm's frank; Marysville postmark containing FREE; also FREE in grid.

Circular Mail

Perhaps because of an ingrained belief that the "printed word" had special intrinsic value, printed circulars and newspapers could always be sent at more favorable postage rates than letters. The charge for circulars was based upon a single sheet, then folded, in a wrapper or an unsealed envelope, the rate being multiplied by the number of sheets enclosed. Prior to 1851 an unsealed circular required 3¢ postage per sheet. From 1 July 1851 to 30 September 1852 the rate for a circular weighing up to one ounce was established as follows:

Up to 500 miles	1¢ prepaid
500 to 1,500 miles	2¢ prepaid
1,500 to 2,500 miles	3¢ prepaid
2,500 to 3,500 miles	4¢ prepaid
over 3,500 miles	5¢ prepaid

Rates were doubled for unpaid circulars.

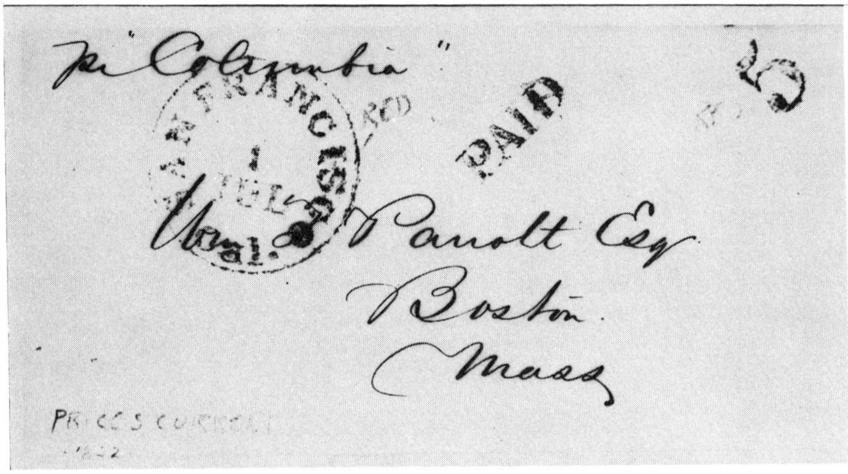

SF to Boston July 1, 1852: the scarce 5¢ prepaid circular rate for over 3,500 miles.

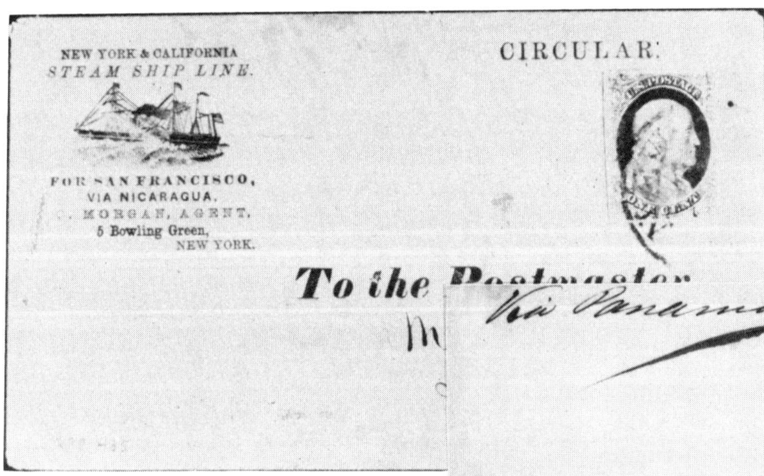

The 1¢ circular rate (after 1 Oct. 1852) paid by 1¢ 1851 on NY & California SS Line corner card.

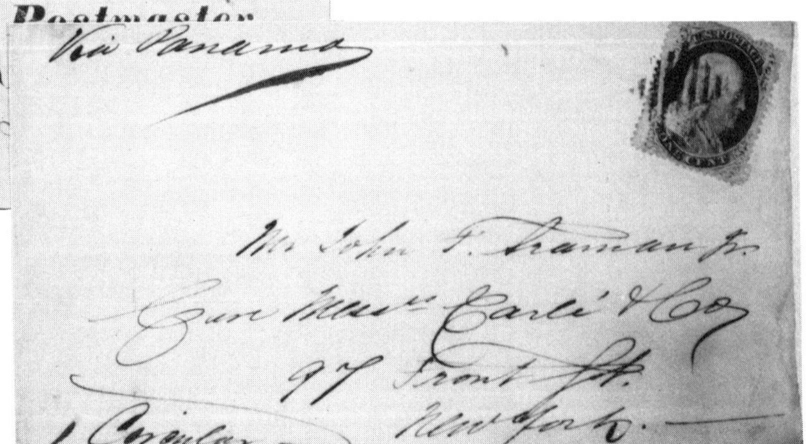

From California to NY "Via Panama," 1¢ 1857 paying "1 Circular."

On 1 October 1852 the factor of distance was eliminated and the postage on all printed circulars not exceeding 3 ounces was set at 1¢ plus 1¢ for each additional ounce, if prepaid. If a circular was sent collect, the rate was doubled. After 1 January 1857 collect mail was eliminated. If a circular was mailed in an unsealed envelope a notation "Circular" was frequently placed on the cover by the sender (often in manuscript) to distinguish it from letter mail.

Drop Mail

The term "drop mail" denotes letters intended for delivery to the addressee at the post office where they were mailed. Drop letters did not enter the mail system and, prior to 3 April 1860, were not delivered from the post office to a street address. From 1 July 1851 to 27 February 1861, the drop rate was 1¢. Drop letters could be deposited in the post office either prepaid or unpaid. After 3 April 1860,

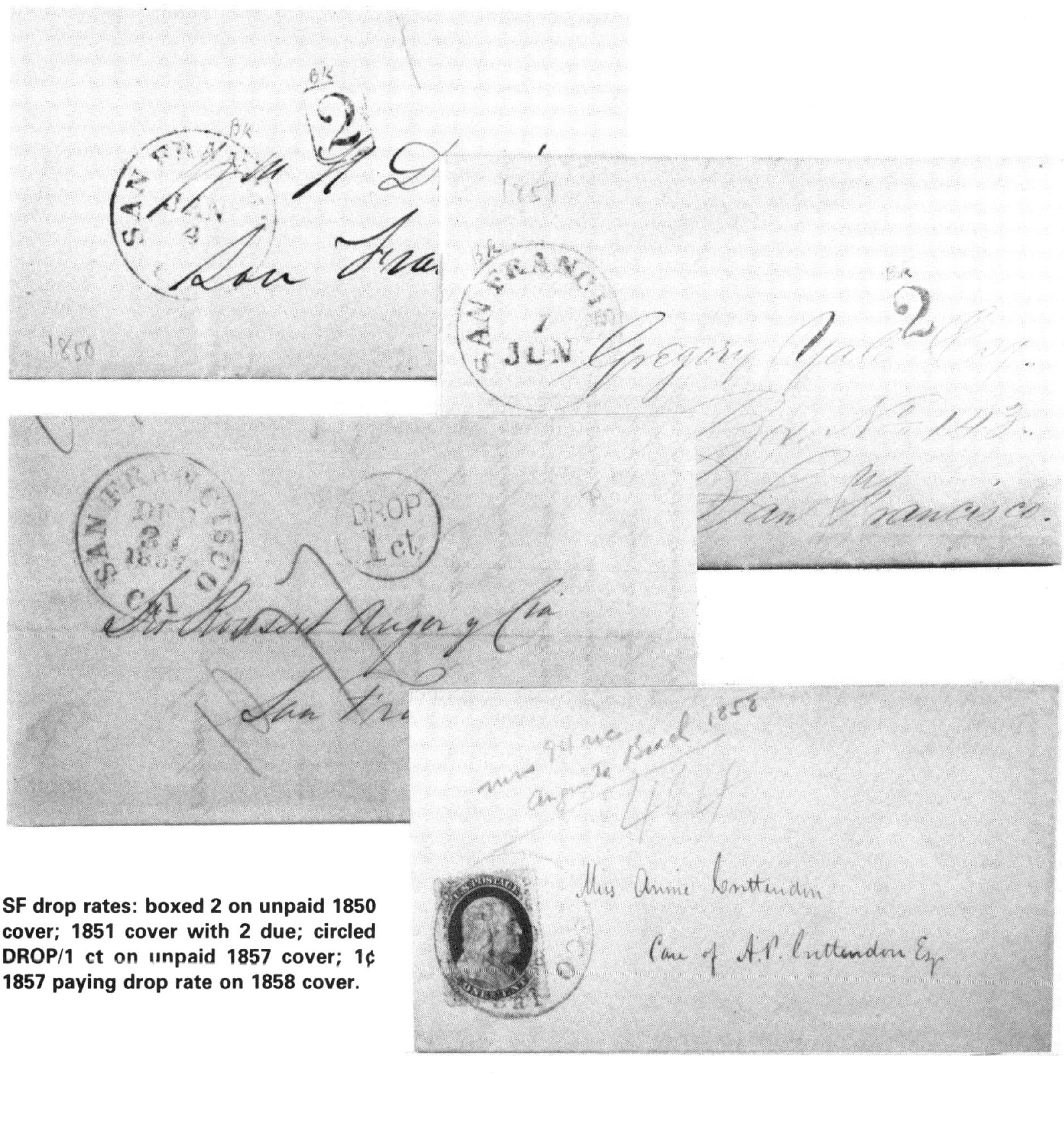

SF drop rates: boxed 2 on unpaid 1850 cover; 1851 cover with 2 due; circled DROP/1 ct on unpaid 1857 cover; 1¢ 1857 paying drop rate on 1858 cover.

The 2¢ drop rate paid by Black-jack on 1864 cover.

the drop letter rate was waived if the letter was delivered to a street address and if the carrier rate was paid. San Francisco was the only California town that had a government carrier service at this time but this service was very limited because of the severe competition of private local carriers through 1864. Prepayment by postage stamp was required by the Act of 27 February 1861. Effective 1 July 1863 the drop letter rate was raised to 2¢. It was this charge that necessitated the printing of the 2¢ "Black Jack."

Registered Mail

The first U.S. Post Office Department system of registering mail was established on 1 July 1855. The registry fee was 5¢ in addition to the regular postage, payable in cash. The use of postage stamps to pay the registry fee was not required until 1 June 1867. Interestingly, in the beginning postmasters were forbidden to indicate by any notation that the letter contained valuable enclosures.

On 3 March 1863 (effective 1 July) the

Ms. registration markings from Rattlesnake with 10¢ 1857 to Me.

Gibsonville ms. "Registered 5, PAID" (in cash) with 10¢ 1857 to Me. and there forwarded DUE/3.

registration fee was raised to 20¢. Not until 1857 did a postal regulation require that all registered letters be numbered, thus establishing an accounting system for such mail, but there was no indemnification in case of loss. Seemingly, the only value of registration was to ensure more careful handling of this mail, and after numbering was instituted, it allowed the tracing of a lost letter to the point of its disappearance. With no indemnification and no system of insuring mail yet instituted, it is not surprising that few registered letters are known from California between 1849 and 1869.

Placerville to Switzerland with straightline REGISTERED; overpaid by 24¢ stamp.

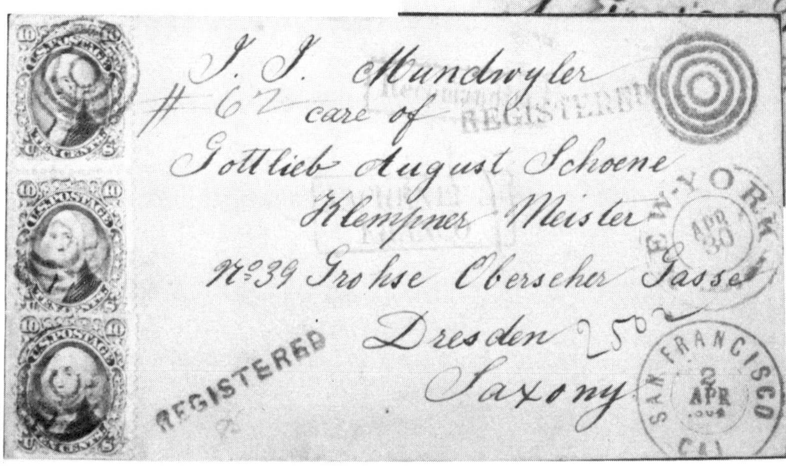

SF to Saxony with straightline REGISTERED (upper is NY) on 1864 cover by Prussian Closed Mail.

SF to NY, 1865, three Blackjacks paying double rate on Pacific Stage & Express Co. corner card.

Stockton ADVERTISED in oval on cover from Missouri, due 40¢; 4¢ added for advertising.

Sacramento boxed ADVERTISED on envelope from Mud Springs. No amount shown.

SF ADVERTISED-2 on letter from Owego, N.Y.; also NOT CALLED FOR.

ADVERTISED 1 and NOT CALLED FOR on envelope from Sacramento to SF.

Advertised Mail

As noted earlier, prior to the initiation of carrier service to street addresses, all mail had to be picked up by an addressee at the post office. If a letter was not called for within a specified period of time, the postmaster was required to advertise the name of the addressee in a local newspaper. Prior to July 1851 the fee was a maximum of 4¢. After that it was lowered to 1¢. The postmaster was authorized to pay the newspaper this fee for each name published. The 1¢ fee was charged to the addressee to reimburse this cost of ad-

Virginia City, March 28, 1864, to SF. On April 6 the cover was backstamped SAN FRANCISCO CAL/ ADVERTISED and "Adv 1" noted on front. Someone supplied the correct address and the letter was forwarded at 3¢ on May 7 to Long Bar, total 4¢ due.

vertising. A notation was made on the cover, in manuscript or handstamp, that the letter had been advertised and this marking was frequently accompanied by the amount of the advertising fee. These markings were accounting marks for the post office and, when the fee was noted, constituted a due marking payable by the addressee. If such a letter was not called for within the required period of time (a period which varied according to the income of the advertising post office, but which rarely exceeded six weeks) the unclaimed letter was sent to the Dead Letter Office. In this event the postmaster received reimbursement for the advertising fees on his accounts with the Post Office Department. Until a Dead Letter Office was established in San Francisco by the Act of 24 July 1854 to serve California, Oregon, and Washington territory, all dead letters were returned to the Dead Letter Office in Washington.

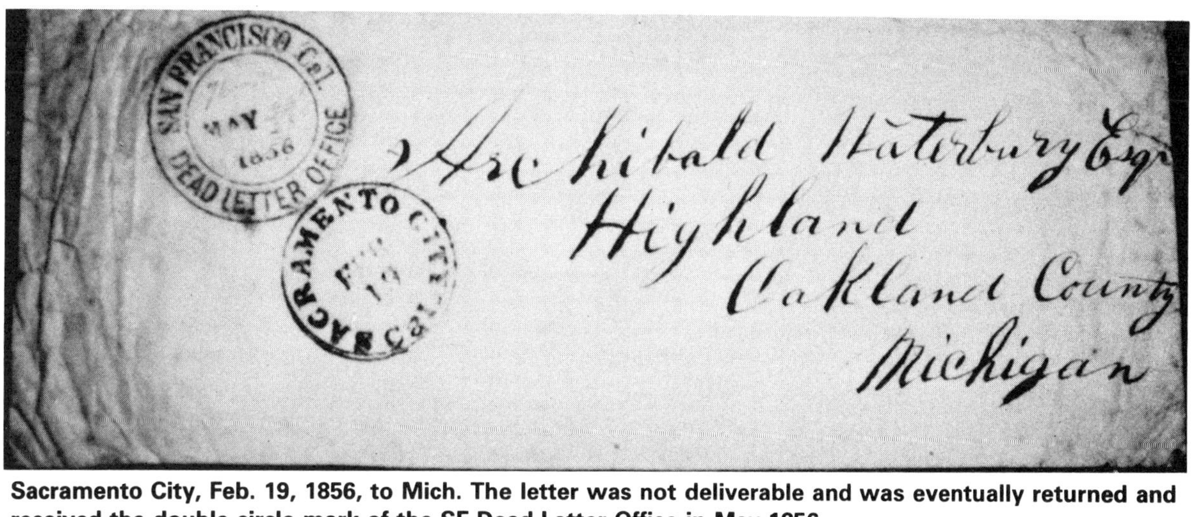

Sacramento City, Feb. 19, 1856, to Mich. The letter was not deliverable and was eventually returned and received the double circle mark of the SF Dead Letter Office in May 1856.

SF to San Jose, unpaid 12½, forwarded 12½ to Vallejo, total 25¢ due.

Wolcottville, Conn., to Sacramento, unpaid 40, forwarded 12½ to Nevada City, total 52½¢ due.

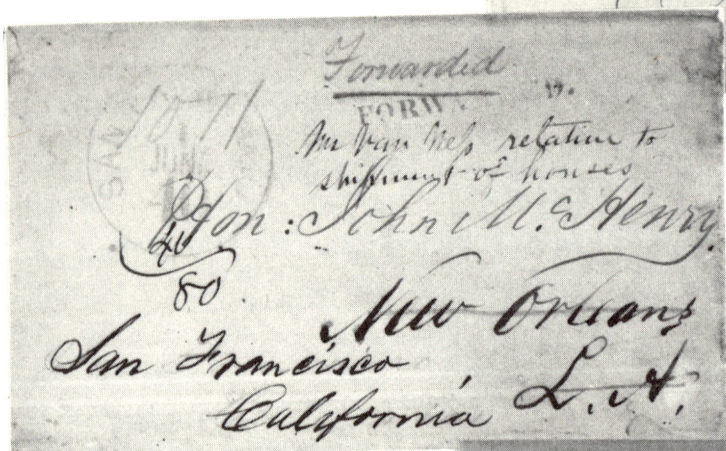

SF (40 in postmark) to New Orleans unpaid and forwarded back to SF for another 40¢, total 80¢ due.

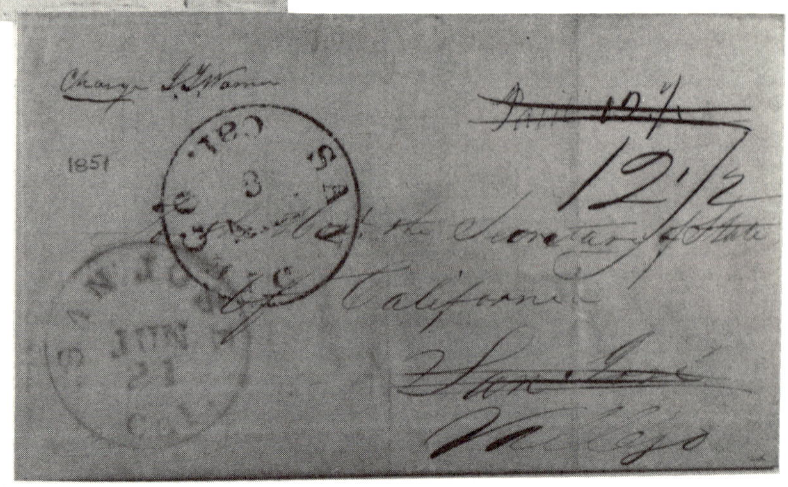

San Diego "Paid 12½" to San Jose, then forwarded to Vallejo with 12½¢ due.

Unpaid double letter from Easton, Md., to SF, rated 80¢ due. Forwarded to Sonoma for 25¢ additional, total $1.05 due.

Munroe, Wis., to Sacramento and there FOR'D 5 to Placerville.

NY to Shasta with 10¢ 1855, then FORWARDED 3 to Callahan's Ranch.

Petra, Mississippi, to Diamond Springs, Dec. 1858, then "ford 3" to Sac City.

SF to Auburn with 3¢ 1851. Forwarded from Auburn to Sacramento PAID/3 in cash.

Topsham, Me., to Yankee Jims, with 10¢ 1857. Forwarded to SF with 3¢ 1857.

Forwarded Mail

If a letter was improperly addressed, or if an addressee had moved and had informed his former post office of his new address, letters would be forwarded to his new location. Forwarded letters were considered "new mail" and were charged the appropriate rate from the original destination to the new one. In the period 1 July 1851 to 1 April 1855, forwarded letters on which the postage for forwarding had not been paid incurred the penalty rates per single letter of 5¢ under and 10¢ over 3,000 miles. From 1 April 1855 the forwarding fee was charged at the regular prepaid domestic postage rate from the original destination to the new address. The fee could be paid in cash or stamps by the addressee. An exception to the forwarding charge was made during the Civil War for letters forwarded to servicemen on active service. Finally on 1 July 1866 forwarding fees were abolished, if the addressee had filed a request and his new address with the post office.

SF to Placerville, with 3¢ 1861, but missent. Undated Onion Valley postmark with ms. "missent" inserted.

Culloma PAID 3 to Santa Clara, but missent to Shasta. Shasta ms. postmark and "Missent & f'd."

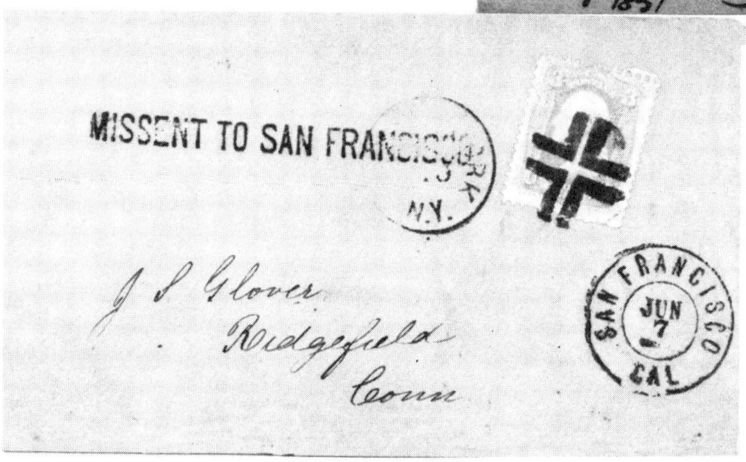

NY to Conn., badly missorted. SF double circle and straightline MISSENT TO SAN FRANCISCO.

Missent Mail

Occasionally, a California cover is seen bearing the notation "Missent" either handstamped or in manuscript. This marking indicates an error by the post office department in routing the letter and there was no charge for forwarding it to its correct destination.

Way Mail

The "Way" marking, either in manuscript or handstamped, is found only on mail carried by contractors employed to carry the U.S. mails. Such contractors carried the mails by steamboat, stagecoach, rail, or ship. Way markings found on mail carried by contract ships were largely limited to California mail. Outgoing mailbags were locked and sealed when placed on these ships. If the ships carried no route agent employed by the POD, any letters picked up at port stops in transit, or written by ships' passengers during the voyage, were placed by a ship's officer or agent in an open mailbag for delivery to the ship's port of entry post office. There they received the townmark and rating mark of that office and entered the mail system. Additionally, the receiving post office handstamped these letters "Way" to distinguish them from letters in the locked mailbags. This was to satisfy the postal regulation which stated that way letters would be charged 1¢ each for the service of the contractor in delivering them to the post office and that this charge was to be collected from the addressee. Collection of the way fee from the addressee was terminated at most post offices by 1 January 1853. The 1855 *PL&R* stated that if the letter was prepaid the carrier would still receive his 1¢ fee, but that the charge would not be added to the postage.

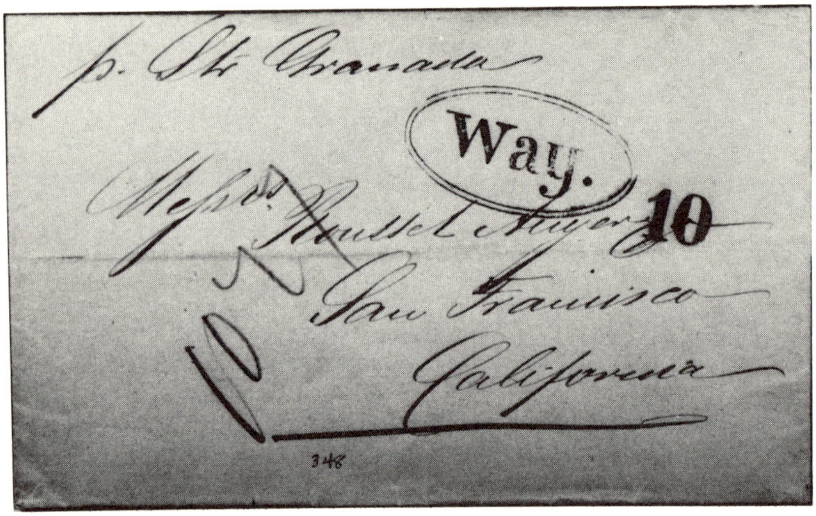

Oval "Way." and 10 rate on letter arriving at SF by contract vessel.

After that date, collect letters, in contrast, were treated as ship letters with 6¢ due if delivered at the port of entry and the regular postage plus 2¢ due if delivered beyond the port of entry. On 27 February 1861 the ship letter fee was changed to 5¢ if delivered to the port of entry. Finally, on 3 March 1863, all way fees were abolished effective 1 July 1863.

Other Markings

The classification of three additional markings used at San Francisco on mail incoming from abroad in the 1860s is still not clear. The markings are all in straightline form and made from heavy sans-serif capitals from the same font. (The CHINA STEAM markings mentioned in the chapter on steamship mail were composed from the same type.) It is likely all were made up locally by or for the San Francisco post office; they presumably reflect interpretation of postal laws and regulations at that office. The marking PURSER is known on incoming covers dated 1865-67 which were subject to the 10¢ "blanket" rate applicable to mail from countries with which the United States had no treaty arrangements, or so it appears from the ratings on these covers. In the last few months of 1864 the marking FOREIGN was used, also, in many cases, on mail to which the blanket rate applied. The scarcest of this group is U S SHIP., used in this period on mail brought in from ships of the U.S. Navy.

PURSER and DUE 7 (to make 10¢ rate) on cover to SF with 3¢ 1861.

Shanghai to SF in 1864, struck FOREIGN and rated 10.

Similar markings on cover from Chile, British postage paid to Panama.

Cover from Hawaii, marked FOREIGN, rated 5 due.

US SHIP. on cover brought in by Navy vessel. No due markings.

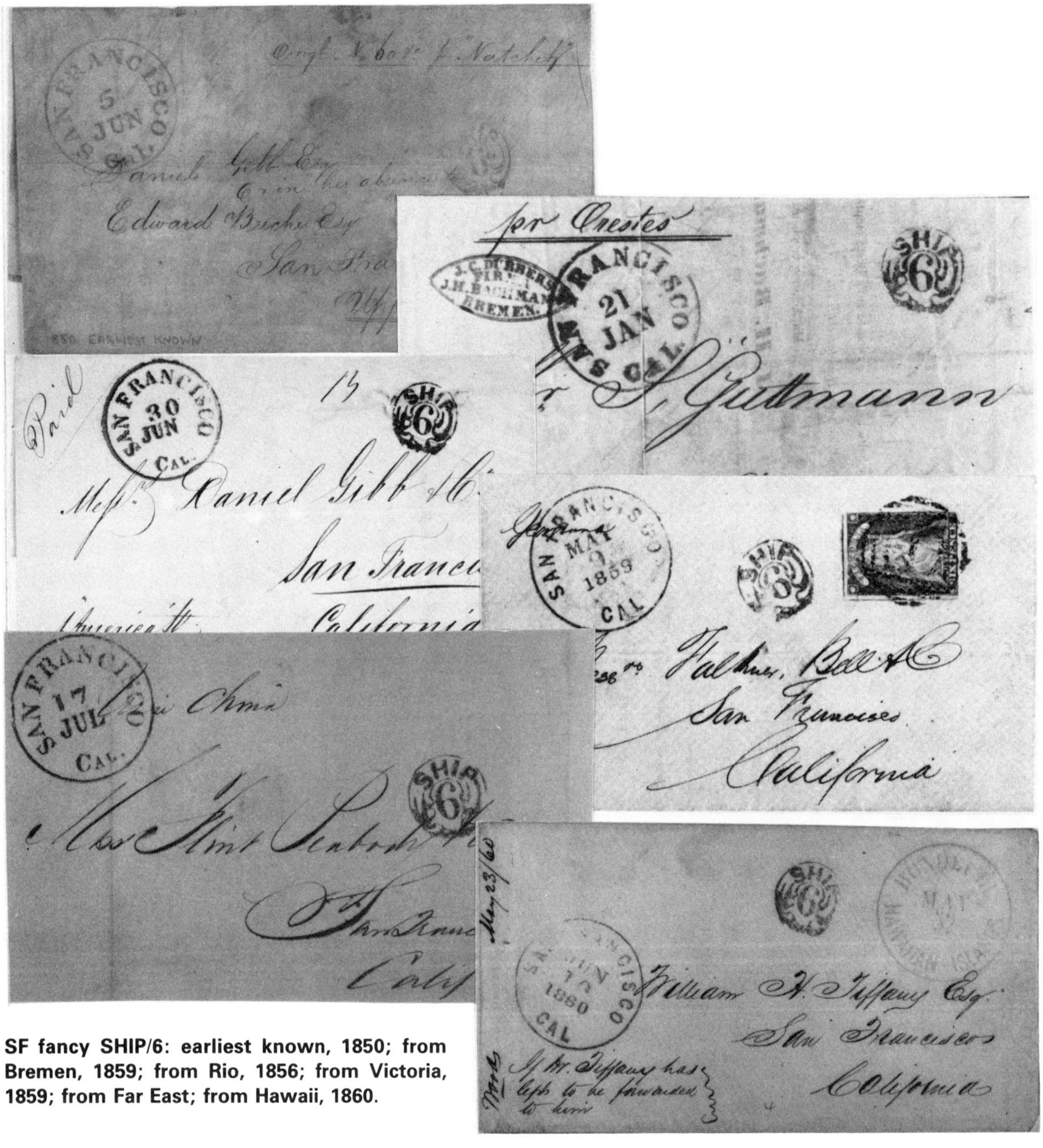

SF fancy SHIP/6: earliest known, 1850; from Bremen, 1859; from Rio, 1856; from Victoria, 1859; from Far East; from Hawaii, 1860.

Ship Letter Mail

Despite the fierce competition of overland mail-carrying stage lines, until the completion of the transcontinental railroad in 1869, almost three-fourths of all mail to and from California was carried by ships. Therefore ship postal markings are of special interest to the postal historian. The bulk of this ship mail traveled on contract steamers

Different style SHIP/6 used at SF on 1860 letter from Hawaii.

which rarely applied markings to the covers. (A detailed discussion and illustrations of markings placed on covers by the contract lines and their competitors is found in Chapter 4.)

Nevertheless, mail did come into California from locations outside the isthmian routes served by the main contract steamer lines, and such mail did receive special postal markings. The Act of 3 March 1825, not amended until 27 February 1861, continued a rate of 6¢ (established March 1799) for each letter delivered by a non-contract ship to a port of entry for local delivery, either prepaid or collect. This rate applied only if the letter was addressed to someone within the port of entry town. If the letter was addressed to a point beyond the port, then the charge was 2¢ plus regular domestic postage to the destination post office. The Act of 27 February 1861 reduced the rate to 5¢ at the port of entry to correct the anomaly that ship letters addressed to such ports were charged more than those addressed beyond after the single letter rate was reduced to 3¢. The Act of 3 March 1863 changed the domestic rate to 3¢ for anywhere in the United States and effective as of that date ship letters were charged double the applicable domestic rate.

The port of entry post office was required

Sidney, New South Wales, to SF in 1861. Rated SHIP/5, the reduced rate for ship letters delivered at port of arrival.

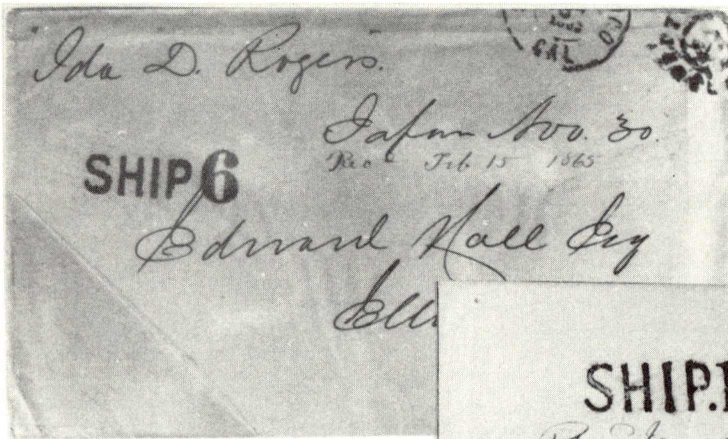

Ship letter originating in Japan in 1865, addressed to Conn. Rated SHIP 6 at SF at double domestic rate.

Hall correspondence letter from Japan in 1862. At SF rated SHIP 12, 10¢ transcontinental and 2¢ ship.

to handstamp all mail received from such non-contract vessels with the word "SHIP" and a numeral indicating the postage rate, thus allowing for the postal accounting reports of moneys (the 2¢ ship fee) paid to the ship captains and to distinguish such mail from that carried by contract steamers.

One of the most interesting ship markings is the well-known fancy scrolled San Francisco "SHIP 6," often erroneously called the "Clipper Ship Marking." This handstamp was used from 1850 to 1860, applied in San Francisco to letters received from private vessels for local delivery. Until about July 1851 the handstamp color was orange-red, then it was black until use of the mark ended sometime in 1860. Often the name of the ship bringing these letters into port would appear in manuscript on the cover. Letters bearing this fancy marking are known to have originated in South and Central America, Cuba, Hawaii, Wales, Scotland, Manila, Australia,

Letter from Hawaii to SF (after July 1, 1863). The 4¢ ship rate at the port prepaid by a pair of Black-jacks.

Hong Kong, Singapore, Bangkok, France, and Germany.

Numerous independently owned ships, many of foreign registry, plied the west coast Pacific waters carrying mail to and from California. Most of this mail entered United States postal distribution through the San Francisco post office and was treated as ship letter mail. Of particular interest to the postal historian, however, are letters carried by a British-owned steamer line, the Pacific Steam Navigation Company. This company had been chartered in London on 17 February 1840 and two steamers, the *Peru* and the *Chile,* had been dispatched from England to South America. Headquartered in Callao,

Oval STEAM markings on letters to SF: STEAM PANAMA rated 10 due in SF postmark; STEAM ACAPULCO on 3¢ envelope, DUE 7 to make 10¢ rate; STEAM MAZATLAN due 10, 1865; STEAM MANZANILLO on double letter rated 20 due, 1865.

Peru, and granted a mail subsidy by the British government, these ships ran from Valparaiso north to western ports of Central America and Mexico. In 1848 they began connecting across the Isthmus of Panama with the British Royal Mail Steam Packet Company, out of Southampton, which carried mail between Europe and the West Indies. When California became an American possession, the Pacific Steam Navigation Company extended its route to make San Francisco a port of call. During the late 1850s and early 1860s each letter picked up in Panama, Acapulco, Mazatlan, and Manzanillo by these vessels was marked with a fancy oval handstamp, probably by a P.S.N.C. ship's officer. The Panama handstamps are the most common, the Acapulco ones are scarce, the Mazatlan ones are rare, and the few Manzanillo ones are very rare. All these covers known to the author are rated at 10¢ postage.

Epilogue

Thus concludes the story of a state which slept in pastoral simplicity for centuries until struck by a golden bolt of lightning, which in 20 years transformed it from a sleepy frontier territory into a flourishing and powerful state.

The transformation was the product of free enterprise, moulded by the necessities of the period. The expressmen, stage lines, and steamers have passed into history. The locals and forwarders are gone. There remains only the debris of their days, preserved like artifacts in libraries, museums, and the collections of the postal historians. But it was an heroic story, and its memories are woven into the tapestry of our nation, written by pioneers who had no awareness that they were inscribing a history that would become part of America's legendary past.

Chiseled into the marble at the entrance to the National Archives Building in Washington, D.C., are the words: "The Past is Prologue." It is a reminder that to understand the present, one must know the past. This book has dealt with a narrow segment of the past, but in its restricted subject matter it has at least hinted at the romance of those days when "God gave us men like our mountains." It has been a story worth the telling.

Culloma forwarding postmark on cover from Va. Post office opened 1849; name changed to Coloma in 1851.

Shaded star cancel from Sacramento on cover by Butterfield route to Washington.

Shield on allover patriotic from Sacramento. This post office opened in 1849.

SF double circle on 1862 cover with pair of red brown 5¢ for transcontinental rate.

Covers from early post offices.

Appendix 1: California Post Offices

Chronological list through 1869, as published in the *Official Register*. Through appointment by the Special Postal Agent, many of the early post offices were in operation prior to the dates appearing in the *Official Register*. Date in italics indicates reopened office.

1848

9 Nov - San Francisco (unofficial)
21 Nov - Monterey (unofficial)

1849

8 Nov - Benicia
8 Nov - Culloma
8 Nov - Sacramento City
8 Nov - San Jose
8 Nov - Sonoma
8 Nov - Stockton
8 Nov - Vernon (closed 3 Nov 1853). *See* 12 Dec 1866.

1850

9 Apr - Fremont (closed 30 Mar 1855). *See* 3 Jul 1855.
9 Apr - Junction
9 Apr - Los Angeles
9 Apr - Mission San Jose
9 Apr - Napa (Nappa City)
9 Apr - Placerville
9 Apr - San Diego
9 Apr - Santa Barbara
9 Apr - Santa Cruz
9 Apr - Weaverville
1 Jul - Monroeville (closed 18 Sept 1862)
7 Jul - Eliza (closed 6 June 1851)
12 Dec - Nevada City

1851

13 Jan - Coloma (spelling of Culloma changed on this date)
21 Jan - Haydensville
10 Jul - Bidwells Bar (closed 27 May 1864). *See* 27 Nov 1865.
10 Jul - Centreville (Nevada)
10 Jul - Double Springs
10 Jul - Hamilton
10 Jul - Jackson
10 Jul - Mokelumne Hill
10 Jul - Shasta
28 Jul - Empire (closed 21 Oct 1851)
28 Jul - Graysonville (closed 21 Oct 1851)

Crowds lined up at the San Francisco post office, 1849.

Stockton to Pa., 40¢ due, 1850.

Stockton 6 PAID to Mass.

Circle PAID/12 on double letter to Del.

Stockton to Me., with pair of 3¢ 1851 cancelled by web.

Stockton double circle used to Canada, with 1861 10¢ and 5¢ for 15¢ rate.

A group of covers from Stockton, established 1849.

Nevada City to Canada with 10¢ 1861. After Feb. 1864, when rate was reduced.

Placerville to Washington, with 10¢ 1857.

Santa Cruz to Me., the rate paid by ten copies of 1¢ 1861.

Some post offices established in 1850.

1851 (cont.)

28 Jul - Green Springs (El Dorado)
28 Jul - Horr's Ranch
28 Jul - Knights Ferry
28 Jul - Livermore Ranch
28 Jul - Louisville
28 Jul - Mariposa
28 Jul - Martinez
28 Jul - Marysville
28 Jul - Mountain Inn
28 Jul - Nicolaus
28 Jul - Oak Spring
28 Jul - Rough and Ready
28 Jul - San Joaquin
28 Jul - San Juan
28 Jul - San Luis Obispo
28 Jul - Santa Clara
28 Jul - Sonora
28 Jul - Trinidad
28 Jul - Tuolumne City (closed 21 October 1851). *See* 5 June 1867.
28 Jul - Vallejo (closed 3 Sep 1853). *See* 19 Jan. 1855.
28 Jul - Wood's Diggings
28 Jul - Yuba
7 Aug - Mormon Island
7 Oct - Agua Fria
7 Oct - Big Bar
7 Oct - Chico
7 Oct - Dobbin's Ranche
7 Oct - Downieville
7 Oct - Georgetown
7 Oct - Goodyears Bar
7 Oct - Hall's Ranch
7 Oct - Jacksonville
7 Oct - Lassens
7 Oct - Park's Bar
7 Oct - Quartzburg
7 Oct - Salmon Falls
7 Oct - Volcano
21 Oct - Antioch (to Marsh's Landing), *again* 18 Jan 1855.
21 Oct - Gilroy
21 Oct - Moon's Ranch
21 Oct - Staple's Ranch
21 Oct - Yuba City
6 Nov - Carsons Creek
6 Nov - Contra Costa
6 Nov - Mud Spring
6 Nov - Murphy's
6 Nov - San Rafael (closed 18 Oct 1953). *See* 31 Mar 1854.
6 Nov - Texas Hill
6 Nov - Union (Santa Clara)
29 Nov - Colusi(a)
29 Nov - Garrotte
29 Nov - Tehama

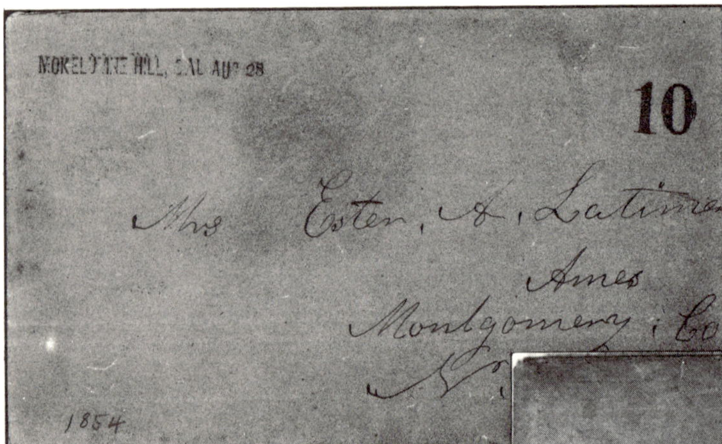

MOKELUMNE HILL CAL AUG 28 postmark on unpaid 1854 cover to N.Y. state.

Marysville PAID BY STAMPS and barred cancel on 6¢ Nesbitt to Ohio.

Two post offices established in 1851.

1852

20 Jan - San Ramon
21 Jan - Big Oak Flat
21 Jan - Cold Spring
21 Jan - Dry Town
21 Jan - Sutter Creek
9 Feb - Petaluma
20 Feb - Bodega
20 Feb - Cottonwood
20 Feb - Ophir
20 Feb - Ringgold
5 Mar - Foster's Bar
5 Mar - Nashville
19 Mar - North Branch
24 Mar - Cache Creek
24 Mar - Green Springs (Tuolumne)
24 Mar - Ophirville
23 Apr - Santa Rosa
7 May - Elk Horn
18 May - Alamo
21 Jun - Consumne
28 Jun - Charleys Rancho
6 Jul - Aurum City
20 Aug - Grass Valley
3 Sep - Jone Valley
3 Sep - San Bernardino
3 Sep - Yankee Jims
15 Sep - Columbia
15 Sep - Marsh's Landing
15 Sep - Washington (Nevada)
2 Oct - Redwood
9 Oct - Greenwood
12 Oct - Horsetown (closed 20 Oct 1853). *See* 20 Dec 1853.
19 Oct - Ashland
19 Oct - Mendocino (Trinity)
19 Oct - Union Town
10 Nov - Kilna
20 Nov - Camp Seco (Tuolumne)
20 Nov - Maxwell's Creek
16 Dec - Garden Valley (closed 20 Dec 1853). *See* 2 Dec 1854.
16 Dec - Third Crossing

1853

6 Jan - Oro City (closed 20 Dec 1853). *See* 2 Mar 1855.
6 Jan - Steinbergers
19 Jan - Curtisville
3 Feb - Shingle Spring (closed 30 Mar 1855). *See* 10 Oct 1865.
3 Feb - Yolo
3 May - San Leandro
27 May - Angel's Camp
21 Jun - Eureka
21 Jun - Round Tent
14 Jul - Merry Oaks (closed 30 Mar 1855). *See* 17 Nov 1859.
21 Jul - Auburn
8 Aug - Alvarado
16 Aug - Jamestown
19 Aug - Yreka
3 Sep - Monroeville
11 Oct - Millerton (closed 7 Oct 1863). *See* 6 Oct 1864.
11 Oct - Punta de los Reyes
11 Oct - Snelling's Ranch (closed 13 Feb 1861). *See* 5 Jul 1861.
12 Oct - Crescent City
12 Oct - Laguna Seca
12 Oct - Woodville
17 Oct - Diamond Spring
17 Oct - Red Bluff
21 Nov - Johnson's Ranche
21 Nov - Spanish Flat
21 Nov - Watsonville
22 Nov - Buckner
22 Nov - Cedarville
22 Nov - Don Pedro's Bar
22 Nov - Illinoistown
22 Nov - Indian Diggings
17 Dec - Monte
20 Dec - Horsetown. *See also* 12 Oct 1852.
31 Dec - Fiddletown

Grass Valley patriotic cover to SF; Diamond Springs with 1851 issue to Conn.

1854

18 Jan - Elliott's Ranch
18 Jan - Mountain View
3 Feb - Campo Seco (Calaveras)
3 Feb - San Padro
4 Feb - Green Valley (to Hitchcock Ranch), *again* 12 Sep 1865.
4 Feb - Mountain Springs
11 Feb - Michigan Bluff
18 Feb - Camptonville
31 Mar - Montezuma
31 Mar - Salinas
31 Mar - San Rafael. *See also* 6 Nov 1851.
4 Apr - San Lorenzo
15 Apr - Woodside
18 Apr - Alameda
18 Apr - Chinese Camp
18 Apr - Dry Creek
18 Apr - Pilot Hill
18 Apr - Suisun
18 Apr - Washington (Yolo)
3 May - Clinton
3 May - French Camp (closed 12 Nov 1862). *See* 22 Mar 1865.
3 May - Grafton
3 May - Oregon House
3 May - Oroville
3 May - Western
13 May - Elk Grove
13 May - Forbestown
13 May - Onisbo
22 May - Shaw's Flat
24 May - Lewiston
24 May - Trinity
30 May - Acelanus
30 May - Haskell's Ranch
30 May - Stringtown (closed 18 Jan 1855). *See* 5 Oct 1857.
1 Jun - Vacaville
16 Jun - Iowa City
17 Jun - McDermott's
17 Jun - Newtown
17 Jun - Putah (closed 2 Dec 1856). *See* 15 Jul 1858.
14 Jul - Alviso (closed 27 Jan 1855). *See* 17 Aug 1859.
14 Jul - Cordelia (to Rockville), *again* 16 Feb 1869.
14 Jul - Grand Island
14 Jul - Yornet (Yeomet)
18 Jul - Belmont
26 Jul - San Gabriel
17 Aug - Cherokee
17 Aug - Forlorn Hope (closed 29 Jan 1859). *See* 25 Apr 1860.
17 Aug - Vallicita
29 Aug - Secret Ravine (to Auburn Station), *again* 31 Jul 1863.
29 Sep - Smith's Ranch
29 Sep - Tomalles (Tomales)
1 Nov - Rattlesnake Bar
1 Nov - Russian River
14 Nov - San Andreas
15 Nov - San Pablo
16 Nov - Forest City
2 Dec - Garden Valley. *See also* 16 Dec 1852.
2 Dec - La Grange
28 Dec - Georgiana
29 Dec - Ottitiewa

1855

3 Jan - Foreman's Ranch
3 Jan - O'Byrnes Ferry
15 Jan - Saint Louis
18 Jan - Antioch. *See also* 21 Oct 1851. (Closed 21 Oct 1862). *See* 5 Mar 1863.
19 Jan - Vallejo. *See also* 28 Jul 1851.
19 Feb - Linden
19 Feb - Michigan Bar
2 Mar - Elizabethtown
2 Mar - Oakland
2 Mar - Oro City. *See also* 6 Jan 1853.
3 Mar - Centreville (Centerville) (Alameda)
16 Mar - Little York (closed 16 Nov 1855). *See* 3 May 1856.
16 Mar - Pleasant Springs
30 Mar - Nelson's Creek
12 Apr - Alpha
18 Apr - Sierra

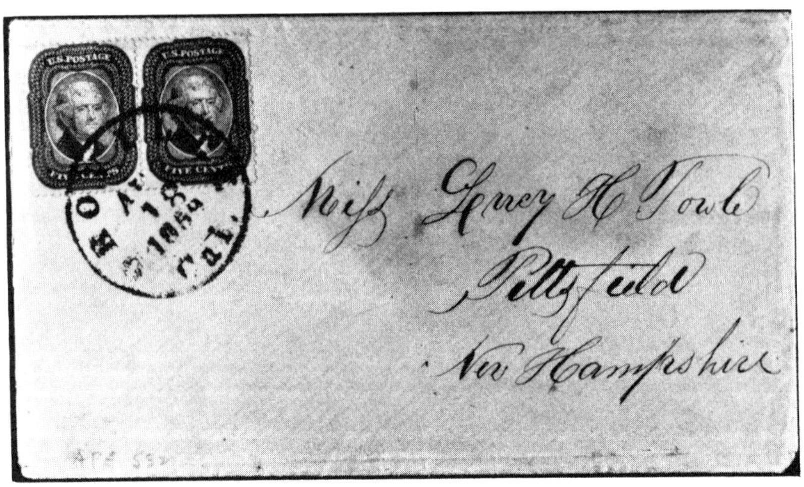

Oroville (est. 1854) to N.H., with pair 1857 5¢.

1855 (cont.)

4 May - Split Rock
17 May - Buckeye
17 May - Sutter
28 May - Bottle Hill
2 Jun - Emery's Crossing
2 Jun - Empire Ranch
2 Jun - Fourth Crossing
2 Jun - Lisbon
2 Jun - Mayfield
3 Jun - Visalia
18 Jun - Gibsonville (closed 25 Oct 1869). *See* 16 Dec 1869.
18 Jun - Patterson
3 Jul - Aquaduct City
3 Jul - Fremont. *See also* 9 Apr 1850.
3 Jul - Hill's Ferry
3 Jul - Indian Gulch
3 Jul - Neilsburgh
6 Jul - Strawberry Valley
13 Jul - Gwin
14 Jul - Clarksville
19 Jul - American Ranch
31 Jul - Loving's Ferry (to Holden's Ferry), *again* 18 Dec 1858.
31 Jul - Oak Grove Farm
31 Jul - Plum Valley
31 Jul - Trinity Centre
18 Aug - Onion Valley
28 Aug - Brooklyn
31 Aug - Grizzly Flats
31 Aug - Natividad
31 Aug - Windsor
13 Sep - Rabbit Town
3 Oct - Meadow Valley (to Spanish Ranch), *again* 23 Mar 1864.
2 Nov - McCartysville
2 Nov - Princeton
3 Nov - Bucksport
17 Nov - Pleasant Valley
17 Nov - Red Dog
19 Nov - Bondville
15 Dec - El Dorado
17 Dec - Quincy
19 Dec - Leach's Store
19 Dec - Owsley's Bar

1856

3 Jan - Coon Creek
3 Jan - Henley
3 Jan - King's River
3 Jan - Pea Vine
2 Feb - Brush Creek
2 Feb - Novato
18 Feb - Folsom City
18 Feb - French Gulch
18 Feb - Whiskey Creek
3 Mar - Kelsey
3 Mar - Scott River
5 Apr - Canon City
5 Apr - Honcut (closed 23 Mar 1868). *See* 21 Jul 1868.
23 Apr - Minersville
3 May - Little York. *See also* 16 Mar 1855.
3 May - Mill Valley
10 May - Sarahville
19 May - Sebastopol (closed 15 Jan 1857). *See* 30 Sep 1858.
19 May - Woodville
23 May - Franklin (closed 19 Oct 1858). *See* 3 Jul 1862.
31 May - Milpitas
17 Jun - Todd's Valley
18 Jun - Brick Post
18 Jun - Hornitas
18 Jun - Middletown
12 Jul - Bloomfield
12 Jul - Pine Grove
31 Jul - West Point
4 Aug - Tarr's Ranch
4 Sep - Merced Falls
16 Sep - Redwood City
2 Oct - Half Way House

Folsom City (est. 1856) on patriotic with 3¢ 1861.

Forwarding postmark of North San Juan (1857) on cover originating at SF.

1856 (cont.)

30 Oct - Antelope
3 Nov - Mount Ophir
4 Nov - Two Mile Bar
15 Nov - Damascus (to Forks House), *again* 25 Jul 1861.
18 Nov - Dutch Flat
3 Dec - Scottsburgh (closed 1 Dec 1858). *See* 30 Jun 1859.
15 Dec - Reynolds Ferry (closed 6 Jul 1860). *See* 3 Nov 1860.
15 Dec - Saint (St.) Helena
15 Dec - Walnut Grove
17 Dec - Hansonville

1857

15 Jan - Fourteen Mile House (closed 28 Jul 1857). *See* 19 Mar 1858.
15 Jan - Oakville (closed 30 Mar 1859). *See* 4 Nov 1867.
17 Jan - Moore's Flat
17 Jan - Rail Road Flat (closed 16 Aug 1858). *See* 17 Mar 1869.
31 Jan - Bangor
3 Feb - Barton's Store
3 Feb - Frenchtown
3 Feb - Suisun City
3 Feb - Thompsons Flat
16 Feb - Ione City
2 Mar - La Fayette (Lafayette)
3 Mar - Montevideo
10 Mar - Rio Seca
11 Mar - Keysville
18 Mar - Chip's Flat
19 Mar - Prairie
13 Apr - Stony Point
14 Apr - Healdsburgh
17 Apr - Table Rock
25 Apr - Fugitt
25 Apr - Wood's Ferry
2 May - Russville
21 May - North San Juan
1 Jun - North Bloomfield
19 Jun - El Dorado Ranch
19 Jun - Omega
5 Jul - Soquel
6 Jul - Greenville
6 Jul - Sweetland
17 Jul - Elk Grove (Sacramento)
17 Jul - Tejon
17 Jul - Two Rocks (closed 17 Sep 1858). *See* 27 Mar 1863.
28 Jul - Jonesville
15 Aug - Butte Mills
15 Aug - Cloverdale
4 Sep - Eel River
5 Oct - Stringtown. *See also* 30 May 1854.
10 Oct - Chili (closed 14 Feb 1860). *See* 31 Aug 1861.
12 Oct - Butte City
12 Oct - San Mateo (closed 19 Oct 1858). *See* 16 Oct 1861.
17 Oct - Phillip's Flat
17 Oct - Rich Gulch
18 Nov - Cottage Grove
18 Nov - Jenny Lind
18 Nov - Star House (closed 17 Sep 1858). *See* 28 Apr 1859.
19 Nov - Alleghany
19 Nov - Eureka North
19 Nov - La Porte
2 Dec - Orleans
8 Dec - Tuttletown
10 Dec - Springfield

1858

15 Jan - Searsville
15 Feb - Callahan's Ranch
15 Feb - Fort Bragg (closed 30 Jun 1858). *See* 15 Jan 1861.
2 Mar - Denverton
5 Mar - Wyatt's Store
19 Mar - Fourteen Mile House. *See also* 15 Jan 1857.
19 Mar - Happy Camp
19 Mar - Indian Springs
19 Mar - Jacinto
19 Mar - Seiad Valley
27 Mar - Poverty Bar
30 Mar - Mountain Well
9 Apr - Long Bar
16 Apr - Kerrick's Ranch

1858 (cont.)

16 Apr - Petersburgh
19 Apr - Bath (to Forest Hill), *again* 17 Aug 1861.
29 Apr - Ferry Point
15 May - Willow Point
18 May - Clairville
2 Jun - Colorado
2 Jun - Musquito
2 Jun - Punta Arenas
3 Jun - Holden's Ferry
3 Jun - Virginia
15 Jun - Pacific Home
15 Jun - Sawyers Bar
15 Jun - Upper Clear Lake
21 Jun - Bear Valley
30 Jun - Grove City
1 Jul - Rockville
3 Jul - Rock Creek
15 Jul - Fort Goff
15 Jul - Putah. *See also* 17 Jun 1854.
17 Jul - Big Valley
17 Jul - Salsbury
30 Jul - Mountain Ranch
31 Jul - Lower Lake
18 Aug - Rio Vista
30 Aug - Charleston
14 Sep - Duroc
18 Sep - Timbuctoo
30 Sep - Forks of Salmon
30 Sep - Oak Point
30 Sep - Poland
30 Sep - Sebastopol. *See also* 19 May 1856.
18 Oct - Anderson
18 Oct - Burnt Ranch
19 Oct - Yankee Hill
1 Nov - Grizzly Bear House
1 Nov - Uncle Sam
15 Nov - Marietta
1 Dec - Mendocino (Mendocino)
16 Dec - San Antonio
16 Dec - Weitchpec
17 Dec - Cascade City
17 Dec - Orr's Ranch
18 Dec - Loving's Ferry. *See also* 31 Jul 1855.
31 Dec - Fairfield
31 Dec - Ukiah

1859

29 Jan - Wyandotte
31 Jan - Elk Camp
31 Jan - Lakeville
31 Jan - San Quentin (closed 13 Oct 1859). *See* 25 April 1862.
28 Feb - Olema (closed 23 Oct 1860). *See* 23 Apr 1864.
3 Mar - Keeneysburgh
15 Mar - Fort Tejon
22 Apr - Temecula (San Diego)
28 Apr - Star House. *See also* 18 Nov 1857.
28 Apr - Tule
2 May - Lancha Plana
11 May - Central House
15 Jun - Pescadero
27 Jun - Forest Hill
29 Jun - Albion
30 Jun - Scottsburgh. *See also* 3 Dec 1856.
15 Jul - Kayote
16 Aug - Pacheco
17 Aug - Alviso. *See also* 14 Jul 1854.
17 Aug - Sink
18 Oct - Warner's Ranch (closed 7 Jun 1860). *See* 5 Jul 1861.
17 Nov - Burwood
17 Nov - Kingston (closed 12 Mar 1862). *See* 10 April 1866.
17 Nov - Merry Oaks. *See also* 14 Jul 1853.
17 Nov - Noyo River
3 Dec - French Corral
14 Dec - Merritt
23 Dec - Douglas City
23 Dec - Six Mile Bar

1860

4 Jan Fresno City
4 Jan - San Miguel (closed 9 Nov 1860). *See* 17 Aug 1861.
6 Jan - Haywood
27 Jan - Millville

Ms. postmark of Tule (opened 1859) on 3¢ envelope to N.Y. state.

Ms. postmark of Haywood (est. 1860) on cover to Me. by Butterfield route with "Star of the Union" four horse stage coach and 10¢ 1857.

1860 (cont.)

15 Feb - Dougherty's Station
13 Mar - Firebaugh's Ferry
2 Apr - Messerville
13 Apr - Fair Play
19 Apr - Plumas
25 Apr - Forlorn Hope. *See also* 17 Aug 1854.
28 Apr - Hicksville
2 May - Hermitage
5 May - Arcata
8 May - Richland
17 May - Forks House
28 May - North Columbia
1 Jun - Bellota
7 Jun - Knights Valley
20 Jun - Ferndale
5 Jul - Liberty
7 Aug - Analy (closed 13 Feb 1861). *See* 12 Nov 1866.
25 Aug - Sheldon
29 Aug - Jay Hawk
31 Aug - Danville
25 Sep - Sanel (closed 29 Jan 1869). *See* 10 Aug 1869.
9 Oct - Hitchcock Ranch
16 Oct - Susanville
17 Oct - Elderton
3 Nov - Reynolds Ferry. *See also* 15 Dec 1856.
19 Nov - Fort Jones
26 Nov - Calpella
19 Dec - Linn's Valley

1861

5 Jan - Pacific
8 Jan - Eight Mile Corners
10 Jan - Etna Mills
15 Jan - Fort Bragg. *See also* 15 Feb 1858.
23 Jan - Butte Valley
23 Jan - Clay's Bar
29 Jan - Humbug Creek
2 Feb - Oro Fino
12 Feb - Temescal
13 Feb - Hay Fork
1 Mar - Little Lake
6 Mar - Unionville
6 Jun - Half Moon Bay (closed 20 Sep 1862). *See* 29 Nov 1862.
6 Jun - Haran
6 Jun - Lexington
19 Jun - Anaheim
19 Jun - Lockford
19 Jun - Longville
19 Jun - Martins Ferry (closed 12 Nov 1862). *See* 22 May 1865.
19 Jun - South Fork
19 Jun - Table Bluff (closed 14 Oct 1862). *See* 18 Apr 1867.
28 Jun - Mokelumne City
3 Jul - Quartz Valley
5 Jul - New Almaden
5 Jul - Sellon's Ranch
5 Jul - Snellings Ranch. *See also* 11 Oct 1853.
5 Jul - Warner's Ranch. *See also* 18 Oct 1859. (Closed 11 Jul 1862). *See* 6 Sep 1867.
5 Jul - Woodland
25 Jul - Damascus. *See also* 15 Nov 1856.
25 Jul - Globe Ranch
25 Jul - Hoopa Valley
25 Jul - Viola
13 Aug - Clipper Mills
17 Aug - Bath. *See also* 19 Apr 1858.
17 Aug - San Miguel. *See also* 4 Jan 1860.
19 Aug - Junction City

1861 (cont.)

19 Aug - Port Wine
27 Aug - Hydesville
27 Aug - Yolo Centre
31 Aug - Chili. *See also* 10 Oct 1857.
9 Sep - Esmeralda
11 Sep - Buck's Ranch
17 Sep - Lake Valley
17 Sep - Taylor's Ranch
4 Oct - Clayton
4 Oct - Placer
4 Oct - Spanish Ranch
7 Oct - Lone Star
16 Oct - San Mateo. *See also* 12 Oct 1857.
12 Nov - Lakeport
14 Nov - Magalia
21 Nov - Corralitos
21 Nov - Maine Prairie
21 Nov - San Luis Rey (closed 20 May 1865). *See* 20 Nov 1865.
21 Nov - Slippery Ford
2 Dec - Calaveras
2 Dec - Ione Valley
19 Dec - Copperopolis
19 Dec - Janesville (Shasta)
23 Dec - Rome

1862

2 Jan - Smithville
3 Feb - Telegraph City
5 Feb - Lincoln
25 Feb - Albany
5 Mar - Albeeville
17 Mar - Farmington
24 Mar - San Buenaventura
25 Apr - San Quentin. *See also* 31 Jan 1859.
30 Apr - Burnett
30 Apr - Millers Ranch
6 Jun - Brownsville
10 Jun - White River (closed 17 Jun 1864). *See* 10 Dec 1866.
18 Jun - Anthony House (closed 22 Mar 1866). *See* 22 Feb 1867.
25 Jun - Mount Eden
26 Jun - Collinsville (closed 6 Oct 1862). *See* 26 Feb 1864.
3 Jul - Franklin. *See also* 23 May 1856.
10 Jul - Mount Bullion
25 Jul - Sierra Valley
23 Aug - Rich Bar
3 Sep - Gualala
4 Sep - Washington
7 Oct - Woodbridge
8 Oct - Forest Home
6 Nov - Sand Rock
12 Nov - Inskip
12 Nov - Solano
29 Nov - Half Moon Bay. *See also* 6 Jun 1861.
1 Dec - Walnut Creek
20 Dec - Duncans Mills

1863

12 Jan - Churntown
26 Jan - West Butte
10 Feb - Preston
19 Feb - Linden
21 Feb - Meridian
24 Feb - Timber Cove
5 Mar - Antioch. *See also* 21 Oct 1851, 18 Jan 1855.
25 Mar - Auburn Station
27 Mar - Greenland
27 Mar - Two Rocks. *See also* 17 Jul 1857.
12 May - Konigsberg (closed 4 Jun 1864). *See* 10 Aug 1864.
3 Jun - Bolinas
3 Jun - Mattole
3 Jun - Round Valley
13 Jun - Santa Ynez
10 Jul - Fisherman's Bay
15 Jul - Gold Run
22 Jul - Cahto
31 Jul - Secret Ravine. *See also* 29 Aug 1854.
12 Aug - Smith River

Arcata postmark on flag patriotic to Me. with 3¢ 1861. Post office opened 1860.

1863 (cont.)

19 Aug - Amador City
19 Aug - Blue Mountain
19 Aug - Somersville
13 Oct - Webster
21 Oct - Markleville
25 Nov - Elliott
25 Nov - Pope Valley
30 Nov - Knoxville
30 Nov - Monitor
22 Dec - Bank Mills
26 Dec - Taho
28 Dec - Carey's Mills
28 Dec - Locust Shade

1864

7 Jan - Taylorsville
8 Jan - Browns Valley
8 Jan - Loyalton
8 Jan - Painsville
8 Jan - Pentz
26 Feb - Camanche
26 Feb - Collinsville. *See also* 26 Jun 1862.
23 Mar - Cucamonga
23 Mar - Meadow Valley. *See also* 3 Oct 1855.
23 Mar - Pleasant Valley
19 Apr - Bridgeport
23 Apr - Brighton
23 Apr - Olema. *See also* 28 Feb 1859.
23 Apr - San Simeon (closed 22 May 1865). *See* 2 Oct 1867.
23 Apr - Wilmington
26 Apr - Trucky River
5 May - Snugville
24 May - Summit
28 May - Copper Vale (Coppervale) (closed 28 Oct 1867). *See* 13 Jul 1868.
1 Jun - San Carlos
15 Jun - Janesville (Lassen)
20 Jun - Milford
20 Jun - Newcastle
23 Jun - Sierra City (closed 20 Sep 1865). *See* 19 Mar 1867.
29 Jun - Hopewell
9 Jul - Roseville
25 Jul - Binghamton
25 Jul - Silveyville
10 Aug - Konigsberg. *See also* 12 May 1863.
22 Aug - Soldier's Bridge
29 Aug - San Marcos (closed 16 May 1865). *See* 29 Jan 1869.
29 Sep - Union (Merced)
6 Oct - Leitch's Ferry
6 Oct - Millerton. *See also* 11 Oct 1853.
10 Oct - Hope Valley
11 Oct - Lake House
11 Oct - Latrobe
11 Oct - Silver King
28 Nov - Freeport
8 Dec - Los Gatos (closed 16 Apr 1867). *See* 7 Aug 1867.
14 Dec - Langworth
14 Dec - Saint (St.) John

1865

11 Jan - Black Point
11 Jan - Martinsburgh (closed 30 Nov 1866). *See* 1 Dec 1868.
13 Jan - Saratoga (closed 29 Jan 1869). *See* 1 Apr 1869.
17 Jan - Calistoga
22 Mar - French Camp. *See also* 3 May 1854.
22 Mar - Pine Creek
23 Mar - Wilsons Ranch
28 Mar - Silver Mountain
7 Apr - Smith's Ferry
17 Apr - Monte Vista
28 Apr - Pattees Ranch
19 May - Firebaugh
22 May - Martin's Ferry. *See also* 19 Jun 1861.
12 Sep - Green Valley. *See also* 4 Feb 1854.
13 Sep - Petrolia(ea)
14 Sep - Battle Creek
14 Sep - Waterloo
19 Sep - Geneseo
10 Oct - Shingle Spring. *See also* 3 Feb 1853.
10 Oct - Sportsmans Hall
13 Oct - Last Chance
25 Oct - Mark West
20 Nov - Little River (Littleriver)
20 Nov - San Luis Rey. *See also* 21 Nov 1861.
27 Nov - Bidwells Bar. *See also* 10 Jul 1851.
12 Dec - Smartville
15 Dec - Harrisburgh
26 Dec - Hermit Valley
27 Dec - Big Trees (Bigtrees)

1866

8 Jan - Rock Springs
9 Jan - Havilah
11 Jan - Benton
25 Jan - Paulinville
30 Jan - Alisal
14 Feb - Colfax
27 Feb - Independence
8 Mar - Millbrae
12 Mar - Fish Springs (Mono)
12 Mar - Owensville (to Glen Mary), *again* 27 Jan 1869.
16 Mar - Sugar Pine (Sugarpine) (Tuolumne)
10 Apr - Kingston. *See also* 17 Nov 1859.
24 Apr - Buena Vista
25 May - Carner Place
28 May - Chrysopolis
28 May - Evan's Ranch
4 Jun - Cleveland
7 Jun - Crystal Lake
7 Jun - Donner Lake
25 Jun - Meadow Lake

Fort Bidwell double circle; post office established 1866.

1866 (cont.)

19 Jul - Bullards Bar
30 Jul - San Isidro
2 Oct - Elbow
2 Oct - Hopeton
3 Oct - Cisco
12 Nov - Analy. *See also* 7 Aug 1860.
21 Nov - Meincke
30 Nov - Clipper Gap (Clippergap)
30 Nov - Wheatland
4 Dec - Kings River
10 Dec - White River. *See also* 10 Jun 1862.
12 Dec - Colegroves Point
12 Dec - Vernon. *See also* 8 Nov 1849.
26 Dec - Klamath

1867

14 Feb - Greenville
22 Feb - Anthony House. *See also* 18 Jun 1862.
14 Mar - Castroville
19 Mar - Carpenteria (Carpinteria)
19 Mar - Sierra City. *See also* 23 Jun 1864.
25 Mar - Guenoc
8 Apr - Fair Oaks (San Mateo)
9 Apr - Santa Margarita
17 Apr - Petroliopolis
18 Apr - Table Bluff. *See also* 19 Jan 1861.
14 May - Yountville
28 May - Sherwood Valley
4 Jun - Pleasanton
5 Jun - Capistrano
5 Jun - Tuolumne City (closed 20 March 1871). *See also* 28 July 1851.
14 Jun - Hot Springs (closed 21 Nov 1867). *See* 7 Feb 1868.
14 Jun - Los Nietos
8 Jul - Monticello
7 Aug - Illinois Mills
7 Aug - Los Gatos. *See also* 8 Dec 1864.
7 Aug - Navarro Ridge
15 Aug - Pleasant Grove Creek
22 Aug - Newport
26 Aug - Grainland
26 Aug - Graniteville
6 Sep - Carneros
6 Sep - Warner's Ranch. *See also* 18 Oct 1859, 5 Jul 1861.
24 Sep - Blue Canyon
2 Oct - San Simeon. *See also* 23 Apr 1864.
2 Oct - Sebastopol
4 Nov - Oakville. *See also* 15 Jan 1857.
25 Nov - Bald Hills
25 Nov - Paradise
2 Dec - Mendon (to Indian Diggings), *again* 15 Jun 1869.
13 Dec - Sherman

1868

3 Jan - Spadra
23 Jan - Truckee
7 Feb - Hot Springs. *See also* 14 Jun 1867.
13 Feb - Camp Grant
12 Mar - Davisville (Solano), *again* 8 Jun 1868.
23 Mar - Emigrant Pass
26 Mar - Fish Springs (Inyo)
14 Apr - Fort Bidwell
21 Apr - Coast Range
21 Apr - Newville
22 Apr - Collegeville
22 Apr - Gas Jet
22 Apr - San Joaquin Valley
23 Apr - Kernville
24 Apr - Atlanta
24 Apr - Rocklin
11 May - You Bet
18 May - Valta
26 May - Purissama (Purisima)
26 May - Suscol
8 Jun - Davisville (Yolo). *See also* 12 Mar 1868.
10 Jun - Milquatay

1868 (cont.)

11 Jun - Buckeye Ranch
16 Jun - Glen Mary
19 Jun - Morgan
29 Jun - San Gorgonio
10 Jul - Eagleville
10 Jul - San Felipe
10 Jul - Sheridan
13 Jul - Copper Vale (Coppervale). *See also* 28 May 1864.
14 Jul - Yorkville
21 Jul - Honcut. *See also* 5 Apr 1856.
10 Aug - Bucks Ranch
26 Aug - Farmersville
26 Aug - Lake City
28 Aug - Bakersfield
13 Oct - Big Meadows
14 Oct - Ravena City
23 Nov - Coleville
1 Dec - Black Diamond
1 Dec - Hites Cove
1 Dec - Martinsburgh. *See also* 11 Jan 1865.
1 Dec - Old Creek
4 Dec - Capay

1869

8 Jan - Gabilan
15 Jan - Arroyo Grande
15 Jan - Hollister
15 Jan - Nottingham
17 Jan - Fir Cap
26 Jan - Dixon
27 Jan - Owensville. *See also* 12 Mar 1866.
29 Jan - San Marcos. *See also* 29 Aug 1864.
29 Jan - Santa Maria
29 Jan - Spring Valley
29 Jan - Tehichipa (Tehachapi)
3 Feb - Adelante
10 Feb - Schoolhouse Station
12 Feb - Bullionae
16 Feb - Cordelia. *See also* 14 Jul 1854.
17 Mar - Rail Road Flat. *See also* 17 Jan 1857.
17 Mar - Soledad
22 Mar - New Idria
22 Mar - Las Cruces
24 Mar - Picacho
26 Mar - Florin
1 Apr - Saratoga. *See also* 13 Mar 1865.
1 Apr - Woodfords
8 Apr - South San Diego
15 Jun - Black Bear
15 Jun - Mendon. *See also* 2 Dec 1867.
15 Jun - Petersburgh
15 Jun - Yocumville
30 Jun - Galt
6 Jul - Compton
6 Jul - Plainsberg
26 Jul - Junction House
9 Aug - Yo Semite
10 Aug - Sanel. *See also* 25 Sep 1860.
4 Oct - Alexandria
4 Nov - Cerro Gordo
5 Nov - Point of Timber
5 Nov - San Benito
16 Nov - Moores ('s) Station
17 Nov - Live Oak
17 Nov - Mokelumne
6 Dec - Pino
14 Dec - Zem Zem
16 Dec - Gibsonville. *See also* 18 Jun 1855.
22 Dec - Cedarville
29 Dec - Long Valley
29 Dec - Melita
29 Dec - National City

Statistical Summary

Year	*New*	*Reopened*	*Total*
1848	2	-	2
1849	7	-	7
1850	11	-	11
1851	63	-	63
1852	40	-	40
1853	33	1	34
1854	58	2	60
1855	54	4	58
1856	41	1	42
1857	50	1	51
1858	55	4	59
1859	25	4	29
1860	30	2	32
1861	52	8	60
1862	26	3	29
1863	30	3	33
1864	37	5	42
1865	23	6	29
1866	30	4	34
1867	28	7	35
1868	39	5	44
1869	40	8	48
	774	68	842

Appendix 2: California Expresses and Stage Lines before 1870 For which no Franks are now Known

Sam **Abbey's Express**
1858 North San Juan to French Corral
Captain **Agers Truckee & Humboldt Express Mail & Passenger Line**
1863 Truckee (Cal.) to Unionville (Nev.)
Alexander's Sierra Stage Line
1863 Eastern side of Sierra, Mono County
Allman & Peck's Express
1860 Napa to Healdsburg
Almy's Oakland San Francisco Express
1851-1854 Contra Costa County; S.F. to Marysville
Amador Stage Line
-1864 Amador to Austin
Angel's Daily Express
1861 North San Juan to Columbia Hill
Angle & Co.'s Express
1850 Sacramento to S.F.; also transcontinental, Europe
Atkerson's Express
1857 Weaverville down Trinity R.
Auburn & North Fork Express
1850 Sacramento and Auburn; along north fork American R.
Bacon's Pony Express
1855 Angels Camp to Robinson's Ferry (Southern Mines)
Baker's Express (George H. Baker)
1850 Marysville to camps on Feather R.
Baker's Fast Freight & Passenger Line (H. W. Baker)
1866-1867 Virginia City (Nev.) via Placerville to Sacramento
Baker's Western Shasta Express
1869 Shasta to Roaring River
Bald Hill Express
1857-1858 Shasta, Bald Hill and Horsetown
William T. **Ballou's California Express**
1849-1850 Stockton to Southern Mines
Barrow & Porter & Co.'s Stockton & Kansas City Mail
1858 Stockton to Kansas City
Baxter & Monroe's U.S. Mail Line
1851-1853 Sacramento to Shasta
Hill **Beachy's Railroad Stage Line**
1867 S.F., Sacramento to Virginia City (Nev.)
Orlando **Bennett's Express**
1856 Weaverville to Cox's Bar and Manzaneta
Bennett & Crandall's Letter Express
1867 Virginia City (Nev.) to Placerville
Billings & Co. Express
1850-1851 Stockton to Merced
Birch & Co.'s Daily Express
1849-1850 Sacramento to Coloma
1850-1852 Sacramento to Nevada City
Blackman & Co.'s Express
1852-1853 Marysville to Slate Ridge
Bodega Empire Stage Co.
1855-1856 Petaluma, Pine Grove and Bodega
Brewster's Stage (Saddle Train)
1860-1871 Oroville to Quincy
Britten's Weekly Express
1861 Weaverville to Shasta
Brown, Knowlton & Co.'s Express
1850-1852 California to all U.S.
Brown & Parish's Mail Line
1852 Sacramento to North & Middle Forks American R.
Brown's Accommodation Line
1852-1854 Sacramento to Auburn
W. C. Brown's Express
1858 Shasta to Watson Gulch
Browne's News Depot & Pony Express
1858 Jackson to Grass Valley
John R. **Buchel's North Fork Feather River Express**
1851-1852 Marysville to camps on North Fork Feather R.
Buckbee's Express
1850 Marysville to Feather R. camps
Francke **Buckingham's Express**
1851 Stockton to Tuolumne and Merced Rs.
William **Bunty Stage**
1854 Sonora, Columbia, Stockton
Burns & Co.'s Express
1850 Stockton to Sonora
Burt & Co.'s Express
1853 Sacramento to Drytown, Sutter Creek
Byrne's Old Line of Stages
1866-1867 Columbia to Sonora
Charles J. **Cady's Express**
1847 San Francisco to Fort Sutter
California & Oregon Stage Line
1862-1865 Portland (Ore.) to Lincoln (Cal.)
California & South Idaho Stage Line
1865-1867 Boise (Idaho) to S.F.
California Stage Co.
1854-1865 California, Oregon, Nevada

Camp & Co.'s Express
1850-1851 Marysville to North Fork Feather R.
James E. **Carr's Mule Team**
1858 Yreka to Scott Valley
Carr's Susanville & Camp Bidwell Line (Joseph D. Carr)
1868-1870 Bidwell to Susanville
Cartwright, Swain & Co.'s Stage Line
1852 Sacramento to Mokelumne Hill
Chain Lightin' Express
1851 Knights Ferry to Columbia
H. Chandler & Co.'s Horsetown & Shasta Express
1856-1863 Shasta to Horsetown
Chandler's Pittsburg Express (F. B. Chandler)
1863 Pittsburg to Shasta
Chapman & Co.
1854 Between Salem & Downieville; into Humboldt (Nev.)
Nelson **Chase's Tri-Weekly Express**
1859-1862 Yreka to Fort Jones via Rough & Ready, Aetna Mills
Cheeseman's Express
1852 Marysville to Feather R. camps
Chico & Susanville Stage Line
1865-1866 S.F. via Marysville, Chico, to Susanville
Dr. **Clark's Coach Line**
1856 Jamestown, Chinese Camp, Montezuma, Don Pedros, Coulterville & La Grange
Clough & Boyer's Express
1854-1855 Sonora to Southern Mines camps
Clough & Theall Express
1854 Sonora to Southern Mines camps
Columbia & Stockton Mail & Express Stage Line
1866 Columbia to Stockton via Sonora & Jamestown
Conger & Dorning's Express
-1867 French Corral to North Bloomfield
Contra Costa Express
-1855 Oakland to towns in Contra Costa Co.
Coover & Co.'s Stage Line
1856-1857 Columbia to Stockton
Copper Mines Express
1861 Stockton to Copperopolis
Copperopolis & Murphy's Accommodation Line
1863 Copperopolis to Murphy's
H. W. Corbett Oregon Stage Co.
1865-1870 Portland (Ore.) to Lincoln (Cal.)
Couch Line
1866 Drytown to Jackson, Mokelumne Hill, San Andreas, Murphy's & Sonora
Crandall & Co.'s Diablo Express
1855 S.F. to San Ramon
Crandall's Pioneer Express
1857 Placerville to Carson City (Nev.)
Crescent City Express
1854 S.F. to Crescent City
Cummings New Line of U.S. Mail Stages
1860 Oakland to Stockton

S. Cutler's Knoxville & Suisun Express
1857-1864 Benecia, Fairfield, Suisun City & Knoxville
Davis & Bagley's Express Co.
1852-1853 Marysville to Feather R. camps
Hamlet **Davis & Co. Express**
1849-1850 S.F., Sacramento to Nevada City
Davis & Hirst's Express
1850-1851 Nevada City to S.F.
Davis Line
1851 Sacramento to Marysville
Davis & Organ Express
1855 El Dorado to Volcano
Dean's Express
1855-1858 Marysville to Bidwells Bar
Dearing & Co.'s Express
1852 Marysville to Feather R. camps
Denig's Express
1857-1863 San Andreas to Southern Mines camps
Dillon, Hedge & Co.
1853-1855 Stockton to Sonora; San Jose to Monterey
M. J. Dooley & Co.
1863 Copperopolis to Shasta
Dornin's Express
1861 North San Juan, Cherokee, Humbug
Douglas & Reany's Eureka Line
1851-1852 Stockton, Angels Camp, Carsons Creek
Dry Creek Opposition Line of Stages
1851 Sacramento to Mokelumne Hill
Dugan & Co.'s Oregon Express
1852-1853 Shasta area
Duke's Express
1864 Los Angeles to Prescott (Ariz.)
Durgan's Express
1850s Nevada City to Dutch Flat via Red Dog
Durkee's Empire Stage Line
1850-1851 Sacramento to Marysville
D. W. Earl's Express
1862-1863 Virginia City (Nev.) to Sacramento
Elliott's Atlantic & Pacific Express
1851-1855 Operated over both Panama, Nicaragua Lines
Ellis' Express
1854 Coloma to Georgetown
Everett's Pedestrian Express
1861-1864 Knights Ferry to Mariposa Mines
Fisher's Express
1857 Weaverville to Canon City (Northern Mines)
Fisher's Express & Stage Line (A. N. Fisher)
1855-1859 Stockton, Columbia, Sonora (Southern Mines)
Fleming's Chowchilla Copper Mines Stage
1866 Mariposa to Haskell's Works
The **Forest Line**
1853-1854 Sacramento to Sonora
1857 Sacramento, Jackson, Mokelumne Hill
Curry **Foster's Express**
1859 Washoe Mines (Nev.) to Nevada City (Cal.)

Fowler's Express
1858 Sacramento, Tehama, Red Bluff
Furman, McIntyre & Co.'s Express
1851 Middle Yuba to Grass Valley
Garland & Dean Express
1854 Oroville to Quincy
Garland Quincy & Crescent Mills Stage Line
1866-1872 Oroville, Quincy, Crescent Mills
Garland's Quincy & Indian Valley Stage Line
1865-1866 Quincy, Crescent Mills, Indian Valley
Gelait, Matteson & Garland Express
1867 Genoa (Nev.) to Murphy's
Gilbert's Express
1866-1871 Shasta to South Fork
Gilkey's Klamath & Salmon Rivers Express
1857-1860 Union, Orleans Bar; to camps on Klamath, Salmon Rs.
M. Goldie's Pony Express
1864 Sonora to Sugar Pine and Mono
George L. & Henry **Greathouse Express**
1854-1855 Shasta to Callahan Ranch
Charles **Greene Express**
1852 Sacramento to Sonora via Ione Valley
Greene & Voghan's Express
1861 Sacramento, Stockton, Sonora
Gregory & Wells Express
1850s Nevada City to Moore's Flat and Eureka
Grip Express
1864 S.F., San Jose, Watsonville, Santa Cruz
Grumont Co.
1864 S.F., San Jose, Santa Clara, Watsonville, Santa Cruz, Almaden Mines
Hall & Crandall Stage Line
1850-1854 S.F., San Jose to Monterey via Gilroy; Sacramento to Marysville; Marysville to Yreka
Halstead's Express
1849 Sacramento, Oroville, Quincy
Hamilton & Kellogg's Express
1850-1851 Stockton to Sonora
Hamilton & Smith's Stage Line
1857 Folsom, Jackson
Hanford & Downes' Express
1853 Volcano to neighboring camps
Hanson's Amador Stage Line
1864 Amador to Austin
J. A. Hardgraves & Son
1864 Reno and Virginia City to Greenville (Cal.)
Harrill's Express
1853 Shasta to Pitt River camps
Harris & Co.'s Express
1854 Placerville to Yankee Jim's
Hart & Co.'s Express
1855 Crescent City to Yreka
Harvey's Tri-Weekly Stage
1866 El Dorado to Drytown

Hatch & Co.'s Express
1854-1866 Placerville to Mud Springs
Haven & Livingston's Express
1849-1850 Stockton, S.F. to New York
Hawley's Express
1849-1850 S.F., Benecia, Sacramento
Haworth & Swift U.S. Mail Line, The Telegraph Line
1851-1860 Sacramento to Nevada City
Haws & Harmon's Express
1854 Sacramento to Iowa Hill
Hay Fork Express Co. (J. S. Hoyt)
1867-1873 Weaverville to Hay Fork
Henkle & Co.'s Express
1852 Marysville to Shasta, Weaverville
Henley, McKnight & Co.'s Express
1850 Sacramento to S.F.
Hice & Thomas Express
1859-1860 Hortinas to Visalia
Hickman's Express
1862 Nevada City, Dutch Flat via You Bet, Red Dog
Hoff's Pescadero & Santa Cruz Express
1867 Pescadero to Santa Cruz
Hoffman & Little's Express
1851 Nevada City to Sacramento
Holdridge & Keyser Opposition Line
1856 Sacramento, Placerville to Diamond Springs
M. H. Hoy's Passenger Express
1857 Sonora to San Jose
Hubb's & Co. Express
1851 Marysville to Downieville
Hughes & Co. Express
1852 Marysville to Downieville
Humboldt & Red Bluff Express
1863 Humboldt to Red Bluff to Eureka
A. C. Hunnewell's Miners Express
1854-1855 Columbia to Southern Mines camps
Husband's Express
1858 Yreka to Scott and Klamath R. camps
Idaho & California Stage Co.
1865-1866 Ruby City (Id.) to Chico (Cal.)
Idaho Stage Co.
1865 Idaho to Chico (Cal.)
Jackson's Express
1857-1858 Mining camps in El Dorado Co.
D. Jacobs' Express
1852 Stockton to Coulterville
Jatunn & Parks Pilot Line
1852 Auburn to Marysville
Jenkins & McConnell's Express
1853-1854 Coloma, Mormon Island, Sacramento
Johnson & Davis Express
1850 Sacramento to Marysville
Johnson's Opposition Line
1850-1852 Sacramento to Marysville

Jones & Co.'s Express
1852-1862 Sacramento to Marysville, Feather R. camps
Joslin's Ridgeville Express
1856 Weaverville to Ridgeville
Keeler, DePeyster & Co.'s Express
1850-1851 Camps on Stanislaus, Tuolumne, Merced, Mariposa Rs.
Keeler's Express
1850 Stockton to Merced
Kelty & Co.
1851 Columbia to Sonora
Kimball's Express
1849 Sacramento to S.F.
Leonard **King's Stage Line**
1854 Grass Valley to Illinoistown
Kistle & Moffet's Yreka & Humbug Express
1858-1859 Yreka to Humbug
C. A. Latimer Pioneer Express
1860 San Andreas to Calaveras
Lebdell's Pioneer Stage Line (J. B. Lebdell)
1855 Nevada City via Grass Valley to Iowa Hill
Craven **Lee's New Express**
1858 Salmon River to Weaverville
W. Linton's Express
1858 Weaverville to Canon City
Lloyd's Express
1862 San Andreas to Calaveritis
W. H. Lobdell & Co. Stage Line
1855 Volcano to Jackson
Long & Kenyon's Shasta & Red Bluff Stage
1857 Shasta to Red Bluff
W. D. Long's Express
1861-1862 Nevada City to Dutch Flat
Loop & Shafer's Weaverville & Uniontown Express
1859 Trinity River towns to Humboldt Bay
Lount & Co.'s Express
1849-1850 Stockton, Sacramento to New York
Love's Pony Express
1865 San Andreas to Robinson's Ferry
Lovett's San Juan & Los Angeles Stage Line
1867 Los Angeles, Ventura to Paso Robles, San Juan
Luther's Package & Letter Express
1850 S.F. to San Jose
Lynde & Co.'s Pony Express
1860 Sonora to Walkers River, Mono Lake
C. McCammon & Co.'s Accommodation Line
1866 Placerville, Diamond Springs to Shingle Springs
George **McCarter Express**
1859 Placerville (Cal.) to Carson City (Nev.)
McCarthy & Cooper Express
1856 Sonora to Columbia
McCombe & Co.
1854 Shasta, Yreka to Jacksonville (Ore.)
McCombe & McLaughlin Express
1854 S.F., San Jose to Santa Clara, Mountain View; also Shasta, Callahans Ranch, Yreka and Portland (Ore.)

McConnell & Co.'s Express
1854 Coloma to Sacramento
McConnell's California Stage Line
1858 Yreka to Shasta
McCullough's Yreka & Humbug Express
1860s Yreka, Humbug, and Klamath R.
McDonald's Express
1862-1869 Shasta to Roaring River, Horsetown, Arbuckle
McDowell & Co.'s Express
1850 S.F., Sacramento, to Nevada City
J. F. McFarlane's Express
1860 Sonora to Mono Lake
McGee's Pitt River & Shasta Express
1853 Shasta to Pitt River diggings
McKenzie & Co.'s Express
1856-1857 Napa City to the Geysers
McKnight's Express
1850 Sacramento to San Francisco
C. McLaughlin & Co.
1858 Oakland to San Jose, Santa Cruz, Monterey
McLaughlin & Mann's Express
1853 Marysville to Downieville
McLaughlin's U.S. Mail Line
1853-1856 Camps on Yuba R.
McLearn's Express
1860 Nevada City to mining camps
Malzeard's Express
1849 Sacramento to Mormon Island
Mann's San Jose & Los Angeles Stage Line
1866 Los Angeles, Santa Barbara, San Jose
Martin's Express
1857 Angel's to Murphy's and French Camp
Marvin's Express
1850-1854 S.F. to Stockton to Southern Mines
John N. **Massey's Yreka & Humbug Express**
1859 Yreka, Humbug and Klamath R.
Maurison & Ackley's United States Mail Line
1850 San Francisco to San Jose
Maurison & Co.'s Express
1849-1850 Stockton to Stanislaus R. mines
Maynard's Hay Fork & Weaverville Express
1866-1868 Weaverville to Hay Fork
Meek & Vera's Express
Meek & Powers' Express
Meek & Aubrey's Express
1856-1881 (chronologically listed) Marysville, Parks Bar, Timbuctoo, Rose's Bar, Wood's Bar
Menner's Express Line
1867 French Corral, North San Juan, North Bloomfield
Messenger & Co.'s Express
1850 Stockton to Sacramento
F. J. Mette & Co.'s Express
1859 Oakland to Martinez via San Pablo
Miller & Hopkins Half Moon Bay & Accommodation Line
-1862 San Francisco to Half Moon Bay

Miller's United States & California Express
1851 California to U.S.
Monnet & Co.'s Humbug Express & Stage Line
1864-1868 Yreka, Humbug to Klamath R. camps
Monson & Co.'s Express
1850 S.F., Sacramento, Nevada City, Greenwood
Monterey & San Luis Obispo U.S. Mail Stage Line
1858 Monterey to San Luis Obispo
W. F. Montgomery's Express
1851 S.F. to Humboldt and Klamath R.
Mooney & Merrick's Yreka & Humbug Express
1860 Yreka to Humbug, Klamath R.
Joseph **Mooney's Express**
1860 Yreka to Humbug, Klamath R.
Moore & Edmundson's Passenger Express
1850 Stockton to Sonora
C. F. Moore Express
-1868 Shasta to Janesville
Morgan's Austin & Ione Stage & Express Line
1864 Austin to Ione (Cal.)
H. W. Morgan's Express
1856 Sonora to Robinson's Ferry
Morrow's Express
1854-1855 Marysville to Bidwells Bar
Muma & Co.'s Express
1856 La Porte to Plumas County camps
Murphys & Carson Valley Express
1858 Murphys to Carson Valley (Nev.)
Murray & Armstrong's Shasta & Arbuckle Express
1854 Shasta to Arbuckle
Sam **Nelly & Co.**
1854 Campo Seco from Stockton via Lancha Plana, Chili Camp
Nevada Stage Co., McCue Line
1857-1867 Sacramento to Carson City (Nev.); Napa, Vallejo, Benecia
Nicholas' Yreka & Humbug Express
1862-1864 Yreka, Humbug to Ridgeville
Noble & Walters Express
1868 San Bernardino to Prescott (Ariz.)
North Fork, Weaverville & Douglas City Express
1866-1867 North Fork, Weaverville to Douglas City
North San Juan & Humbug City Express
1858-1859 North San Juan to Humbug City
Old Line Stage
1859-1863 Columbia to Sonora
Opposition Stage Line
1856-1857 Stockton to Columbia via Knight's Ferry, Jamestown, Springfield
Ormsby's Coloma Stage Line
1851-1853 Sacramento to El Dorado via Coloma, Placerville
Orr's Forest Stage Line
1863 Michigan Bluff, Forest Hill, Yankee Jims, Todd's Valley
Page & Hoffman Express
1849-1850 S.F., Sacramento to Mud Springs
Page's Opposition Line
1857 Stockton to Columbia via Knight's Ferry, Jamestown, Springfield
Palache & Strange's Express
1855 Murphys to Calaveras Co. camps
Parker's Express
1866 Indian Creek to Weaverville
Patterson's U.S. Mail Line
1852 Sacramento, Coloma to Placerville
N. Pauly & Co.'s Express
1864 Gibsonville to camps on Gibsonville Ridge
Peel & Robertson's Pioneer Express
1859 Quincy to Honey Lake
The **Peoples Opposition Line**
1852-1853 Ophir, Auburn, Yankee Diggings
Pierce & Francis' Chico & Idaho Stage Line
1865 Chico via Susanville to Ruby City (Id.)
Pioneer U.S. Mail Line
1858 Sacramento to Stockton
Piper's Ridgeville Express
1856 Weaverville to Ridgeville
Pittsburg & Copper City Express
1864-1865 Shasta to Pittsburg and Copper City
Pleninger & Co.'s Humbug Express
1868-1872 Yreka, Humbug to Klamath R. camps
Putnam & Shuey's Daily Express
1850s Oakland, San Leandro, San Lorenzo, Hayward
R. H. Rand's Honey Lake Express
1858 Quincy to Honey Lake
W. C. Randolph Express
1849-1851 Stockton to Mariposa Co. mines
Raphael's Express
1852 Stockton to Calaveras Co. mines
Rawling's Express
1852-1853 Marysville to Foster's Bar
Ray & Danforth Express
1853 Marysville to Oregon towns
A. E. Raynes & Co.'s Express
1851 Trinidad to Bestville on Salmon R.
Read's Express
1857 Placerville to Indian Diggings
Red Bluff & Humboldt Express
1862 Red Bluff to Humboldt mines
Reed & Kendall's Express
1856 San Francisco to San Jose
Rice's Express Co.
1850 S.F. to Middle and South Yuba R.
William **Rightmire's Stages**
1850 Sacramento to Auburn, and North, Middle Forks Yuba and American Rs.
Riley's Express
1856 Folsom to Michigan Bar
Rinewald's Express
1866-1868 Yreka to Scott's Bar via Fort Jones, Indian Creek

Rodgers Express Co.
1856 Yreka to Scott R. mines
Rogers & Co.'s Stage
1860 Marysville, Downieville, Virginia City (Nev.)
Rollins Express
1852 Marysville to Fosters Bar
Rosencrants & Cutts U.S. Mail Line
1852 Sacramento, Ophir and Auburn
E. Rowe & Co.'s Express
1857 Shasta to Weaverville
Ruby's Peoples Line
1851 Marysville to Nevada City via Rough & Ready, Grass Valley
Sacramento & Coloma Stage Line
1850 Sacramento to Coloma
Sacramento & Nevada City Express Line
1850 Sacramento, Rough & Ready, Nevada City
Sale & Johnson's Express Train
1850s or 1860s Nevada City to Grass Valley
Salmon River Express Co.
1857 Yreka to Aetna Mills
John G. **Sanborn's Express Co.**
1859 Weaverville to Taylor's Flat
San Ramon Valley Line
1866 Walnut Creek and Moraga Valley
Sawyers Bar & Scott Valley Express
1867-1869 Yreka, Rough & Ready, Aetna Mills, Unionville
Sawyers Express
1856-1857 Weaverville to Canon City
Scott River, Sawyers Bar & Scott Valley Express
1869-1871 Yreka to Scott River mines
Scott Valley Express
1862-1864 Yreka, Fort Jones, Rough & Ready, Unionville
John **Sharp Express**
1850-1851 Sacramento to Marysville
Shasta & Weaverville U.S. Mail & Express
1868-1871 Shasta to Weaverville
Silman & Carter's Express
1857 Slough City to Visalia via Millerton
Singer, Dean & Co.'s Express
1852-1855 Marysville to Feather R. camps
Singer & Fargo's Express
1852 Marysville to Feather R. camps
Sissons, Dolly & Co.'s Express
1866 Stockton to San Andreas, Mokelumne Hill
Sissons' Mokelumne Hill & Sacramento Stages
1867 Sonora, Mokelumne and Sacramento
Slocum & Morse's Pilot Line of Stages
1851 Sacramento to Marysville
Smith's Camptonville & Downieville Stage (T. G. Smith)
1859 Camptonville to Downieville
Smith's Express
1860-1861 North San Juan to Columbia Hill
Smith's Pioneer Stage Line (G. Canning Smith)
1863 S.F., Petaluma, Ukiah, Greenwood
Smith's U.S. Mail Line (John Smith)
1851-1855 Sacramento, Stockton, Sonora, Mariposa
1855-1856 Benecia, Vallejo, Napa City
Smith's U.S. Mail Line (William Smith)
1864-1865 Oroville, Indian Valley to Quincy
A. J. Snow Express
1854 Sonora to Southern Mines camps
Sonora & Aurora Express
1864 Sonora to Sugar Pine
Sonora & Coulterville Stage Line
1864 Sonora to Coulterville
South Coast Express
1851-1857 S.F., Monterey, San Pedro to San Diego
Southern Accommodation Line
1853 San Francisco to San Diego
Stark's California Combination Stage Line
1854 Yreka to Crescent City
F. S. Stevens Pioneer Stage Line
1851-1854 Sacramento to Placerville
Stimson's New York & California Express
1852-1854 San Francisco to New York
Stone & Co.'s Express
1854 Santa Cruz, Watsonville, Gilroy, San Jose, S.F.
Sulloway & Co.
1857 Soda Springs to Yreka
Sunderland's Stage Line
1857-1859 Sacramento to Placerville; Folsom to Coloma, Georgetown
E. B. Sutton & Co.'s Express
1850 S.F., Panama, New York
Swain & Bros. Express
1862 Salmon River mines
Swift & Sink's Stage Line
1859 Cloverdale, Ukiah and Capella
Talmadge's Pioneer Stage Line
1854-1855 S.F., Oakland to San Jose
Terry's Pioneer Stage Line
1859 Mariposa to Bear Valley
Theall's Express
1858 North San Juan to Minnesota
Thede's Canon City & Weaverville Express
1866-1868 Canon City to Weaverville
Thede's Trinity River Express
1861 Weaverville to North Fork
Thomas Stage Line
1855 Coloma to Diamond Springs
Thompson's Oroville & Indian Valley Stage Lines
1866 Oroville, Indian Valley to Goodyears Bar
Thrall's Stage Line
1866 Mokelumne Hill to West Point
Tibbetts U.S. Mail Line
1859-1864 Petaluma to Tomales, Bodega
Tiffany & Co.'s Express
1849 Sacramento to Vernon, Coloma, Mormon Island

Tinsley's North Fork & Weaverville Freight & Passenger Express
1866 Weaverville to North Fork Trinity R.
Titus & Barton's Express
1854 San Jose to Monterey
John S. **Tolles Express**
1849-1852 S.F. to Feather River district
Tomlinson & Co. Express Line
1867 Los Angeles, San Bernardino, Fort Yuma (Ariz.)
E. W. Tracy & Co.'s Express
1855 Shasta to S.F., Weaverville; Yreka to Jacksonville (Ore.)
Tracy & Freeman's Express
1857 Carson Valley (Nev.) to Placerville (Cal.)
Tracy & Harmon Express
1863 Shasta, Copper City and Pittsburg
Trescott's Daily Stage Line
1866 Placerville to Auburn
Turner's Express
1860 Humbug to Yreka
T'vault & Co.'s Oregon & Shasta Express
1852 Oregon City to Shasta City
Uhl's Express
1850 Shasta to Dog Creek
Upman Express
1849 Sacramento to San Francisco
W. G. Van Arsdale Express
1851 Sonora to Columbia
Van Brunt & Johnson's Stage Line
1850 Marysville to Sacramento
Van Wyck's Express
1860 Yreka, Indian Creek, Salmon River
Walker's River Express
1859 Sonora to Walker's River diggings
Walsh & Spaulding Express
1864 Virginia City (Nev.) to Susanville
Ward & Ashe's Eureka Stage Line
1851 Mokelumne Hill to Stockton
Ward's Stage Line
1850 Sacramento to Marysville
G. S. Wardle's Special Express to New York
1849 San Francisco to New York
Watkins Express & Passenger Line
1865-1866 Shasta to Red Bluff
Waugaman's Express
1857 Placerville to Diamond Springs, Grizzly Flats
Weaverville & Hay Fork Express
1867 Weaverville to Hay Fork
Weaverville & Lewiston Express
1867 Weaverville to Lewiston
Weaverville & Lower Trinity Express
1865-1866 Weaverville to Lower Trinity R.
Weaverville & Shasta U.S. Mail & Express
1866-1867 Weaverville to Shasta
Weaverville & Uniontown Express
1859 Weaverville to Uniontown
Weld & Co.'s Inland Express
1849 S.F., Benecia, Sacramento
Wesson & Co. Express
1850 Marysville to all Butte Co. & Plumas Co.
West & Co.'s North and South American Express
1850 S.F. to New York and South America
Westley & Wilder Express
1861 Columbia to Gold Hill
Wheeler, Baxter & Co.'s Express
1851 San Francisco to Sacramento
Wheeler Brothers Mountain Express
1864 Marysville to Feather R. and Gibsonville Ridge
Whistman Express
1849-1850 San Francisco to San Jose
White & Crooks Scott River Express
1856 Scott River mines
Whitney & Ely's Atlantic & Pacific Express
1849 S.F., New York, South American ports
R. Wiley's Express
-1857 Points on Klamath and Salmon Rs.
Willis Pioneer Stage Line
1860 Nevada City to Red Dog
Wilson's Stage Line
1866 Austin to Ione
Wilson's Telegraph Line of Stages
1857 Nevada City to San Juan
Wood's Yuba Express
1851 Downieville to Sacramento
Granville **Woods Express**
1866-1867 Virginia City (Nev.) to Susanville
Woodward & Chandler Express
1854-1856 Shasta to Horsetown
Richard **Yarnold's Carson Valley Express**
1857 Placerville via Carson Valley to Genoa (Nev.)
A. C. Yoacham's Independent Express
1856 Sonora to Coulterville
Yreka & Sawyers Bar Express
1864-1867 Yreka to Sawyers Bar

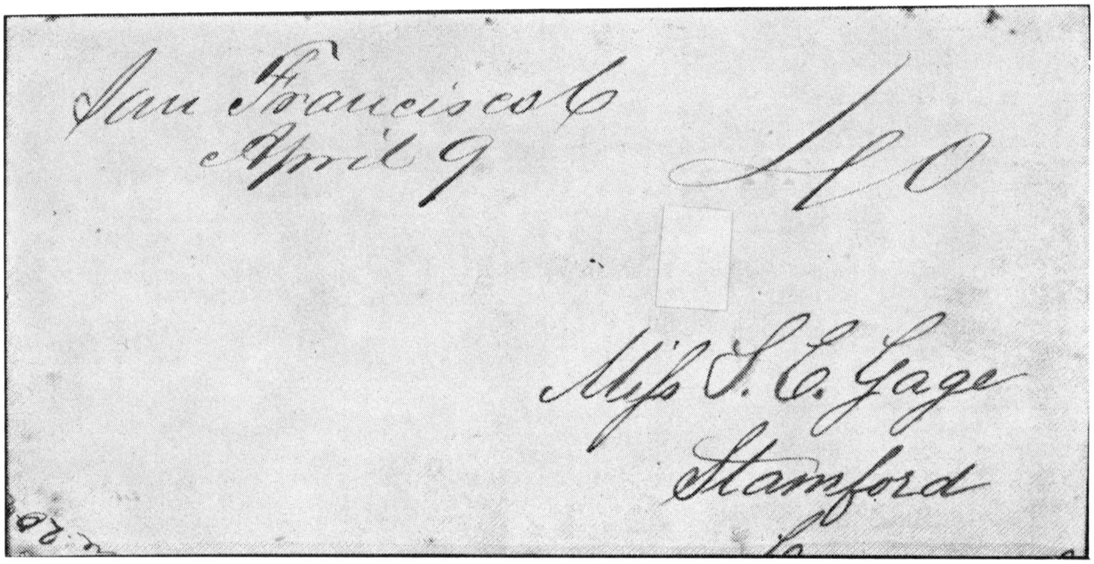

Manuscript "San FranciscoC/April 9" (1849) on unpaid letter to Conn. Headed "South Soleda bay." Carried east on the first contract sailing of *Oregon* from San Francisco April 12.

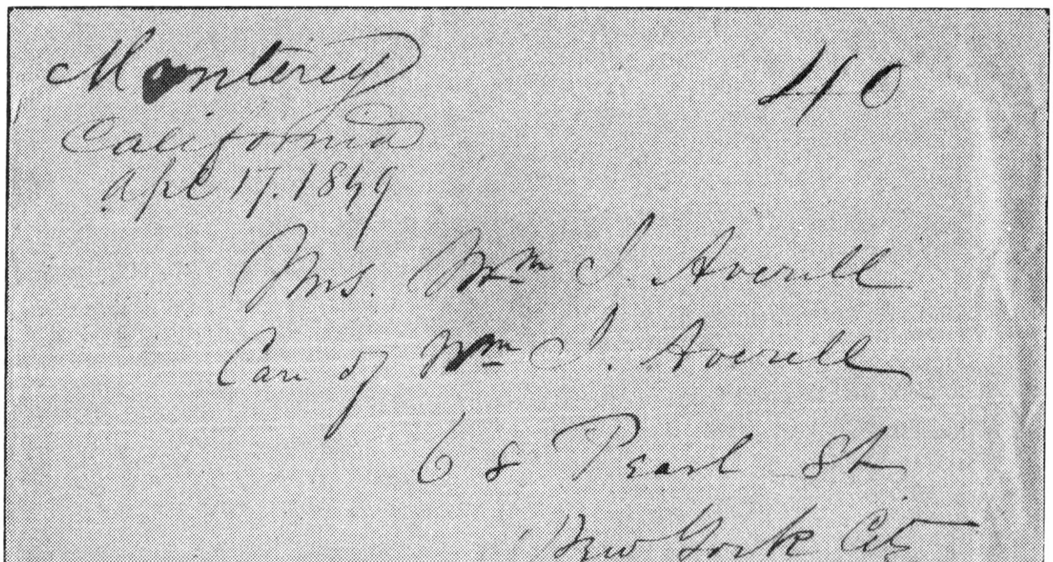

Manuscript postmark of "Monterey/Apl 17, 1849", on folded letter to New York, due 40. The earliest Monterey postmark yet recorded.

Appendix 3: Manuscript Postmarks Through 1869

Town	*Date*
Agua Fria	1851
Alameda	1856
Alamo	1858, 1867, 1868
Alleghany	1861, 1864, 1866
Alpha	1855
Alvarado	1853, 1855
Alviso	1859, 1860, 1861
Amador City	1863
American Ranch	1860
Angels	1855
Antelope	1856
Anthony House	1868
Antioch	1851
Auburn	1852, 1859
Aukum	1859
Aurora	1861
Aurum City	1852
Bangor	1858, 1866
Bath	1858, 1866
Bear Valley	1861
Belmont	1857, 1859, 1860, 1862
Benicia	1849, 1850
Bidwells Bar	1851
Big Bar	1852
Big Meadows	1869
Binghamton	1865
Bloomfield	1856.
Bodega	1852
Bondville	1855
Bridgeport	1867
Brighton	1865
Brooklyn	1855, 1856, 1857
Browns Valley	1864, 1865
Brownsville	1865
Brush Creek	1856
Buckeye (Shasta)	1855
Buckeye (Yolo)	1855, 1862
Buckners P.O.	1854
Buckner City	1857
Bucksport	1853, 1855, 1860
Buck's Ranch	1865
Buena Vista	1867, 1869
Bullionae	1869
Burnt Ranch	1863
Burwood	1859
Butte City	1857
Butte Mills	1857, 1859, 1860
Butte Valley	1861
Butteville	1863
Cache Creek	1854
Cahto	1864
Callahan's Ranch	1859

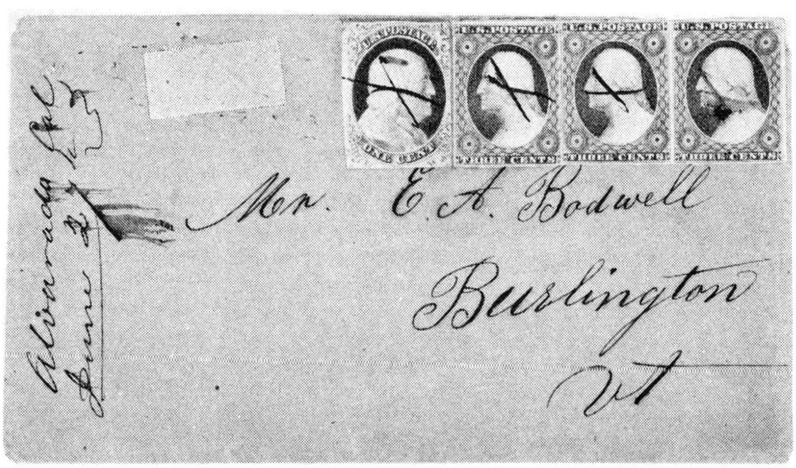

Alvarado postmark on envelope to Vt., with stamps of 1851 issue paying postage.

Buckners on 10¢ Nesbitt to Vt. in 1857.

Town	Date
Camp Seco	1854
Capay	1868
Carsons Creek	1853
Cascade City	1858
Castroville	1859
Centerville	1851, 1852
Central House	1861
Charleston	1859
Chico	1856
Chinese Camp	1854
Clairville	1858
Clarkesville	1856
Clifton	1866
Clinton	1856
Clipper Gap	1866
Cloverdale	1859
Cold Spring	1852
Coleville	1869
Collinsville	1862
Coloma	1851
Colusa	1852, 1854, 1855
Contra Costa	1851
Coon Creek	1859, 1860

Town	Date
Copperopolis	1861
Corbets Ranch	1856
Cordelia	1855
Consumnes	1852
Cottage Grove	1862
Cottonwood	1855, 1856, 1857, 1863
Crescent City	1854, 1857
Crystal Lake	1866
Damascus	1862
Danville	1861, 1862
Davisville	1868
Denverton	1864
Diamond Springs	1853, 1855
Dobbin's Ranch(e)	1851, 1852
Don Pedros Bar	1860
Double Springs	1851
Dougherty Station	1867
Downieville	1852, 1857
Dry Creek	1854
Drytown	1851, 1854
Duncan Mills	1862

Clarksville on 1856 cover to Vt. with scene at post office when the steamer arrives.

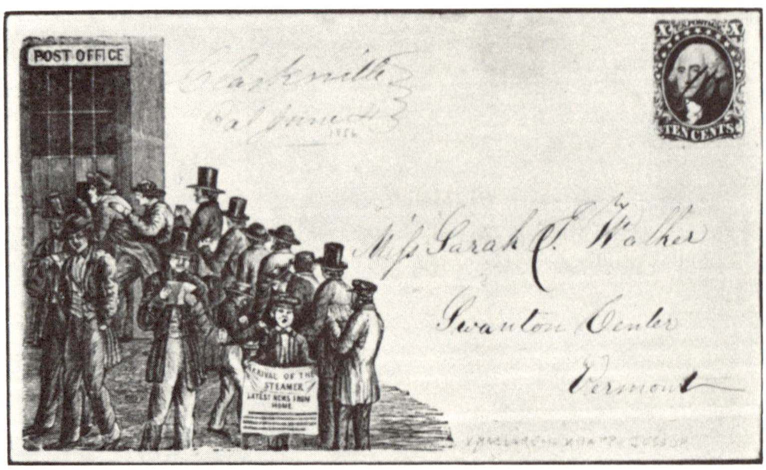

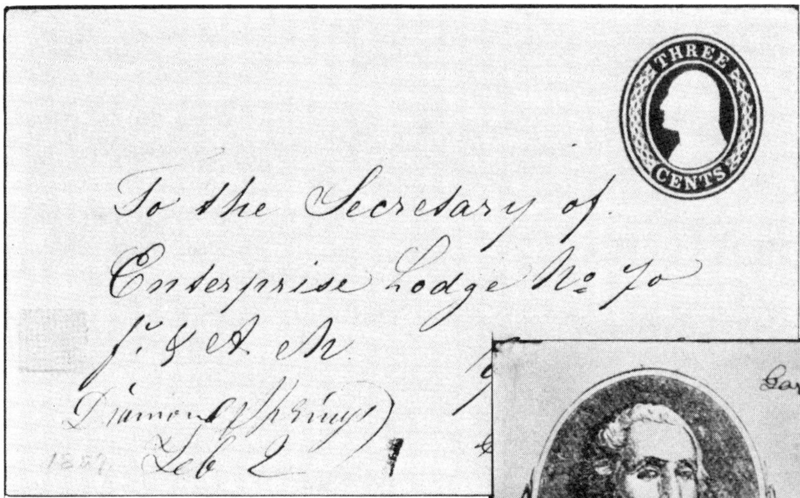

Diamond Springs on 3¢ Nesbitt to Yuba City.

Garden Valley on double letter to Me. on Washington patriotic.

Town	*Date*
Eel River	1857
El Dorado	1861
Elizabethtown	1855
Elk Grove	1858, 1865, 1867
Emerys Crossing	1856, 1858
Emerys Xing	1856
Empire Ranch	1855, 1856
Esmeralda	1862, 1863
Etna Mills	1863
Eureka North	1858
Evan's Ranch	1868
Fairfield	1858
Fair Play	1868
Farmington	1868
Fiddletown	1853
Fir Cap	1869
Forbestown	1854, 1856
Forest Bar	1853
Forest City	1861
Forest Hill	1860
Forest Home	1868
Forks of Salmon	1860, 1862
Fort Bidwell	1869
Fort Jones	1861
Fort Tejon	1860
Foster's Bar	1853
Franklin	1869

Town	*Date*
Fremont	1851
French Camp	1854
French Corral	1860
French Gulch	1858
French Town	1853, 1858
Fresno City	1860
Fugitt	1857, 1860
Galt	1869
Garden Valley	1852, 1859
Garrotte	1855, 1861, 1867
Georgetown	1851
Gibsonville	1855
Gilroy	1851
Gold Run	1864, 1865
Goodyear Bar	1853
Grafton	1854
Graniteville	1867
Grass Valley	1853
Green Springs	1851, 1858
Green Valley	1854
Greenwood	1853, 1854
Grizzly Bear House	1861, 1864
Grizzly Flats	1857
Grove City	1861
Hamilton	1851, 1852, 1861
Hansonville	1857

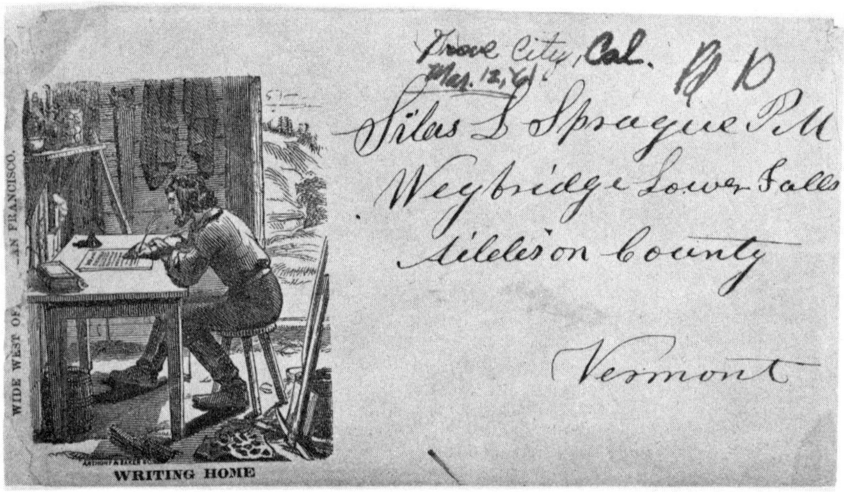

Grove City on miner's envelope "Writing Home," published by Wide West.

Town	*Date*
Happy Camp	1858
Harrisburgh	1865
Havilah	1866
Hayfork	1861
Haywood	1860
Henley	1857, 1860, 1861, 1867
Hicksville	1860
Hitchcock's Ranch	1863
Hopeton	1867
Hopewell	1860
Hornitas	1856
Horr's Ranch	1851
Horse Town	1852
Howe	1867
Hydesville	1861
Illinoistown	1855, 1861
Independence	1867
Indian Diggings	1853
Indian Gulch	1855
Indian Springs	1861
Ione	1856
Ione City	1861
Ione Valley	1862
Iowa City	1854, 1855
Jacinto	1864
Jackson (Amador)	1855
Jacksonville	1851
Jamestown	1853, 1854
Janesville	1862
Jenny Lind	1863, 1864, 1865
Johnsons Ranch	1864, 1869
Junction New York (On the Pacific)	1849, 1850
Keysville	1857, 1862
Knights Ferry	1851
Konigsburg	1863
Lafayette	1857
LaGrange	1857

Iowa City to Mich., unpaid 10 on miner's envelope, "Miner's Coat of Arms," Barber & Baker.

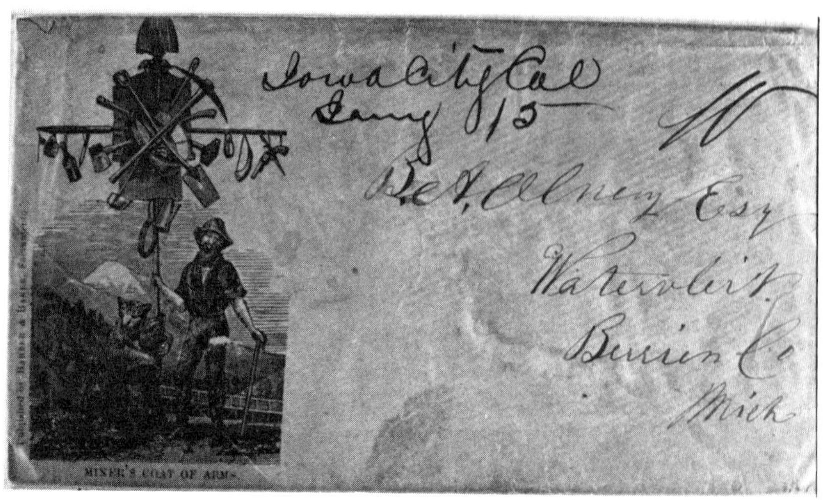

Knight's Ferry as depicted on lettersheet.

Town	*Date*
Lake Bigler	1863
Lancha Plana	1859, 1860, 1863
LaPorte	1857
Lassen's	1857
Last Chance	1867
Lawrence	1867
Lench's Store	1859
Lewiston	1860
Lexington	1861
Liberty	1861
Lisbon	1861
Little York	1856, 1861, 1862
Lockford	1861, 1864
Long Bar	1858
Long Valley	1869
Longville	1863, 1866
Los Angeles	1850, 1851
Louisville	1851, 1852
Loving's Ferry	1856, 1859
Lower Lake	1868
Lynchburg, Martinas Valley	1849
Maine Prairie	1861, 1863, 1864
Mariposa	1852
Mark West	1865
Marsh's Landing	1853
Martinez	1851
Martinsburg	1869
Marysville	1850
Maxwell Creek	1858
McCartysville	1855
M. Valley (prob. Meadow Valley)	1855
Meadow Valley	1856
Merry Oaks	1860
Millerton	1855, 1858
Milpitas	1858, 1859, 1862
Minersville	1857
Mission San Jose	1850
Mokelumne Hill	1852
Mok Hill	1851
Monitor	1863
Monoville P.O.	1861
Monroeville	1851, 1852
Monte	1867
Monterey	1849
Montezuma	1854
Moons Ranch	1853, 1854, 1855, 1860
Moore's Flat	1857
Mormon Island	1851
Morris Bar	1851
Mosquito	1861
Mountain Ranch	1862
Mountain Springs	1860
Mountain View	1867
Mountain Well	1861
Mount Bullion	1863
Mt. Ophir	1853
Mt. Springs	1857
Mud Spring	1851
Murphy's	1857
Napa City	1851
Nappa	1850
Natividad	1858, 1866
Neilsburgh	1855
Nelson's Creek	1855
Nevada	1851
Nevada City	1850, 1851
Newcastle	1865, 1866
Newtown	1854
Nicolaus	1853, 1854
North Bloomfield	1857
North Columbia	1860
Oak Grove	1868
Oak Grove Farm	1855
Oakland	1856
Oak Springs	1851, 1852
Omega	1859, 1863

Three cent Nesbitt to Gold Hill with ms. postmark of La Porte.

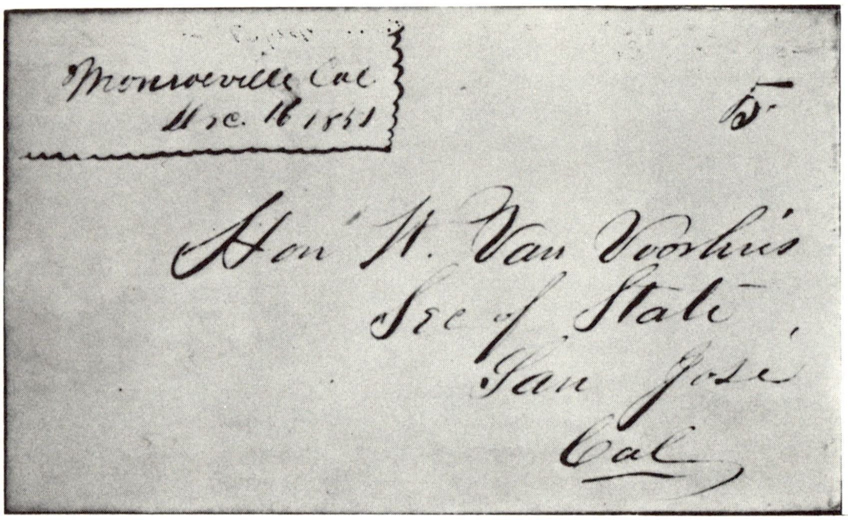

Monroeville on unpaid 1851 cover to San Jose.

Moon's Ranch on 1855 cover to Ill. with Noisy Carriers corner card.

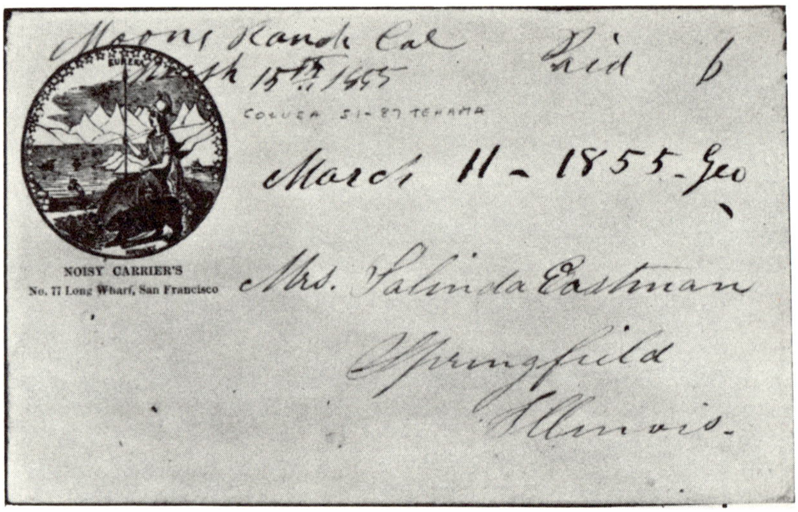

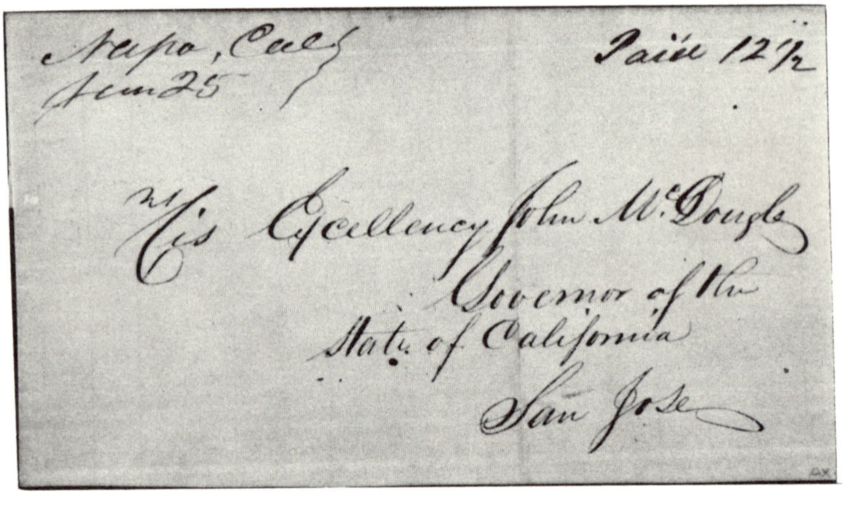

Napa folded letter to San Jose "Paid 12½."

Miners improvise a post office.

Town	*Date*
Onisbo	1863
Ophirville	1857
Oregon House	1851, 1861
Orleans	1865
Oro City	1855
Oro Fino	1862
Oroville	1856
Orr's Ranch	1858
Ottitiewa	1856
Owsleys Bar	1864
Pacheco	1859, 1860
Painesville	1866
Paradise	1869
Park's Bar	1854, 1855
Patterson	1856
Pea Vine	1856
Pescadero	1866
Petaluma	1853
Petrolia	1866
Phillips Flat	1858
Pilot Hill	1856
Pine Grove	1861
Pino	1869
Placerville	1850
Plainsburg	1869
Pleasant Grove Creek	1867
Pleasanton	1867
Plum Valley	1856
Poland	1861, 1867
Port Wine	1861
Poverty Bar	1858
Prairie	1860
Princeton	1857
Quartzburg	1853, 1854
Quartz Valley	1861
Quincy	1865
Rabbittown	1855
Rattlesnake Bar	1854, 1861
Red Bluff	1856
Red Bluffs	1859
Red Dog	1855
Redwood	1853
Redwood City	1856, 1858, 1861, 1862
Reynolds Ferry	1859
Rich Gulch	1860
Rio Vista	1858
R. Views (English for Rio Vista)	1862
Rock Creek	1867
Rocklin	1868
Rockville	1861, 1867

Sacramento "Paid 40" to NY, 1849.

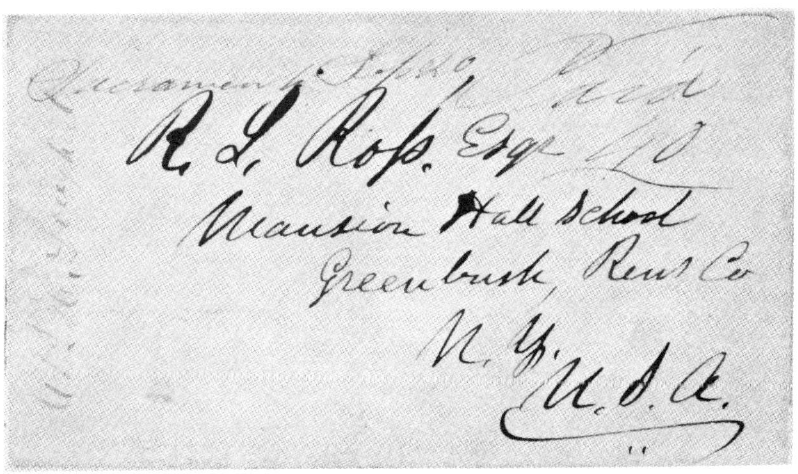

Salmon Falls on envelope to Ill. with 10¢ 1861.

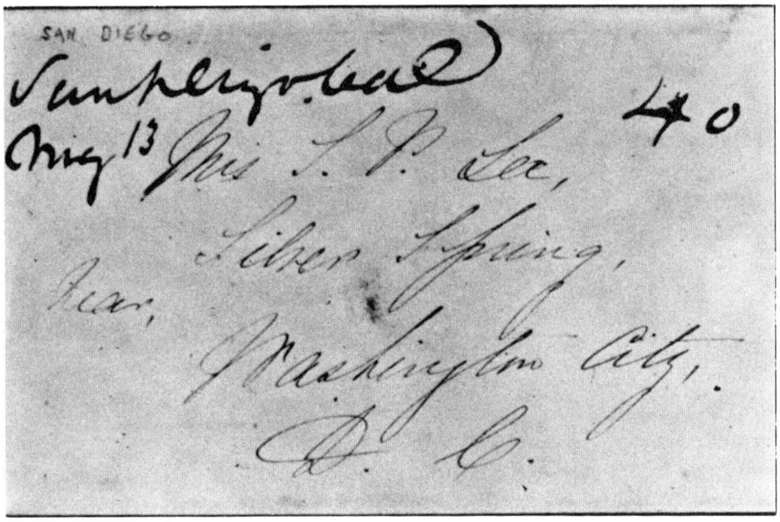

San Diego to Washington, D.C., unpaid, rated 40.

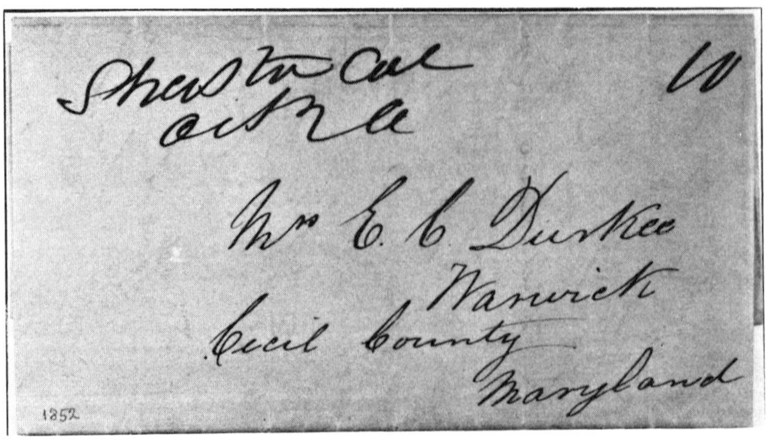

Shasta to Maryland in 1852. Unpaid and rated 10 due.

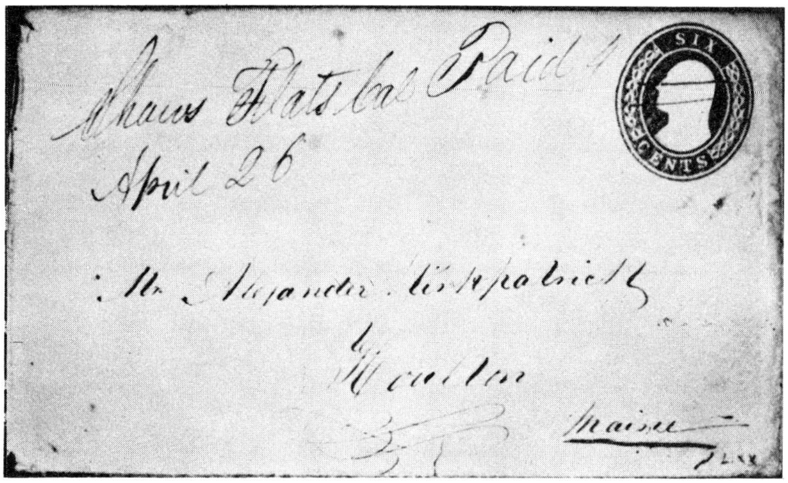

Shaws Flat to Me. with 6¢ Nesbitt.

Town	*Date*
Rough and Ready	1851, 1853, 1854, 1856
Round Tent	1857, 1859
Rush Creek, Cal.	1853
Russian River	1857
Sacramento	1849
Saint Louis	1855
Salmon Falls	1854, 1862
San Andreas	1854
San Antonio	1859
San Buenaventura	1866
San Carlos	1864
San Diego	1849
Sanel	1866
San Francisco	1849
San Jose	1849, 1850, 1851
San Juan	1853, 1858
San Leandro	1856
San Lorenzo	1854
San Luis Obispo	1851, 1852
San Pablo	1860
San Quentin	1866
San Rafael	1852
San Ramon	1852
San Simeon	1868
Santa Clara	1851, 1853, 1856, 1868
Santa Cruz	1851, 1853
Santa Rosa	1856
Sawyers Bar	1861
School House Station	1869
Searsville	1858, 1860, 1861
Seiad Valley	1858
Secret Ravine	1855
Sellon's Ranch	1864
Shasta	1851, 1852
Shaws Flat	1854
Shingle Spring	1867
Silveyville	1864
Sink's of Tejon	1859
Slippery Ford	1864
Slippery Springs	1865
Smith River	1865

Snelling to SF, 3¢ Nesbitt.

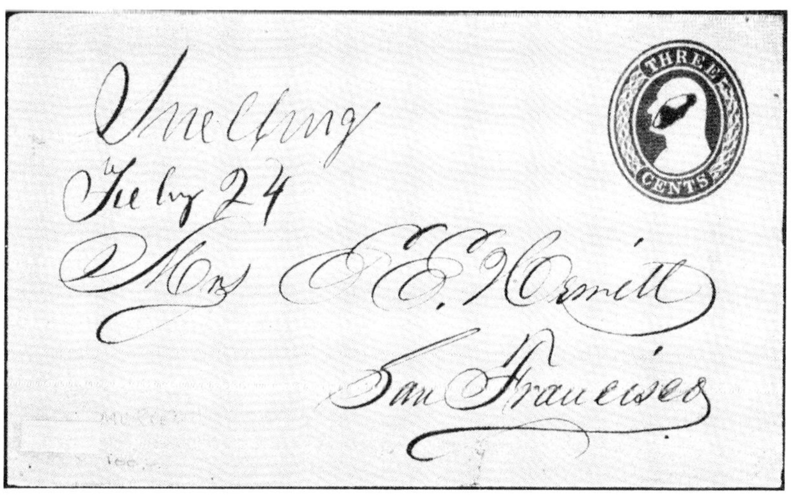

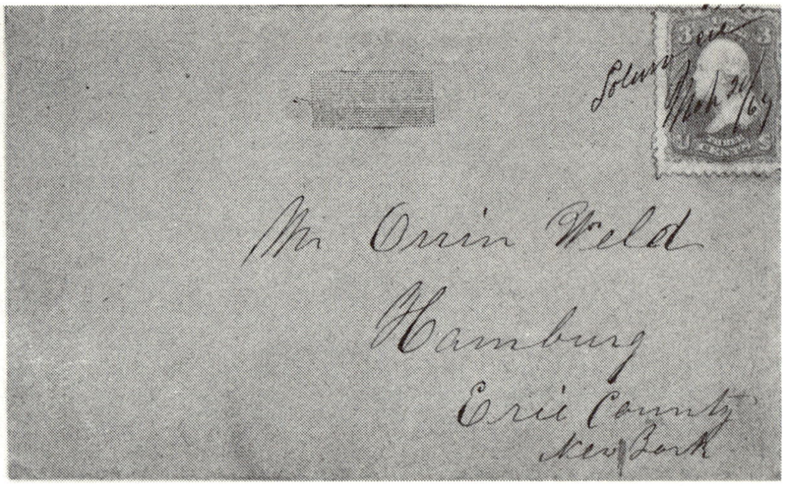

Solano, Cal Mch 20/69 ms. tying 3¢ 1861 to cover to NY state.

Town	Date
Smith's Ranch	1865
Smithville	1862
Snellings Ranch	1861, 1862
Solano	1867
Somersville	1863
Sonoma	1850, 1851, 1854
Sonora	1850, 1852
Soquel	1867
Spanish Flat	1854
Spanish Ranch	1852, 1862
Split Rock	1865
Staples Ranch	1852
Steinbergers	1853, 1856
St. Helena	1858
Stockton	1849
Stony Point	1857
Strawberry Valley	1855
Stringtown P.O.	1855
Suisun	1853, 1856
Summit	1867
Sutter Creek	1852
Sweetland	1857
Table Rock	1857, 1858
Taylors Ranch	1862, 1863
Tehama	1853
Tejon	1857
Texas Hill	1853, 1854
Thompson's Flat	1860
Timbuctoo	1858
Todds Valley	1868, 1869
Tomales	1854
Trinidad	1851
Trinity	1856
Truckee	1868
Two Rocks	1863
Ukiah	1861
Union (Alameda)	1853
Union (Merced)	1866, 1869
Unionville	1861
Upper Clear Lake	1862
Vacaville	1854
Vallicita	1854
Vallejo	1851

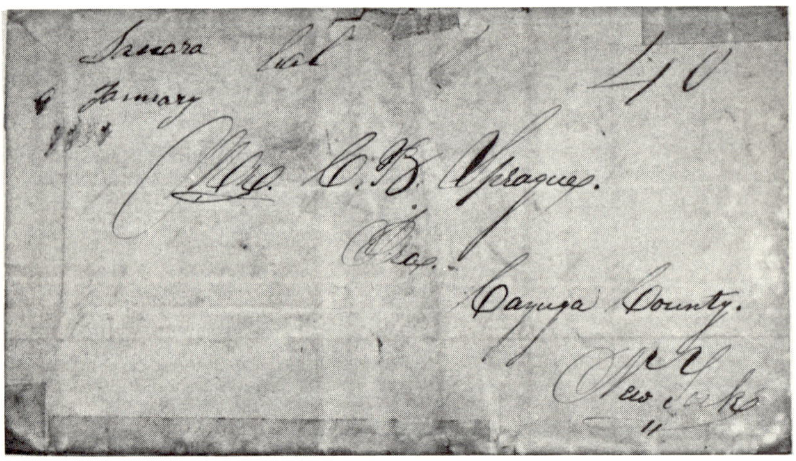

Sonora, Jan. 6, 1851, to N.Y. state at 40¢ unpaid.

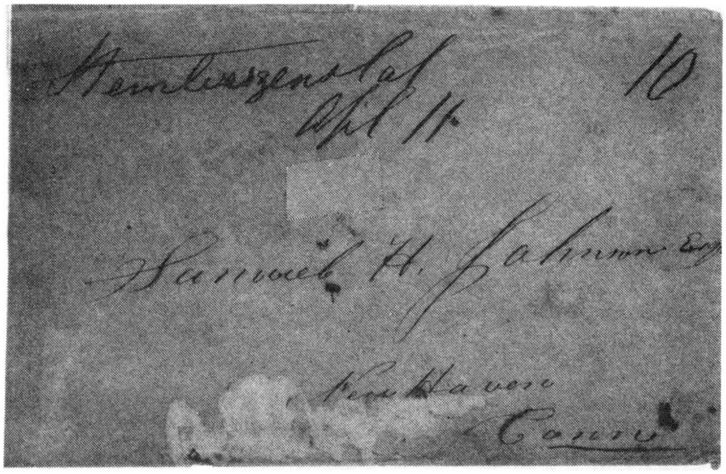

Steinbergers Cal Apl 11 on unpaid letter at 10¢ rate to New Haven, Conn.

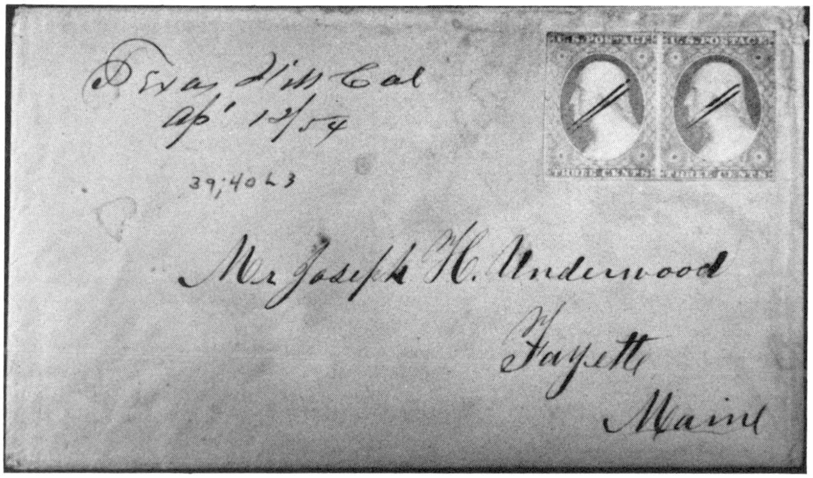

Texas Hill to Me., April 12, 1854, paid with pair of 3¢ 1851.

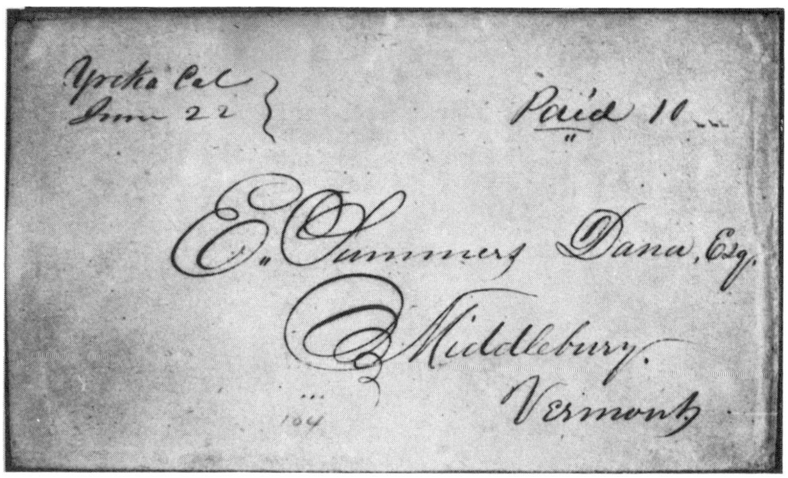

Yreka "Paid 10" to Middlebury, Vt.

Town	*Date*
Vernon	1849
Virginia	1859, 1862
Volcano	1852
Walnut Creek	1865
Washington, South Yuba	1853
Watsonville	1853
Weaverville	1851
Western	1854
West Point	1856
Wheatland	1866
Whiskey Creek	1857, 1859
Wilmington	1864
Woodbridge	1865
Woodland	1862
Woods Diggings	1850, 1851, 1852
Woods Ferry	1858
Woodside	1855, 1856
Wyatt's Store	1861
Yankee Jims	1852
Yeomet (Yornet)	1854
Yolo	1856, 1857
Yolo City	1858
You Bet	1868
Yreka	1855
Yuba City	1855

Yuba City, April 5, 1855, to Illinois on 10¢ Nesbitt.

Appendix 4: California Post Offices Established and Permanently Discontinued, 1849-69

Town	*County*
Acelanus (5/30/54-7/3/55)	Contra Costa
Agua Fria (10/7/51-10/14/62)	Mariposa
Albany (2/25/62-4/13/64)	Sonoma
Albeeville (3/5/62-4/2/63)	Klamath
Alpha (4/12/55-9/4/62)	Nevada
Analy (8/7/60-2/13/61)	Sonoma
Acqueduct City (7/3/55-11/17/55)	Amador
Ashland (10/19/52-12/20/53)	Butte
Aurum City (7/6/52-11/22/53)	El Dorado
Bank Mills (12/22/63-3/13/65)	Santa Clara
Barton's Store (2/3/57-1/18/58)	Solano
Bidwell's Bar (5/1/51-5/27/64)	Butte
Big Valley (7/17/58-11/12/61)	Lake
Blue Mountain (8/19/63-8/8/64)	Calaveras
Bottle Hill (5/28/55-8/17/59)	El Dorado
Brick P.O. (6/18/56-5/28/58)	Sacramento
Buckner (11/22/53-7/17/57)	Sacramento
Buck's Ranch (9/11/61-6/11/68)	Plumas
Bucksport (9/1/52-1/21/53)	Humboldt
Burntranch (10/18/58-11/18/63)	Trinity
Butte City (10/12/57-12/1/58)	Amador
Butte Mills (8/15/57-11/14/61)	Butte
Camp Seco (11/20/52-8/16/53)	Tuolumne
Canon City (4/5/56-11/6/57)	Trinity
Carey's Mills (7/1/58-4/1/69)	Amador
Carneros (9/6/67-4/11/68)	Napa
Carner Place (5/25/66-8/6/67)	Plumas
Carsons Creek (11/6/51-5/27/53)	Calaveras
Cascade City (12/17/58-9/10/60)	Yuba
Cedarville (11/22/53-11/12/63)	El Dorado
Centreville (5/1/51-8/20/52)	Nevada
Charley's Ranch (6/28/52-8/28/58)	Butte
Chip's Flat (3/8/57-11/19/57)	Sierra
Chrysopolis (5/28/66-3/25/67)	Inyo
Churntown (1/12/63-12/12/66)	Shasta
Clay's Bar (1/23/61-2/26/64)	Calaveras
Clinton (5/3/54-1/17/57)	Nevada
Colorado (6/2/58-3/26/60)	Mariposa
Coon Creek (1/3/56-6/13-60)	Placer
Cordelia (7/14/54-7/1/58)	Solano
Curtisville (1/19/53-8/16/56)	Tuolumne
Dobbin's Ranche (7/29/51-5/3/54)	Yuba
Donner Lake (6/7/66-2/18/68)	Nevada
Don Pedro's Bar (11/22/53-4/3/66)	Tuolumne
Double Springs (5/1/51-3/31/60)	Calaveras
Dry Creek (4/18/54-4/9/58)	Yuba
Duroc (9/14/58-11/23/64)	El Dorado
Eight Mile Corners (1/8/61-9/29/63)	San Joaquin
Elbow (10/2/66-4/29/68)	Fresno
Elderton (10/17/60-9/20/62)	Shasta
El Dorado Ranch (6/19/57-9/14/58)	El Dorado
Eliza (known in 1850-51)	Yuba
Elizabethtown (3/2/55-12/17/55)	Plumas
Elk Camp (1/31/59-12/19/62)	Klamath
Elk Grove (5/13/54-12/9/54)	Sacramento
Elk Horn (5/7/52-12/17/53)	Contra Costa
Elliott's Ranch (1/18/54-7/28/57)	Sacramento
Emery's Crossing (6/2/53-7/15/58)	Yuba
Empire City (7/28/51-10/21/51)	Tuolumne
Empire Ranch (6/2/55-12/12/65)	Yuba
Eureka North (11/19/57-12/20/61)	Sierra
Evan's Ranch (5/28/66-6/24/68)	Lassen
Ferry Point (4/29/58-7/16/67)	Del Norte
Foreman's Ranch (1/3/55-2/19/63)	San Joaquin
Forks House (5/17/60-7/25/61)	Placer
Forlorn Hope (8/17/54-2/14/61)	Mariposa
Fort Tejon (3/15/59-9/20/62)	Kern
Foster's Bar (7/29/51-7/19/66)	Yuba
Fourteen Mile House (1/15/57-11/15/58)	San Joaquin
Fremont (4/9/50-11/23/64)	Yolo
Frenchtown (2/3/57-10/19/58)	Butte
Fresno City (1/4/60-6/8/63)	Fresno
Fugitt (4/25/57-7/5/60)	San Joaquin
Georgiana (12/28/54-8/18/58)	Sacramento
Glen Mary (6/16/68-1/27/69)	Mono
Globe Ranch (7/25/61-1/20/64)	Nevada
Graysonville (7/28/51-10/21/51)	Tuolumne
Green Springs (7/28/51-1/21/52)	El Dorado
Green Springs (3/24/52-1/20/69)	Tuolumne
Greenville (7/6/57-3/30/60)	Yuba
Grove City (6/30/58-8/25/64)	Tehama
Gwin (7/13/55-1/28/64)	Mariposa
Halfway House (10/2/58-3/30/60)	San Joaquin
Hall's Ranch (10/7/51-11/29/51)	Colusa
Hamilton (7/10/51-6/26/55)	Butte
Hansonville (12/17/56-1/2/62)	Butte

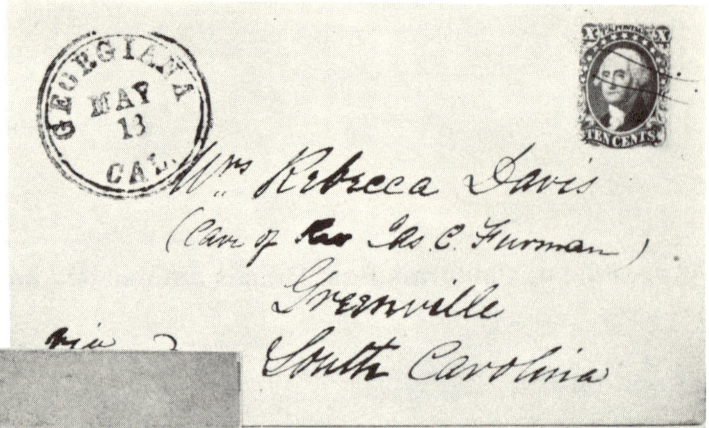

Georgiana to South Carolina with 10¢ 1857. Collin style postmark.

Dry Creek to Ohio "PAID 10." Collin type.

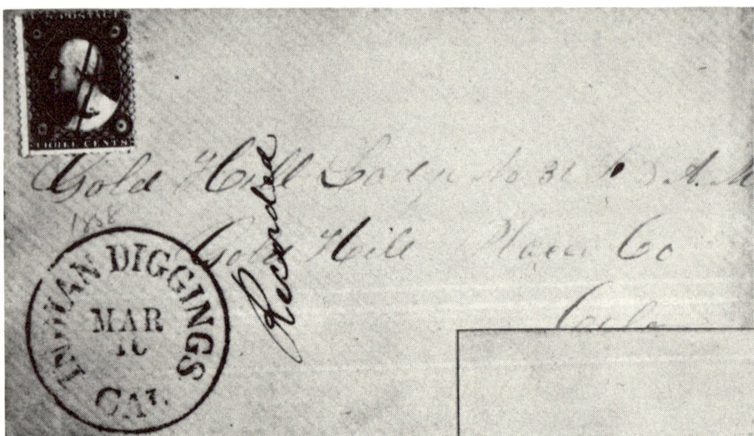

Indian Diggings with 3¢ 1857 to Gold Hill.

Johnson's Ranch to Me. with two 5¢ 1857.

Town	*County*
Haran (6/6/61-3/24/62)	Mendocino
Haskell's Ranch (5/30/54-5/9/56)	Sutter
Haydensville (1/21/51-8/20/52)	Mariposa
Hermit Valley (12/26/65-3/19/67)	Calaveras
Hill's Ferry (7/3/55-5/19/56)	Stanislaus
Hitchcock Ranch (10/9/60-9/12/65)	El Dorado
Holden's Ferry (6/3/58-12/18/58)	Stanislaus
Hope Valley (10/11/64-2/10/68)	Alpine
Hopewell (6/29/64-9/11/65)	Colusa
Humbug Creek (1/29/61-12/18/62)	Siskiyou
Illinois Mills (8/7/67-9/28/68)	Tulare
Illinoistown (11/22/53-2/14/66)	Placer
Indian Diggings (11/22/53-6/15/69)	El Dorado
Jacksonville (10/7/51-3/24/68)	Tuolumne
Jay Hawk (8/29/60-12/26/63)	El Dorado
Johnson's Ranch(e) (11/21/53-11/30/66)	Yuba
Jonesville (7/28/57-9/13/58)	Plumas
Junction (4/9/50-12/20/53)	Contra Costa
Kayote (7/15/59-5/20/61)	Napa
Keeneysburgh (3/3/59-10/1/60)	Tulare
Kerrick's Ranch (4/16/58-9/30/58)	San Joaquin
Keysville (3/11/57-4/23/68)	Tulare
Kilna (11/10/52-10/27/53)	Shasta
Knights Valley (6/7/60-2/25/62)	Napa
Konigsberg (5/12/63-3/28/65)	Alpine
Laguna Seca (10/12/53-1/13/55)	Santa Clara
Lake House (10/11/64-1/21/69)	Sacramento
Lake Valley (9/17/61-12/26/63)	El Dorado
Last Chance (10/13/65-1/29/69)	Placer
Leach's Store (12/19/55-3/26/59)	Mariposa
Leitch's Ferry (10/6/64-5/14/66)	Stanislaus
Lisbon (6/2/55-11/30-66)	Placer
Livermore Ranch (7/28/51-1/6/53)	Contra Costa
Locust Shade (12/28/63-1/23/68)	San Joaquin
Long Bar (4/9/58-6/22/64)	Yuba
Louisville (7/28/51-10/9/52)	El Dorado
Loving's Ferry (7/31/55-6/3/58)	San Joaquin
Marietta (11/15/58-3/17/62)	San Joaquin
Marsh's Landing (9/15/52-5/3/54)	Contra Costa
Martinsburgh (1/11/65-9/22/70)	Butte
McCarthysville (11/2/55-12/22/63)	Santa Clara
McDermott's Bridge (6/17/54-10/2/56)	San Joaquin
Meadow Lake (6/25/66-9/20/69)	Nevada
Mendocino (10/19/52-12/20/53)	Trinity
Meritt (12/14/59-9/14/60)	Yolo
Merry Oaks (7/14/53-11/6/60)	San Joaquin
Messerville (4/2/60-8/19/61)	Trinity
Middletown (6/18/56-3/19/58)	Shasta
Miller's Ranch (1/2/62-1/25/66)	Butte
Mill Valley (5/3/56-1/3/61)	Calaveras
Minersville (4/23/56-12/12/64)	Trinity
Mokelumne City (6/28/61-5/4/64)	San Joaquin
Monoville (12/3/59-4/16/62)	Mono
Monte Vista (4/17/65-1/2/68)	Alameda

Mountain Springs to N.J. with 1855 10¢.

Mount Ophir to Vt. with 1857 10¢.

Mud Springs double circle on unpaid cover to SF.

Ophirville to Me. on 6¢ Nesbitt.

Park's Bar to Mass. with 10¢ 1857.

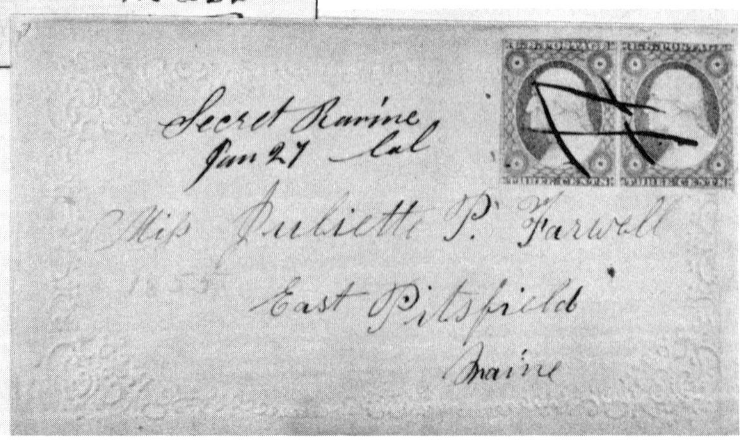

Secret Ravine on fancy embossed envelope to Me. with pair 3¢ 1851.

Town	County
Mountain Inn (7/28/51-1/21/52)	Tuolumne
Mountain Springs (2/4/54-7/15/63)	Placer
Mountain Well (3/30/58-7/26/66)	Nevada
Mount Opher (Ophir) (11/3/56-6/1/68)	Mariposa
Mud Springs (11/6/51-12/15/55)	El Dorado
Musquito (6/2/58-3/29/69)	Calaveras
Nashville (3/5/52-6/20/54)	El Dorado
Neilsburgh (7/30/55-11/30/66)	Placer
Nelson's Creek (3/30/55-10/10/59)	Plumas
Newport (2/22/67-11/10/68)	Solano
Noyo River (11/17/59-3/3/60)	Mendocino
Oak Grove Farm (7/31/55-5/19/56)	Solano
Oak Point (9/30/58-3/3/59)	San Joaquin
Oak Springs (7/28/51-3/31/54)	Tuolumne
O'Byrnes Ferry (1/3/55-4/18/60)	Calaveras
Ophir (2/20/52-11/3/56)	Mariposa
Ophirville (3/24/52-5/11/56)	Placer
Oro City (1/6/53-6/3/58)	Placer
Orr's Ranch (12/17/58-11/25/63)	San Joaquin
Ottitiewa (12/29/54-11/19/60)	Siskiyou
Owensville (3/12/66-6/16/68)	Mono
Owsley's Bar (12/19/55-5/29/65)	Yuba
Pacific (1/15/61-6/7/64)	Humboldt
Pacific Home (6/15/58-4/19/60)	Sonoma
Painesville (1/8/64-9/29/69)	Nevada
Parks Bar (10/7/51-9/18/58)	Yuba
Pea Vine (1/3/56-6/6/64)	Butte
Petersburgh (4/16/58-11/25/63)	Tulare
Phillip's Flat (10/17/57-11/17/58)	Mariposa
Placer (10/4/61-1/2/62)	Placer
Pleasant Springs (3/16/55-10/17/57)	Calaveras
Pleasant Valley (11/17/55-9/16/56)	Mariposa
Plumas (4/19/60-12/4/62)	Yuba
Preston (2/10/63-11/12/66)	Marin
Poland (9/30/58-3/23/67)	San Joaquin
Poverty Bar (3/27/58-2/9/64)	Calaveras
Punta De Los Reyes (10/11/53-3/30/55)	Marin
Quartzburg (6/30/51-2/2/61)	Mariposa
Quartz Valley (7/3/61-6/27/62)	Siskiyou
Rabbit Town (9/13/55-11/19/57)	Sierra
Rattlesnake Bar (11/1/54-10/4/69)	Placer
Red Dog (11/17/55-11/17/69)	Nevada
Redwood (10/9/52-1/6/53)	San Mateo
Reynolds Ferry (12/15/56-2/11/68)	Calaveras
Rich Bar (8/23/62-5/22/63)	Plumas
Ringgold (2/20/52-11/22/53)	El Dorado
Rock Springs (1/8/66-12/6/66)	San Bernardino
Rome (12/23/61-11/25/64)	Sutter

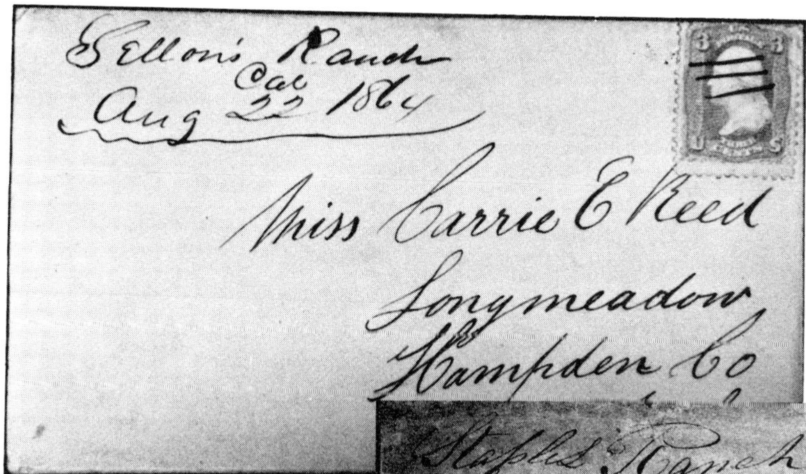

Sellon's Ranch in 1861 cover to Mass., with 3¢ 1861.

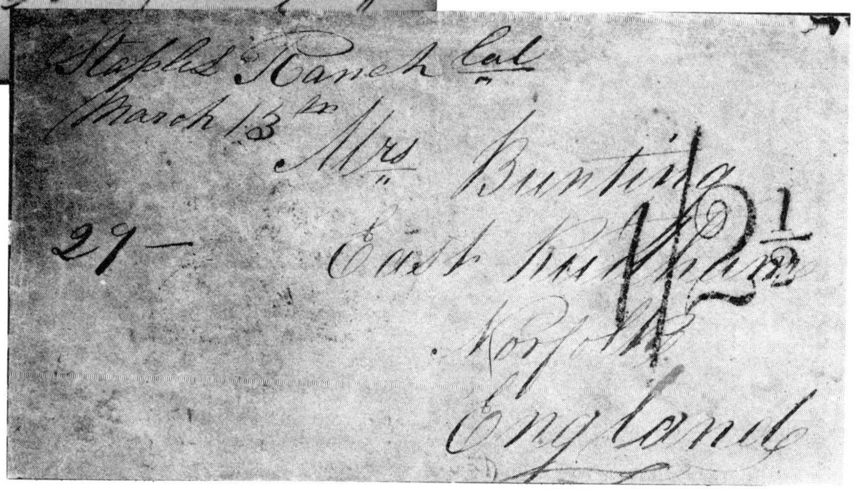

Staples Ranch to England, unpaid and rated 29. At Liverpool struck 1/2½, the sterling equivalent of 29¢.

Town	County
Round Tent (6/21/53-11/12/60)	Yuba
Russian River (11/1/54-4/14/57)	Sonoma
Russville (5/2/57-1/4/58)	Sacramento
Salsbury (7/17/58-10/22/69)	Sacramento
San Carlos (6/1/64-2/27/66)	Tulare
Sand Rock (11/6/62-3/18/64)	Butte
Sarahville (5/10/56-11/16/59)	Amador
Scottsburgh (12/3/56-1/19/64)	Fresno
Secret Ravine (8/29/54-5/15/68)	Placer
Sellon's Ranch (7/5/61-7/12/66)	Yuba
Silver King (10/11/64-4/17/66)	Alpine
Six Mile Bar (12/23/59-4/25/60)	Calaveras
Smith's Ferry (4/7/65-12/4/66)	Fresno
Smithville (1/2/62-12/6/69)	Placer
Snugville (5/15/64-1/22/69)	San Joaquin
Solano (11/12/62-6/8/68)	Solano
Soldiers Bridge (8/22/64-9/24/67)	Lassen
Sportsmans Hall (10/10/65-5/28/67)	El Dorado
Springfield (12/10/57-2/10/68)	Tuolumne
Staples Ranch (10/21/51-12/28/63)	San Joaquin
Star House (11/18/57-10/31/62)	Sacramento
Steinbergers (1/6/53-9/16/56)	San Mateo
Stringtown (5/30/54-5/1/58)	Butte
Sutter (5/17/55-1/13/60)	Sacramento
Tarr's Ranch (8/4/56-1/29/59)	Butte
Taylors Ranch (9/17/61-1/7/64)	Plumas
Texas Hill (11/6/51-12/18/56)	Sacramento
Third Crossing (12/16/52-11/14/54)	Calaveras
Trucky River (4/26/64-11/1/64)	Placer
Two Mile Bar (11/14/56-2/25/57)	Calaveras
Union (11/6/51-8/8/53)	Alameda
Union Town (10/19/52-5/5/60)	Humboldt
Unionville (3/6/61-2/2/62)	El Dorado
Vernon (6/30/49-11/3/53)	Sutter
Viola (7/25/61-1/22/66)	Sacramento
Virginia (6/3/58-4/2/66)	Placer
Washington (4/18/54-7/14/56)	Yolo
Webster (10/13/65-6/14/67)	Placer
Western (5/3/54-4/3/56)	Sacramento
Whiskey Creek (2/18/56-6/28/64)	Shasta
Willow Point (5/15/58-1/13/60)	Yolo
Wilsons Ranch (3/23/65-3/23/68)	Placer
Wood's Diggings (7/28/51-8/16/53)	Tuolumne
Wood's Ferry (4/25/57-10/7/62)	San Joaquin
Woodville (5/19/56-6/19/58)	Butte
Woodville (10/12/53-6/2/55)	Tulare
Yolo Centre (8/27/61-10/22/62)	Yolo
Yornet (Yeomet) (7/14/54-6/8/61)	Amador

(Total: 236)

Texas Hill to Me. with 10¢ 1855.

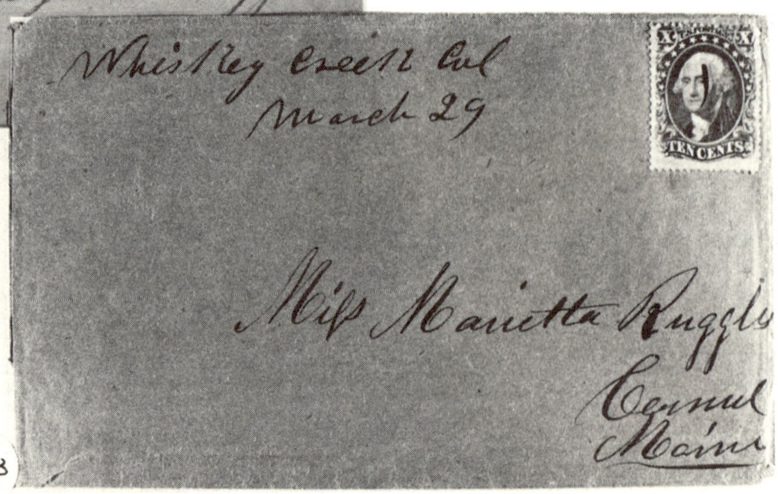

Whiskey Creek to Me. with 10¢ 1857.

Selected Bibliography

General Works

Angel, Myron. *History of San Luis Obispo County, 1883.* Howell-North Books, 1966.

Axon, G. V. *The California Gold Rush.* Mason-Charter, 1976.

Bancroft, H. H. *California Inter Pocula.* The History Co., 1888.

__________. *History of California.* 7 vol. The History Co., 1882-87.

Banning, Wm. and Geo. *Six Horses.* Century Co., 1930.

Bateson, Charles. *Gold Fleet for California.* Michigan State University Press, 1964.

Bean, W. E. *California: An Interpretive History.* McGraw Hill, 1973.

Beck, W. A. and Haase, Y. D. *Historical Atlas of California.* University of Oklahoma Press, 1974.

Beebe, L. M. and Clegg, C. *U.S. West, The Saga of Wells Fargo.* Dutton & Co., 1949.

Bidwell, John. *Trip to California, 1841.* St. Louis, 1842.

Boggs, Mae. *My Playhouse Was a Concord Coach.* Howell-North Press, 1942.

Bolton, H. E. *Padre on Horseback.* Sonora Press, 1932.

Borthwick, J. D. *Three Years in California.* Edinburgh and London, 1857.

Brown, Mackensie. *China Trade Days in California.* University of California Press, 1947.

Caughey, J. W. *California.* Prentice-Hall, 1960.

__________. *Gold is the Cornerstone.* University of California Press, 1948.

Chaitin, Peter M. *Great American West.* Reader's Digest, 1978.

Christman, Enos. *One Man's Gold.* Whittlesey House, 1930.

Clappe, L. A. K. S. *The Letters of Dame Shirley, California in 1851 & 1852.* Rare Americana Series, 1933.

Cleland, R. G. *Cattle on a Thousand Hills.* Huntington Library, 1951.

__________. *A History of California.* Knopf, 1962.

Clemens, Samuel. *Roughing It.* New American Library, 1924.

Colton, Walter. *Three Years in California.* S. A. Rollo & Co., 1850.

Coy, Owen C. *In the Diggings in Forty-Nine.* California State Historical Society, 1948.

Dana, R. H. *Two Years Before the Mast.* Random House, 1936.

Davis, W. H. *Seventy-Five Years in California.* Howell Books, 1967.

Delano, Alonzo. *Life on the Plains and among the Diggings.* Wilson-Erickson reprint, 1936.

__________. *Pen Knife Sketches — Chips from the Old Block.* San Francisco, 1853.

Eaton, Herbert. *The Overland Trail to California in 1852.* Putnam, 1974.

Eggenhofer, Nick. *Wagons, Mules and Men.* Hastings House, 1961.

Engenhoff, E. L. *The Elephant as they Saw It.* San Francisco, 1949.

Farquhar, F. R. *Up and Down California in 1860-1864.* Yale University Press, 1930.

Fehrenbacher, D. E. and Tutorow, N. E. *California, An Illustrated History.* Van Nostrand Co., 1968.

Forbes, Alexander. *California, A History of Upper and Lower California.* London, 1839.

Foster, G. G. *The Gold Regions of California.* New York, 1848.

Frederick, J. V. *Ben Holladay, the Stagecoach King* A. H. Clark Co., 1940.

Greeley, Horace. *An Overland Journey from New York to San Francisco.* Knopf, 1963.

Gudde, E. G. *California Gold Camps.* University of California Press, 1975.

Hinckley, Helen. *Rails from the West.* Golden West Books, 1969.

Hittell, J. S. *Mining in the Pacific States.* San Francisco, 1861.

Hittell, T. H. *History of California.* N. J. Stone & Co., 1885-97.

Hollon, W. E. *Southwest, Old and New.* Knopf, 1961.

Hulbert, A. B. *Forty-Niners.* Little, Brown & Co., 1931.

Hungerford, E. *Wells Fargo.* Random House, 1949.

Hutchinson, W. H. *California, Two Centuries.* American West Publishing Co., 1969.

Jackson, J. H. *Anybody's Gold.* Appleton Century, 1941.

Jackson, W. Turrentine. "A New Look at Wells Fargo, Stagecoaches and the Pony Express," *California Historical Society Quarterly.*

__________. *Wagon Roads West.* Yale University Press, 1965.

Kemble, J. H. *San Francisco Bay — A Pictorial Maritime History.* Cornell Maritime Press, 1957.

Kirsch, R. R. and Murphey, W. S. *West of the West.* Dutton, 1967.

Lewis, Oscar. *The Big Four.* Knopf, 1938.

__________. *Sea Routes to the Gold Fields, 1849-52.* Knopf, 1949.

Lyman, G. D. *Saga of the Comstock Lode.* New York, 1934.

McCarthy, J. R. *California.* 9 vol. Powell Publishing Co., 1929-30.

McMullen, Jerry. *Paddle-Wheel Days in California.* Stanford University Press, 1944.

Marryat, Frank. *Mountains and Molehills.* J. B. Lippincott Co., 1962.

Moody, Ralph. *Stage Coaches West.* Crowell Co., 1967.

Norris, R. N. and Webb, R. W. *Geology of California.* Wiley & Sons, 1976.

Ogden, Adele. *The California Sea Otter Trade, 1784-1848.* Berkeley, 1941.

Pattie, J. O. *The Personal Narrative of James O. Pattie.* Lippincott Co., 1962.

Paul, Rodman W. *California Gold.* Harvard University Press, 1947.

Riesenberg, Felix, Jr. *Golden Gate: The Story of San Francisco Harbor.* New York, 1940.

__________. *The Golden Road.* McGraw Hill, 1962.

Robinson, Alfred. *Life in California.* New York, 1846.

Settle, R. W. and M. L. *Empire on Wheels.* Stanford University Press, 1949.

Smith, Waddell. *Stagelines & Express Companies in California.* Pony Express History Gallery, 1965.

Stimson, A. L. *History of the Express Companies.* New York, 1858.

Strahorn, C. A. *Fifteen Thousand Miles by Stage.* Putnam's Sons, 1911.

Taylor, Bayard. *El Dorado, or Adventures in the Path of Empire.* Knopf, 1949.

Time-Life. *The Expressmen.* Time-Life Books, 1974.

__________. *The Forty-Niners.* Time-Life Books, 1973.

Tucker, T. W. *Waifs from the Waybills of an Old Expressman.* Lathrop, 1872.

Underhill, R. L. *From Cowhides to Golden Fleece.* Stanford University Press, 1939.

Wilson, N. C. *Treasure Express.* Macmillan, 1936.

Winther, Oscar O. *Express and Stagecoach Days in California.* Stanford University Press, 1936.

__________. *Via Western Express and Stagecoach.* Stanford University Press, 1945.

Zollinger, J. P. *Sutter, the Man and his Empire.* New York, 1939.

Postal Markings and Postal History

Alexander, Thomas J. *Simpson's U.S. Postal Markings 1851-61.* U.S. Philatelic Classics Society, 1979.

Ashbrook, Stanley B. *Special Service.* Published serially by the author, Fort Thomas, Ky., 1951-57.

__________. *The United States One Cent Stamp of 1851-1857.* 2 vol. H. L. Lindquist, 1938.

Berthold, Victor M. *Handbook of the Wells, Fargo & Co.'s Handstamps and Franks.* Scott Stamp and Coin Co., 1926.

Bradley, G. D. *The Story of the Pony Express.* McClurg & Co., 1913.

Chapman, Arthur. *The Pony Express.* Putnam & Sons, 1932.

Conkling, R. P. and M. B. *The Butterfield Overland Mail, 1857-1869.* A. H. Clark Co., 1947.

Driggs, H. R. *The Pony Express Goes Through.* Lippincott Co., 1936.

Folkman, David I. *The Nicaragua Route.* University of Utah Press, 1972.

Frickstad, Walter N. *A Century of California Post Offices.* Philatelic Research Society, 1955.

Hafen, Leroy R. *The Overland Mail 1849-1869.* Quarterman Publications, reprint, 1976.

Hargest, George E. *History of Letter Post Communication Between the United States and Europe 1845-1875.* Quarterman Publications, reprint, 1975.

Harlow, Alvin F. *Old Post Bags.* Appleton Century, 1934.

__________. *Old Waybills.* Appleton Century, 1934.

Kemble, John H. *The Panama Route 1848-1869.* Cornell Maritime Press, 1957.

Leutzinger, John F. *The Handstamps of Wells, Fargo & Co.* Published by the author, 1968.

Nathan, M. C. *Franks of Western Expresses.* Collectors Club of Chicago, 1973.

Nathan, M. C. and Boggs, W. S. *The Pony Express.* The Collectors Club of New York, 1962.

Salley, H. E. *History of California Post Offices, 1849-1976.* Postal History Associates, 1977.

Sampson, E. N., Wishnietsky, Benjamin, and Skinner, Hubert. *American Stampless Cover Catalog.* David G. Phillips Publishing Co., 1978.

Skinner, Hubert C. and Eno, Amos. *U.S. Cancellations 1845-1869.* American Philatelic Society, 1980.

Starnes, Charles J. *U.S. Letter Rates to Foreign Destinations 1847 to GPU-UPU.* L. H. Hartmann, 1982.

Thorpe, P. H. *Catalogue of United States Stamped Envelopes.* P. H. Thorpe, 1954.

Wierenga, Theron. *United States Incoming Steamship Mail 1847-1875.* T. Wierenga, 1983.

Wiltsee, E. A. *Gold Rush Steamers of the Pacific.* Grabhorn Press, 1938.

________. *The Pioneer Miner and the Pack Mule Express.* California Historical Society, 1931.

Official Sources

Annual Reports of the Postmaster General, 1847-1869.

Postal Laws and Regulations and/or *Lists of Post Offices,* 1847, 1852, 1854, 1855, 1857, 1859, 1862, 1866.

Register of Officers and Agents, Civil, Military and Naval in the Service of the United States (Federal Register), 1843, 1845, 1847, 1851, 1853, 1855, 1857, 1865.

Periodicals

The American Philatelist. Published monthly by the American Philatelic Society.

The Collectors Club Philatelist. Published bimonthly by the Collectors Club of New York.

Postal History Journal. Published three times a year by the Postal History Society.

The Chronicle of the U.S. Classic Postal Issues. Published quarterly by the U.S. Philatelic Classics Society.

Selected articles from the *Chronicle.*

Alexander, Thomas J. "Bisects of the 12¢ 1851 Issue," 75:121-129 (Aug. 1972).

________. "The Crittenden Correspondence," 111: 164-180 (Aug. 1981).

________. "The Crittenden Correspondence: A Footnote," 116:225-226 (Nov. 1982),

________. "Ocean Mail Route Agents," 115:188-191 (Aug. 1982).

________. "Sonora, California, Straight Line," 105:26-29 (Feb. 1980).

Alexander, Thomas J., and Beals, David T., III. "Gateway to the West," 92:250-257 (Nov. 1976); 93:18-28 (Feb. 1977); 94:84-89 (May 1977); 96:239-243 (Nov. 1977); 97:30-34 (Feb 1978).

Ashbrook, Stanley B. "New York to Panama Ship Sailings," 104:252-256 (Nov. 1979); 105:35-37 (Feb. 1980); 107:174-177 (Aug. 1980); 108:238-242 (Nov. 1980); 109:26-40 (Feb. 1981).

Beals, David T., III. "Military Markings 1851-1861," 64:142-149 (Nov. 1969); 65:12-16 (Feb. 1970); 66:58-62 (May 1970); 67:108-111 (Aug. 1970); 68:150-154 (Nov. 1970); 70:72-76 (May 1971); 79:149-155 (Aug. 1973); 100:256-259 (Nov. 1978).

Christian, C. W. " 'The Shortest Distance'," 84-217-219 (Nov. 1974).

de Voss, James T. "A Remarkable Cover," 91:176-180 (Aug. 1976).

Gallagher, Scott. "The Cover Corner," 111:214-215 (Aug. 1981); 112:284-286 (Nov. 1981).

Garrett, Duane. "Domestic Postal Rates for the 1847 Issue Period: Part II [California]," 109:20-22 (Feb. 1981).

Gilbart, Kenneth. "Hawaiian Steam Service Contract Sailings," 99:210-215 (Aug. 1978).

________. "Addenda: the Hawaiian Steam Service," 106:131-133 (May 1980).

Graham, Richard B. "China and Japan Steam Service," 73:21-30 (Feb. 1972); 75:133-143 (Aug. 1972); 76:206-207 (Nov. 1972); 111:182-193 (Aug. 1981).

________. "Covers from the Column from California," 63:112-113 (Aug. 1969).

________. "A Pair of Pony Covers," 88:222-229 (Nov. 1975).

Hargest, George E. "California to Australia, via Panama and London, in 1869," 74:93-96 (May 1972).

Hubbard, Walter. "Closed Mails from California to Great Britain 1849-1860," 109:60-61 (Feb. 1981).

Laurence, Michael. "Ten Cent 1869 Covers from the Shanghai Consular Post Office," 80:214-222 (Nov. 1973).

________. "The 10¢ Rate: Transpacific Covers via U.S. Mails during the 1869 Period," 100:270-273 (Nov. 1978); 101:42-52 (Feb. 1979); 102:128-133 (May 1979); 104:266-271 (Nov. 1979); 109:52-54 (Feb. 1981).

Reed, Eugene C. "U.S. Mail Steamship Company Sailings to Panama: 1849-1859," 102:98-116 (May 1979).

Schwimmer, Jerome. "Los Angeles: Prestamp and Stampless Period," 106:86-92 (May 1980).

________. "Correction," 110:109 (May 1980).

________. "Los Angeles: the 1851-1861 Period," 110:96-101 (May 1981).

Starnes, Charles J. "British Mail to California via West India Packet," 121:64-67 (Feb. 1984).

________. "Hawaii to Switzerland via the United States," 93:64-66 (Feb. 1977).

________. "The San Francisco Purser Marking," 78:106 (May 1973).

Western Express. Published quarterly by the Western Cover Society.

Selected articles from *Western Express*

Beals, David T., III. "The Leavenworth and Pikes Peak Express Companies," Oct. 1975, 3-13.

Friend, Clifford L. "The 26mm Double Circle Postal Markings of San Francisco," Jan. 1975, 11-18.

Hanson, Robert A. "Notes on the Butterfield Overland Mail Route," Jan. 1969, 4-7.

________. "Notes on the Trail of the Pony Express," Jan. 1967, 5-8.

________. "The Story of the Holy Road," Jan. 1966, 4-8.

McCumber, Norma L. "Pack Mules to Stage Coaches," Jan. 1981, 9-12.

Nathan, M. C. "Difficulties in Establishing Western Post Offices," Jan. 1963, 4-20.

Pearce, Basil C. "California's First Mail Service," Oct. 1970, 11-12.

____________. "Charles P. Kimball, San Francisco's 'Noisy Carrier'," Apr. 1963, 4-19.

____________. "Overland to the Pacific — The San Antonio and San Diego Mail Line," Apr. 1961, 4-10.

Polland, W. Scott, M.D. "California Postmarks 1849-1851, Earlier than Officially Recorded," Oct. 1973, 9-12.

____________. "Early Mail from San Francisco," Apr. 1970, 4-8.

____________. "Mail Routes to California before the Gold Rush," Jul. 1964, 17-21.

____________. "More about 'Mail Routes to California before the Gold Rush'," Jul. 1965, 8-10.

____________. "More about Unusual Usages of the New York Type III Ocean Mail Cancellation," Jul. 1965, 19-21.

____________. "New York and California Steam Ship Lines," Oct. 1963, 4.

____________. "The New York Ocean Mail Postmarks via Panama to California and Oregon 1851-1861," Jan. 1964, 5-20.

____________. "The San Francisco Ship 6 'Clipper Ship Cancellation'," Jan. 1968, 10-14; Jul. 1968, 15-16.

____________. "The San Francisco 'Way' Mark," Oct. 1969, 18-19.

____________. "Stampless Mail from New York to California or Oregon 1848-1856," Oct. 1966, 4-11.

____________. "The Story of the San Francisco Cog Cancellation and its Imitators," Apr. 1966, 4-14.

____________. "The Tehuantepec Route," Apr. 1965, 5-9.

Schwimmer, Jerome. "Los Angeles Covers," Oct. 1979, 3-5.

____________. "Los Angeles Covers — Butterfield Carries the Mail — Via Los Angeles," Jul. 1980, 11-16.

____________. "Los Angeles Covers — Early Foreign Usage," Jan. 1981, 23-26.

____________. "Los Angeles Covers — The Express Companies — Early History," Oct. 1980, 19-22.

____________. "Los Angeles Covers — Overland via Los Angeles," Apr. 1980, 7-12.

____________. "Los Angeles Covers — Wells Fargo Overland Advertising Covers," Apr. 1981, 13-18.

____________. "Los Angeles Covers — Wells Fargo Returns — Odds & Ends," Oct. 1981, 11-16.

____________. "Los Angeles — Rare and Unique," Apr. 1979, 5-6; Jul. 1979, 25-27.

Townley, John M. "Stalking Horse for the Pony Express — The Chorpenning Mail Contracts between California and Utah, 1851-1860," Jul. 1983, 13-23.

"The Capitals of California," Oct. 1963, 3.

"George Chorpenning," Apr. 1957, 15-19.

"Hinckley and Company's Express," Jan. 1963, 30-32.

"Ocean Mail from the West," Oct. 1958, 17-18.

"The Pony Express," Apr. 1960, 2-17; Jul. 1960, 2-12; Oct. 1960, 3-14; Jan. 1961, 6-15; Apr. 1961, 3-17; Jul. 1961, 4-18; Oct. 1961, 2-8.

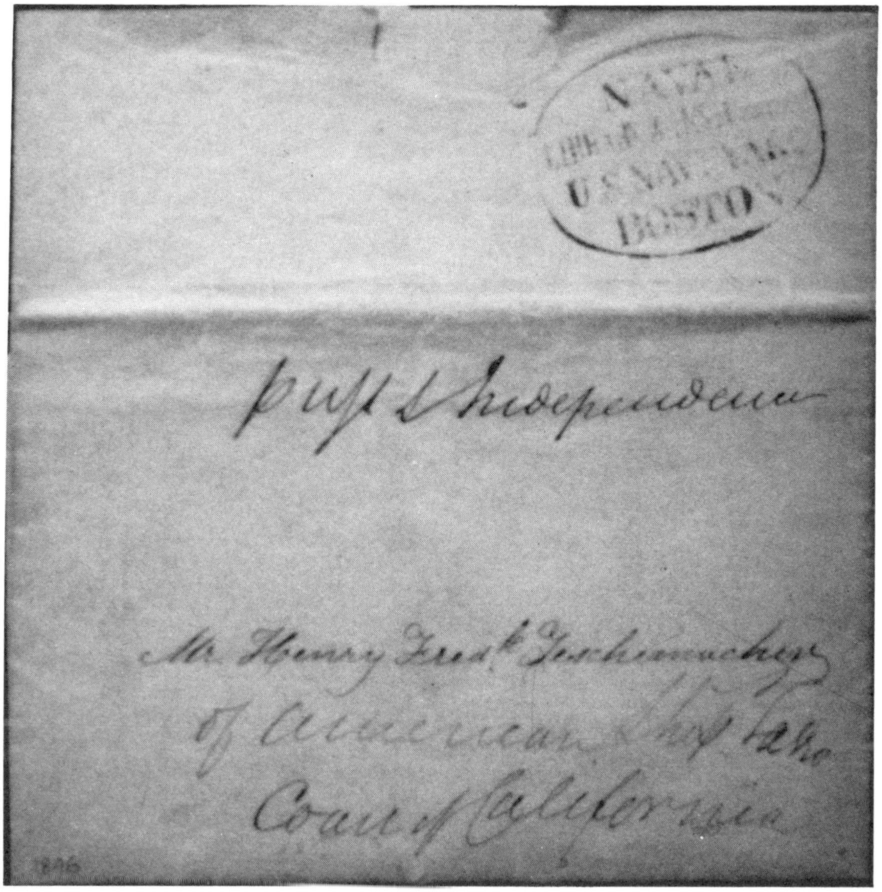

A folded letter from Massachusetts in 1846, from a mother to her son of "American Ship Tasso/Coast of California." On reverse is the oval handstamp of NAVAL/LIBRARY & INSTITUTE/U.S. NAVY YARD/BOSTON, an agency which forwarded mail to naval personnel. This letter was sent "p. U.S.S. Independence."